61st International Scientific Conference on Experimental Stress Analysis (EAN 2023)

AA005105

Kosice, Slovakia
6 - 8 June 2023

Editors:

Martin Hagara
Robert Hunady

ISBN: 978-1-7138-9339-4

Printed from e-media with permission by:

Curran Associates, Inc.
57 Morehouse Lane
Red Hook, NY 12571

Some format issues inherent in the e-media version may also appear in this print version.

Copyright© (2023) by Czech Society for Mechanics (CSM)
All rights reserved.

Printed with permission by Curran Associates, Inc. (2024)

For permission requests, please contact Czech Society for Mechanics (CSM)
at the address below.

Czech Society for Mechanics (CSM)
Dolejskova 5
Praha 8
182 00
Czech Republic

Phone: +420 266 053 045
Fax: +420 286 587 784

csm@csm.cz

Additional copies of this publication are available from:

Curran Associates, Inc.
57 Morehouse Lane
Red Hook, NY 12571 USA
Phone: 845-758-0400
Fax: 845-758-2633
Email: curran@proceedings.com
Web: www.proceedings.com

EXPERIMENTAL STRESS ANALYSIS 2023

61st International Scientific Conference

June 6th – 8th, 2023, Košice, Slovakia

EAN 2023
61st conference on
experimental stress analysis

Proceedings of Full Papers

Edited by:

Martin Hagara, Róbert Huňady

www.ean61.kamasi.sk
ean@kamasi.sk

Czech Society for Mechanics

Technical University of Košice
Faculty of Mechanical Engineering

All rights reserved. No parts of this publication may be reproduced, stored in retrieval system or transmitted in any form or by means of: electronic, magnetic tape mechanical, photocopying, recording or otherwise without permission in writing from the publication.

Title:	Experimental Stress Analysis 2023 Proceedings of Full Papers
Editors :	Martin Hagara Róbert Huňady
Publisher:	Technical University of Košice, Slovakia

The proceedings contain papers that were submitted and presented at the EAN 2023 conference. The aim of the conference is to understand the state-of-art, to exchange experience, to establish and build up cooperation among academic workers, company staff and scientific and research institutions active in the field of experimental mechanics and numerical modelling in mechanics.

All papers in this book have been peer-reviewed by two independent reviewers after consideration by the editors. However, the authors take responsibility for the content and for the correctness of their texts, tables, graphs and figures.

The Contents and Index of Authors are sorted in Slovak alphabetical order.

The complete set of EAN 2023 Conference consists of two parts:

- Book of Extended Abstracts (printed)
- Proceedings of Full Papers (e-book)

Košice, March 2024, first edition

Conference organized by:

Technical University of Košice
Faculty of Mechanical Engineering
Department of Applied Mechanics and Mechanical Engineering

www.sjf.tuke.sk | www.kamasi.sk

Conference Partners:

LABTECH s.r.o. | www.labtech.eu

HBP měřicí technika s.r.o. | www.hbm.cz

SHIMADZU Slovakia | www.shimadzu.sk

IDIADA CZ a.s. | www.idiada.cz | www.simulia.cz .

X-Sight s.r.o. | www.xsight.eu

DEWESoft | www.dewesoft.com

Conference topics:

- Development of experimental methods in mechanics, biomechanics and materials engineering;
- New methods and applications of deformation, stress and vibration analysis in mechanical, civil and other structures;
- Experiment as a tool for the verification of analytical and numerical methods;
- Experimental research and prediction of strength, life and operational reliability of structures and devices;
- Monitoring of operational loadings and operational states of structures and devices;
- Methods and means of teaching experimental methods;

Conference Program Committee:

Ing. Karel Doubrava, Ph.D.
doc. Ing. Peter Frankovský, PhD.
prof. Ing. Nikolaj Ganev, CSc.
prof. Ing. Radim Halama, Ph.D.
prof. Dr. Ing. František Holešovský
prof. Ing. Stanislav Holý, CSc.
Ing. Lubomír Houfek, Ph.D.
prof. RNDr. Miroslav Hrabovský, DrSc.
doc. Ing. Róbert Huňady, PhD.
prof. Ing. Josef Jíra, CSc.
Ing. Kamil Kolařík, Ph.D.
Ing. Aleš Lufinka, Ph.D.
doc. Ing. Tomáš Návrat, Ph.D.
doc. Ing. Pavel Padevět, Ph.D.
doc. Ing. Iva Petríková, Ph.D.
Dr. Ing. Roman Růžek
prof. Ing. Milan Růžička, CSc.
Ing. Jaroslav Václavík
doc. Ing. Miloš Vlk, CSc.
doc. Ing. Robert Zemčík, Ph.D.

Organizing Committee:

prof. Ing. Jozef Bocko, CSc.
doc. Ing. Róbert Huňady, PhD.
doc. Ing. Peter Frankovský, PhD.
doc. Ing. Martin Hagara, PhD.
Ing. Pavol Lengvarský, PhD.
Ing. Ján Kostka, PhD.
Ing. Peter Palička

Contents

Measurement of Moments of Forces on the Control Mechanism of Gate Valves of Hydraulic Power Station

Jozef Bocko[1,a], Miroslav Pástor[1,b], Martin Hagara[1,c] and Pavol Lengvarský[1,d]

[1]*Technical University of Košice, Faculty of Mechanical Engineering, Department of Applied Mechanics and Mechanical Engineering, Letná 1/9 Košice-Sever 042 00, Slovakia;*

[a]*jozef.bocko@tuke.sk*, [b]*miroslav.pastor@tukle.sk*, [c]*martin.hagara@tuke.sk*, [d]*pavol.lengvarsky@tuke.sk*

Abstract: The article discusses the torque measurements on gate valves in the water dam of hydraulic power station. The gate valves serve for regulation of water flow between the pressure chamber and other operational spaces in dam. To open and close the gates, a system of bars (spindles, tubes) connected by universal joints (cardan joints) is employed, guided by a longitudinal guide. Approximately ten years ago, the equipment underwent reconstruction, replacing manual operation with hydraulic actuation for gate manipulation. However, issues arose with opening and closing the gates once the hydraulic actuator and control system were implemented. Furthermore, water entered the hydraulic system. Consequently, manual manipulation became necessary for the gate valves. The article presents an analysis of the causes behind the equipment malfunction. The authors conducted operational measurements of the driving torques of the closing mechanism using their designed dynamometers. These measurements revealed the reason of gate valves malfunction. The analysis of the collected operational data concluded that employing driving units with higher torque is essential for safe and reliable operation. From a strength point of view, the mechanical parts of the linkage are properly designed.

Keywords: Gate valves; Torques; Operational measurement

1 Introduction

Dam structures or water reservoirs are the essential infrastructure of society, used primarily for hydropower, flood control, and irrigation. A decade ago, the number of reservoirs larger than 0.01 ha was estimated to be approximately 16.7 million worldwide, and the number was increasing significantly every year [1]. Information provided in the World Register of Dams database [2], updated in April 2020, considers 58,713 dams classified as large, i.e., dams with a height of at least 15 m measured from the lowest base to the crest, or smaller dams (of height between 5 m and 15 m) that hold more than 3 million cubic meters of water [2]. Using the latest publication of the World Register of Dams, irrigation is the most common purpose of dams. Of single-purpose dams, 48% serve for irrigation, 17% provide electricity generation, 13% ensure water supply, 10% are used for flood control, 5% for recreation, and less than 1% for navigation and fish farming. At present, the number of mainly multi-purpose dams is increasing [3, 4].

The dam described in the article was built between 1963 and 1972. Its purpose is to provide water for the domestic metallurgical industry, electricity supply, flood protection, and recreational purposes. The reservoir has a storage capacity of 45.3 million cubic meters and a flooded area of approximately 4.5 square kilometres. The length of the dam is 11 kilometres. The reservoir also includes a pumped storage hydroelectric power plant with reversible turbines, completed in 1973. It consists of two parts, the second of which functionally forms a balancing reservoir to the pumped storage power station. In 2011, the dam was renovated, with the gates being manipulated to hydraulic actuation. Since the hydraulic actuator and controls were implemented, there have been problems with opening and closing the gates. This manipulation has subsequently had to be done manually. In addition, during operation, the chamber in which the hydraulic equipment is located was flooded for short time. The operator's preliminary information on the condition of the equipment already indicated that the driving torques of the hydraulic mechanism were insufficient for handling the gate valve. The methodology proposed by the authors for the operational measurement of the driving torques of the closing mechanism not only confirmed the above factual situation but also provided important information leading to the identification of possible causes of the limited functionality.

2 Methods

As mentioned in the introduction, after several months of operation, problems with manipulating the gate valves using servomotors with a tripping torque of 160-400 Nm and a breakaway torque of 500 Nm were already detected. According to preliminary information from the operator, the torque value of 500 Nm was insufficient to operate the gate valves. Two dynamometers were designed and manufactured to determine the torque at the gate valves operating load and thus to find out the possible cause of problem. Their design was based on data provided by the operator. However, this procedure also required strength control of the other mechanical parts of the opening mechanism, such as the tubes and connecting elements (pivots/pins). The tube material's chemical composition and mechanical properties were determined to establish the actual status. After evaluating the material's mechanical properties, operational measurements were carried out to determine the torque required for manipulating the gate valves.

A prototype dynamometer was designed for torque determination, which allowed us to obtain the torque value considering actual operating conditions such as friction, corrosion, shape, and wear of mechanical parts. The essential part of the designed dynamometer was a seamless steel tube of the annular cross-section with an outer diameter of ø50 mm and wall thickness of 5 mm mounted in radial bearings pressed in steel discs. The tube material is S235JRH, with a yield strength of 235 MPa and a strength of 360-510 MPa [5, 6]. When designing the dynamometer dimensions, it was necessary to consider the winding cable length of 25 m for strain gauging. Fig. 1 shows a 3D model of the designed dynamometer.

Since the designed dynamometer served to identify the actual value of the torque required for opening and closing the gate valves, it was necessary to verify not only its functionality but also to perform its calibration in laboratory conditions before the operational measurement. For the calibration measurement, two different measuring apparatuses were used. The values were directly read from the display for the measurement with the P3 strain indicator and recorder (Vishay, USA). The Ouantum MX840 strain gauge apparatus (HBM, Germany) was used with the CatmanEasy measurement and evaluation software. Based on the calibration measurements performed, it was found that a torque value of 400 Nm corresponds to a strain of 319-320 $\mu\varepsilon$ for dynamometer 1 and a strain of 322-323 $\mu\varepsilon$ for dynamometer 2. The calibration measurements confirmed a very good agreement so that their possible substitution in the operational measurement would not cause significant deviations (i.e., the deviation of the measurement results would be less than 1%). The measuring chain consisted of a Quantum MX840 strain gauge apparatus and the CatmanEasy evaluation software. In some cases, it was also necessary to use an extender attached to the wheel when the valve gate mechanism was first opened (to apply higher torque).

Fig. 1 Model of the designed dynamometer.

As can be seen from the diagram in Fig. 2, six gate valves (three on the right and three on the left side) are used to fill and discharge water into and out of the pressure chamber, and gate valve 4 is used to discharge water from the pressure chamber into the pump chamber. The operational measurement procedure has been chosen so that the operational modes considered correspond to the worst-case loading mode in a critical situation.

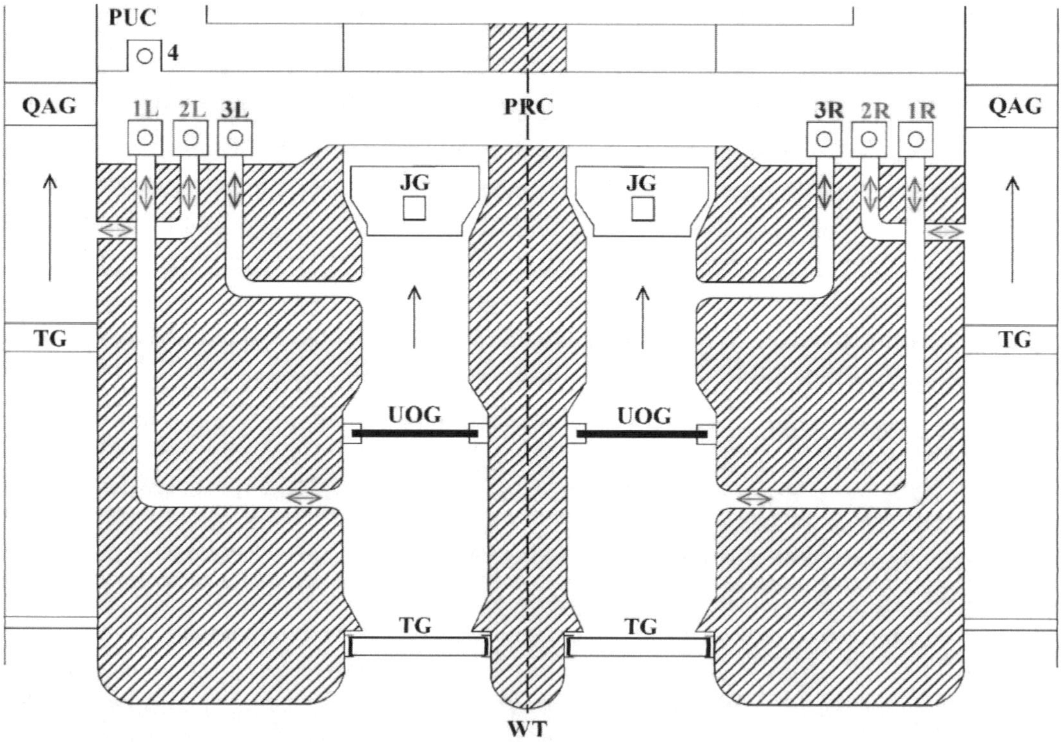

Abbreviations: QAG - Quick-acting gate, PUC - Pump chamber, PRC - Pressure chamber, JG - Johnson gate, TG - Temporary gate, UOG - Upstream operational gate, WT - Water tank

Fig. 2 View of the location of the gate valves under consideration.

For this reason, only the following modes were performed out of all possible modes:

Mode 1: 1L and 1R gate valve handling operation:

- Filling of water from the tank (T) into the pressure chamber (PRC).

Mode 2: 2L and 2R gate valve handling operation:

- Discharge of water from the pressure chamber (PRC) into the zone between the power plant quick-acting gate (QAG) and the temporary gate (TG).

Mode 3: 3L and 3R gate valve handling operation:

- Discharge of water from the zone between the Johnson gate (JG) and the upstream operational gate (UOG) into the pressure chamber (PRC).

Mode 4: gate valve no. 4 handling operation:

- Transfer of water from the pressure chamber (PRC) to the pump chamber (PUC).

In addition to the above modes, others are used in real operation, but in each of the modes tested, the most unfavourable handling case was when the water column pressure was applied from only one side of the gate valve. Each pair of caps 1L/1R, 2L/2R, and 3L/3R has a different function, so each mode was implemented separately.

Before each measurement, the dynamometers were sequentially attached to individual flanges and their corresponding connecting bars, through which the gates were manually manipulated using a wheel located at the top of the dynamometer. The operator found that 60 turns of the wheel are required to open the gate valve fully. As the opening of the gate valves was manual and in some of the modes tested, the pressure chamber was filled (opening of the gate valve at the bottom of the water reservoir directly below the measuring station), it was necessary to pay special attention to ensure the safety of the operating staff during the operational measurements. Therefore, to increase safety, it was agreed after mutual consultation with the operator in charge that the operational measurements would be carried out at 15 turns of the wheel. This number of turns is related to the fact that the most significant resistance to movement occurs when moving from a stationary position, i.e., when static friction is overcome. During subsequent handling, the increase in the torque value required to operate the gate was no longer anticipated. Another of the proposed safety features for the operational measurement was the replacement of the original 10 mm diameter steel pin with an M8 bolt of material grade 8.8. The ultimate strength of the bolt material used is 800 MPa, and the yield strength is 640 MPa.

At the assumed maximum torque value (500 Nm), the shear stress in the used M8 bolt is approximately 250 MPa, which according to Guest's strength theory, is strength-compliant. As it later turned out, strain values corresponding to higher torque values were registered in almost every operational measurement. This phenomenon was confirmed by the deformation of the M8 bolts used. In some cases, these bolts failed by shear on the 1R and 2R gates. In standard operation with 1R and 2R gates, the operator uses a 10 mm diameter pin. As no mechanical (steel) parts have failed, the authors used an M10 bolt of material grade 10.9 for further measurements. This bolt has an ultimate strength of 1000 MPa and a yield strength of 900 MPa.

Fig. 3 to Fig. 9 show documented time-dependent records of the torques determined by recalculation from the measured strains based on the calibration measurements. As can be seen from Figs. 3 to 5, much higher levels of torques, which exceeded 1000 Nm in two cases, were determined on the right side of the water reservoir. After consultation with the operator, it was determined that no critical situation requiring simultaneous manipulation of the pair of corresponding gate valves (on the right and left sides) had occurred during the previous operation history. In regular operation, service personnel use left-hand gate valves. This fact was confirmed by the actual handling during the operational measurements, as documented by the time-dependent torque records in Figs. 6 to 9. Gate valve 4 was the most commonly used, as confirmed by the lowest torque levels, which did not exceed 255 Nm. In Figs. 3 to 9, the areas where it was necessary to release the gate mechanism are marked in red. The green area corresponds to the opening of the gate valve (approximately 15 turns). The closing process is indicated in blue. The purple arrow identifies the instant the gate valve has been fully closed. The yellow arrow identifies the moment the maximum torque value has been reached. When using the M8 bolt, the bolt failed by shear at that instant. When using the M10 bolt, the required release of the locking mechanism was achieved.

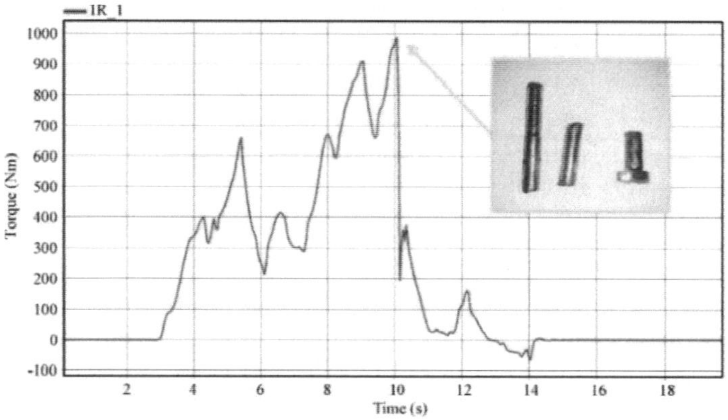

a)

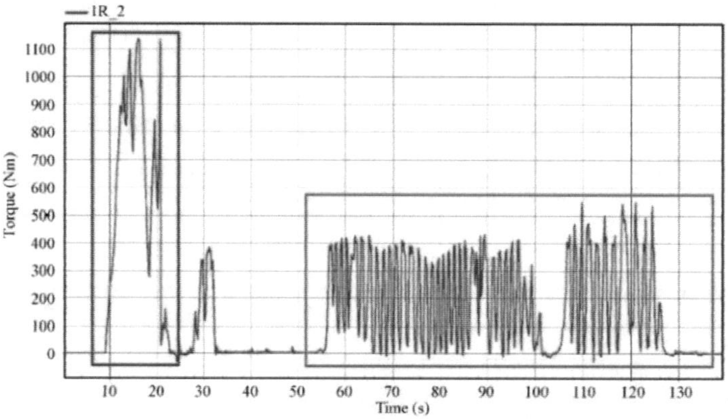

b)

Fig. 3. Time record of torque during manipulation of the 1R gate valve mechanism: (a) failure of the M8 bolt; (b) opening of the gate valve using the M10 bolt.

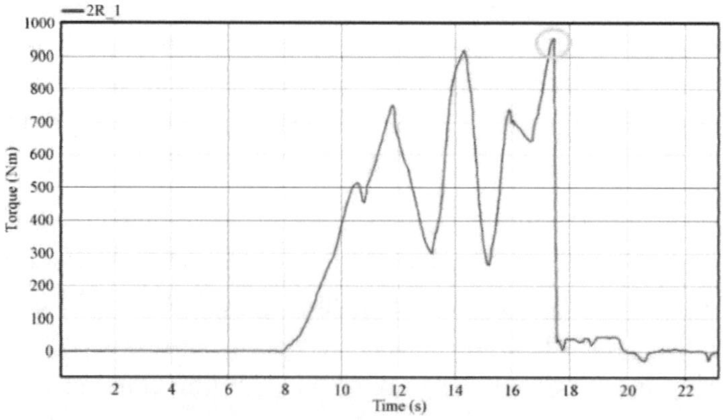

a)

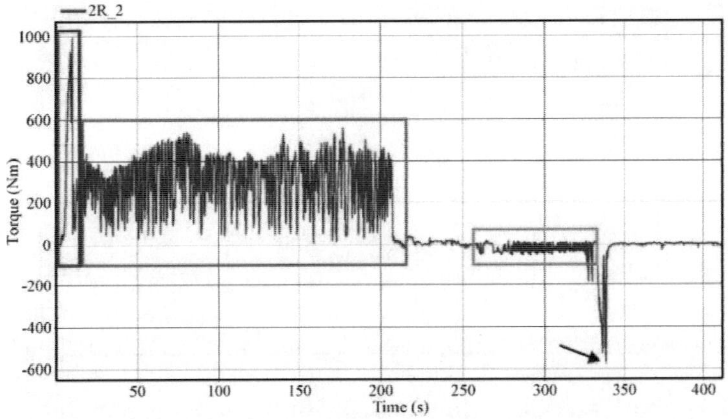

b)

Fig. 4. Time record of torque during manipulation of the 2R gate valve mechanism: (a) failure of the M8 bolt; (b) opening of the gate valve using the M10 bolt.

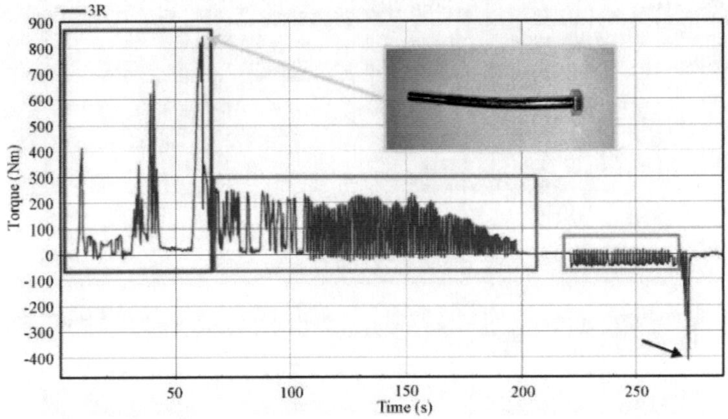

Fig. 5. Time record of torque during manipulation of the 3R gate valve mechanism.

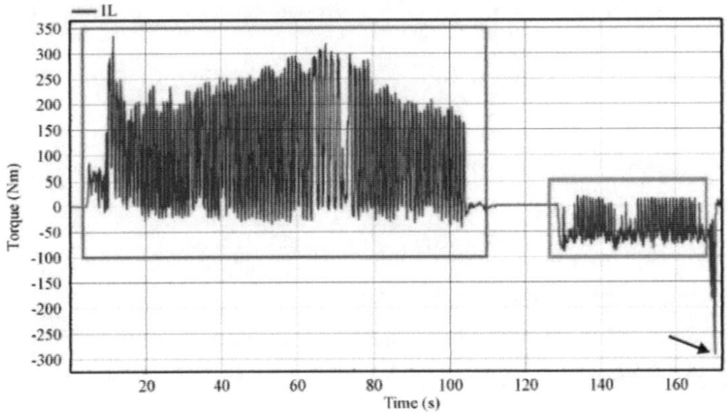

Fig. 6. Time record of torque during manipulation of the 1L gate valve mechanism.

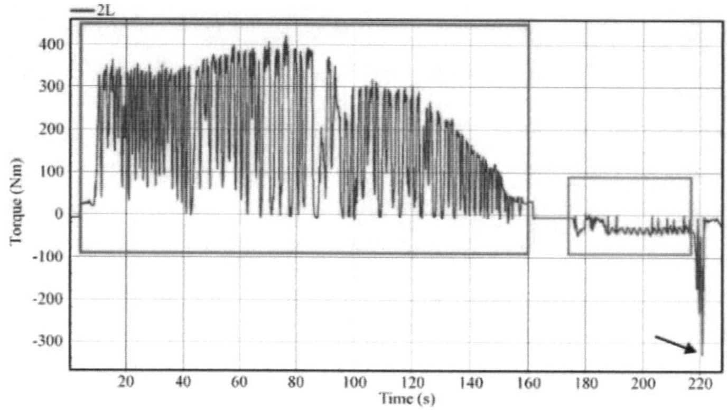

Fig. 7. Time record of torque during manipulation of the 2L gate valve mechanism.

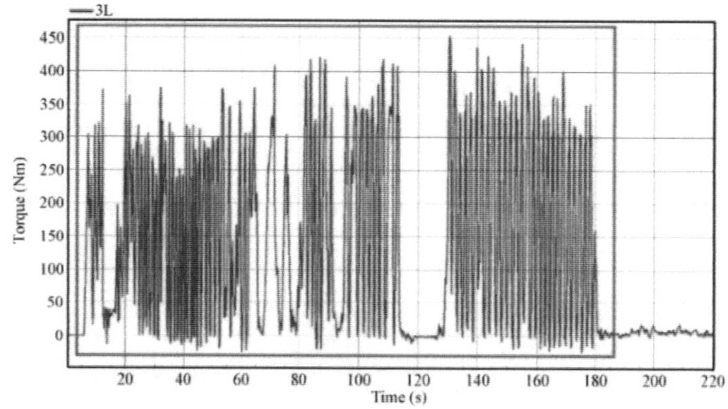

Fig. 8. Time record of torque during manipulation of the 3L gate valve mechanism.

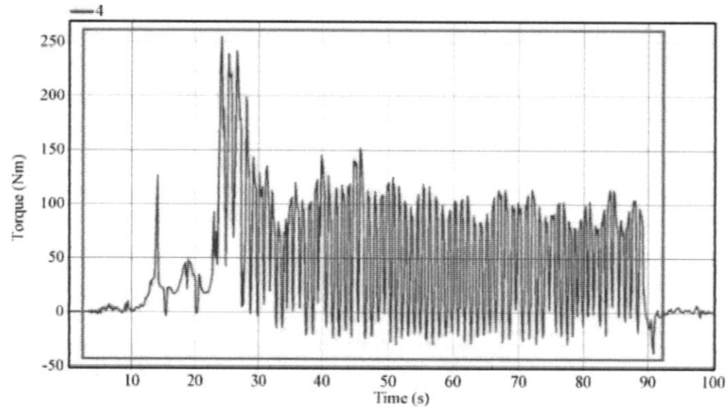

Fig. 9. Time record of torque during manipulation of the 4 gate valve mechanism.

Table 1 shows the maximum values of the torques obtained when opening or closing the closures under consideration. The torques confirm the influence of the manipulation of the gate, as seen by gate 4, which is the most frequently (regularly) used one.

Table 1. Maximum torque values observed when handling gate valves.

Mode	**Obtained value of maximum torque (Nm)**						
	1L	**2L**	**3L**	**1R**	**2R**	**3R**	**4**
Opening	334	420	450	1135*	995*	840	255

**M10 bolt of material grade 10.9 was used*

Discussion and conclusion

Based on the operational measurements carried out, the analysis of the material used, and the subsequent analysis, the following facts can be stated:

- The torque value of the hydraulic actuators of 400 Nm is insufficient for opening or closing the gate valves.
- The maximum torque value obtained during the operation was 1135 Nm. It is the peak value registered in the initial (release) phase of the gate opening when it was necessary to overcome mainly the resistances in the corroded universal joints. The torques were much lower in the subsequent phases after the mechanism started.
- The bar (tube) is made of stainless high-alloy Cr-Ni low carbon steel of likely grade W. Nr. 1.4306 (AISI 304L) with strength characteristics Rp0.2 = 299 MPa, Rm = 607 MPa. When loaded with a torque of 1135 Nm, shear stress is formed in the bar, which exceeds the value of the allowable shear stress but is significantly below the ultimate strength. Thus, the bar is loaded in the elastoplastic region at this stress. When loaded with a torque of 1000 Nm, shear stress is formed in the bar, slightly below the allowable shear stress. This load can be considered limiting when it only occurs in the first stages of the gate valve handling process.
- A visual inspection indicated that the universal joints were significantly corroded, which greatly increases the resistance to movement and requires high driving torques. It should be noted that for vertical manipulation of the gate valve, it is mainly necessary to overcome the friction forces in the mechanism. The experimental measurements' results confirm that the highest torque was only required when the gate valve was opened. Subsequently, the torque values decreased to more than half. From this point of view, paying due attention to the condition of the universal joints and regularly moving them as part of scheduled repairs or maintenance is crucial. Regular opening and closing of the gate valves during maintenance should be carried out in a mode where the gates are not loaded with water so that torque peaks do not occur in the initial stages. It will lead to a reduction in the initial resistance to movement and a reduction in the load on the individual structural elements in the motion chain of the gate valve.

Acknowledgements

This research was funded by Scientific Grant Agency of the Ministry of Education and Science of the Slovak Republic and the Slovak Academy of Sciences, grant numbers VEGA 1/0500/20 and VEGA 1/0516/22.

References

[1] Nilsson, C. "Global distribution of reservoirs," in *Encyclopedia of Inland Waters* (Elsevier, 2009), pp. 625–633.

[2] ICOLD (2020). World Register of Dams-General Synthesis. *Available from* https://www.icold-cigb.org/GB/world_register/world_register_of_dams.asp *Accessed:* 2023-04-12.

[3] Aman, S., Rezk, M.A. and Nasr, R. "Effect of tailwater depth on non-cohesive earth dam failure due to overtopping," *Ain Shams Engineering Journal* (2022), Preprint.

[4] Tabrizi, A.A., Elalfy, E., Elkholy, M., Chaudhry, M.H. and Imran, J. "Effects of compaction on embankment breach due to overtopping," *J Hydraul Res*, **55**, pp. 236-247 (2017).

[5] EN 10027-1:2005, Designation systems for steels – Part 1: Steel names.

[6] ISO 6892-1:2019, Metallic materials – Tensile testing – Part 1: Method of test at room temperature.

A Study of Strain Rate Effects on the Mechanical Response of PolyJet Printed Elastomer Composites

Abdelmadjid Boualleg[1,a] and David Cirkl[1,b]

[1]*Technical University of Liberec; Studentská 1402/2, Liberec, Czech Republic;*

[a]*abdelmadjid.boualleg@tul.cz*, [b]*david.cirkl@tul.cz*

Abstract: This article presents the experiments and finite element analysis on PolyJet Digital Material. The aim of this work is to demonstrate the mechanical properties of a mixture of two materials Agilus30 and VeroClear at various strain rates. The identification of the visco-hyperelastic material model is based on the shear experiment. The calibration model that has been used is the Parallel Rheological Framework model (PRF). One branch of PRF represented the hyperelastic material; the other branch of PRF defines nonlinear viscoelastic properties, known as the Bergstrom-Boyce (B-B) model. The material parameters were used as material input in Marc Mentat software for finite element analysis (FEA) to predict the compressive stress response of cylindrical samples under various strains. Subsequently, these FEA results were compared with the experimental data.

Keywords: PolyJet composite elastomer; Parameter calibration; Bergstrom-Boyce (B-B); Mechanical properties

1 Introduction

The PolyJet Matrix technology is a 3D printing technology that uses a jetting process to deposit droplets of photopolymer materials layer by layer to create a 3D object. One of the advantages of this technology is the ability to use two model photopolymers simultaneously, which can be mixed to create new materials with varying properties known as composite materials [1,2]. Polymeric materials are widely used in 3D printing and can be produced using various 3D printers with a broad range of properties. Soft polymer is widely utilized in additive manufacturing digital material robotics. These materials can be used in various industries, including robotics, as they enable the creation of soft and compliant robotic components [3,4].

Recent researches are focused on understanding and optimizing the mechanical properties of additive manufacturing materials including also computational modeling. This plays a crucial role in predicting and simulating the mechanical behavior of additively manufactured materials. Computational methods such as finite element analysis (FEA) were used to simulate the mechanical response, thermal properties, and other behavior of these materials [5,6].

For hyperelastic materials, the most commonly used constitutive model is the strain energy density function as a relation of the stress and strain of the material. The material parameters that define this function can be determined through a variety of experimental techniques, such as uniaxial tension or compression tests, biaxial tension tests, or shear tests [7-9]. The results of these experiments can be used to fit the strain energy density function to the experimental data, allowing the determination of the material parameters.

The aim of this work is to determine the mechanical properties of 3D printed Digital Material DM40 (Agilus30 and VeroClear). The material exhibits a remarkable nonlinear visco-hyperelastic response that depends on the strain and strain rates. The Bergstrom-Boyce (B-B) model was used to describe the nonlinear behavior of the material [10]. The identification of the visco-hyperelastic material parameters is based on the cyclic loading shear test experimental data of 3D-printed samples.

2 Experimental methods

The Stratasys500 was used to print the specimen, the printed digital material referred to as DM40 is a mixture of two different materials, Agilus30 and VeroClear. The print direction is the vertical z-direction and the Infill percentage of the material is 100%.

2.2 Shear cyclic test

The specimen dimension (length 20 mm, width 20 mm, and height 5 mm). The INSTRON Electropuls E3000 testing machine with a load cell of 1 kN was utilized for the shear test. The load-unload shear tests were performed on the specimen and then the last cycle was held for relaxation. The applied different strains as followed 0.3, 0.4, and 0.5 with various strain rates: 0.2 s^{-1}, 0.4 s^{-1} and 0.6 s^{-1}, respectively, see Fig. 1a. The experiments were repeated three times by allowing the sample to rest for 10 minutes before the next test. The temperature chamber during the test was set to 30° C, see Fig. 1b.

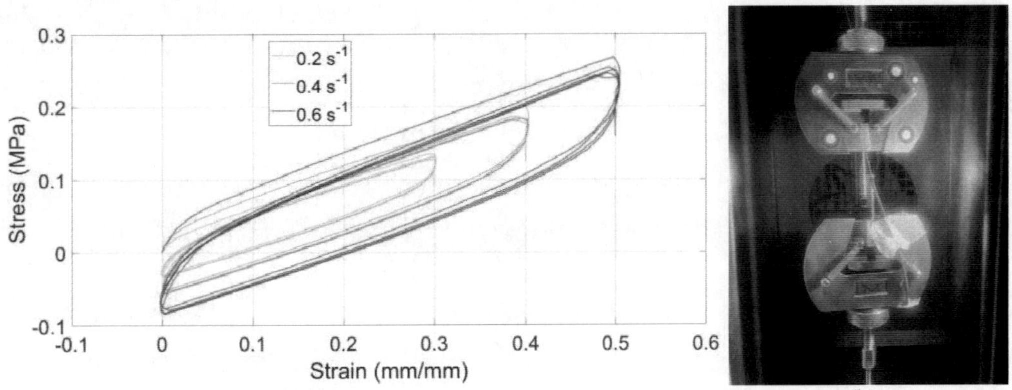

(a) (b)

Fig. 1 (a) Shear cyclic test under various strain and strain rates, (b) shear test.

2.3 Uniaxial compression test

The uniaxial compression tests were performed on the cylinder sample (diameter 17.8 mm, height 25 mm). The samples were compressed to 50% of the hight with different strain rates of 0.4 10^{-2} s^{-1}, 0.2 s^{-1}, 0.4 s^{-1}, and 0.6 s^{-1}. The experiments were repeated three times and the average was calculated. The results are presented in Fig. 2. The temperature chamber during the test was set to 30° C because the material stiffness is affected by the small change in the temperature and also the need to investigate the material to be softer (flexible). The upper compression plate and lower compression plate were lubricated to reduce the barrelling effect of the samples.

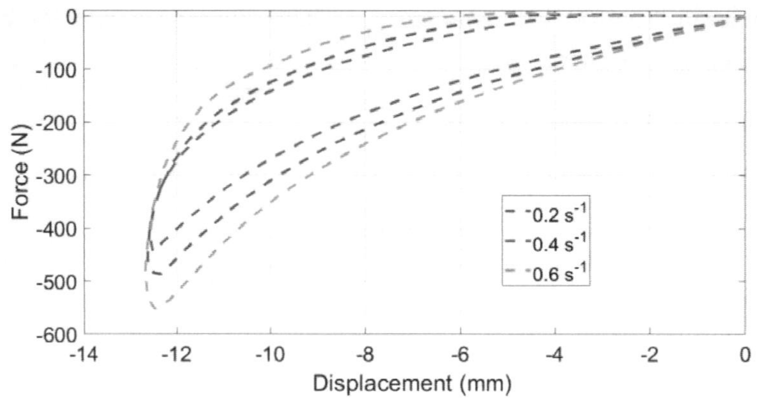

Fig. 2 Uniaxial compression at different strain rates.

3 Constitutive modelling and equation

The Parallel Rheological Framework was used to model the response of the material. The constitutive material modelling includes decoupling a material's behavior into different parallel chains, each with a given individual constitutive behavior, see Fig. 3.

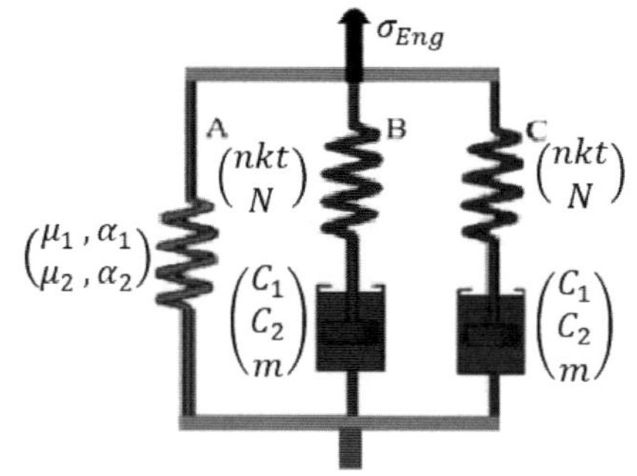

Fig. 3 Three-unit Parallel Rheological Frameworks (Bergstrom 2015).

3.1 Hyperelastic model

The Ogden model that has been selected in this study was fitted by shear experimental data. The strain energy is presented in Eq 1.

$$W = \sum_{i=1}^{N} \frac{\mu_i}{\alpha_i}\left(\lambda_1^{\alpha_i} + \lambda_2^{\alpha_i} + \lambda_3^{\alpha_i} - 3\right) + \frac{K}{2}(J-1), \tag{1}$$

where λ_1, λ_2, and λ_3 are principal stretches, μ_i are moduli and α_i are non-dimensional material constants. J and K are the elastic volume ratio and the material bulk modulus respectively.

3.2 Eight-Chain hyperelastic model

The Eight Chain model considers eight orientations of chains in space, which can be visualized using the eight-chain network system. This system is a representation of how polymer chains are arranged in a specific configuration, the function is expressed as a series of appropriate terms in expansion Eq. 2. [11]

$$\mathcal{W}_{EC} = nkt\left[\frac{1}{2}(I_1 - 3) + \frac{1}{20N}(I_1^2 - 9) + \frac{11}{1050N^2}(I_1^3 - 27) + \frac{19}{7000N^3}(I_1^4 - 81) + \frac{519}{673750N^4}(I_1^5 - 243)\right], \tag{2}$$

where *nkt* represents initial modulus and *N* is limiting chain extensibility are the model parameters, I_1 also corresponds to the first invariant of the right Cauchy-Green deformation tensor.

3.3 Nonlinear viscoelastic element of Bergstrom-Boyce model

The B-B model is used to describe the time-depending behavior of elastomeric materials for large strains. The flow rate rule of the nonlinear viscous element is presented in Eq. 3.

$$\dot{\gamma}=c_1[\lambda^i_{chain}-1]^{c_2} \parallel \tau^V_{ISO} \parallel^m , \quad (3)$$

where C_1, C_2 and m are material parameters, λ^i_{chain} is the inelastic network stretch, and τ^V_{ISO} is the deviatoric Kirchhoff stress representing viscous behavior.

4. Material parameter calibration results

The shear cycle test results were used for fitting Bergstrom-Boyce (B-B) model and the hyperelastic and viscoelastic material constants were determined. The fitting results are presented in Tables 1 and 2. The total parameters are 14. The two-terms Ogden model, represented by chain A, is identified by constants μ_1, μ_2, and α_1, α_2. The constants of nonlinear viscous elements are then *nkt*, *N*, C_1, C_2, m for chain B and C respectively.

Table 1. Ogden Model constants.

Term *i*	Moduli μ_i	Exponents α_i
1	0.0566	0.274
2	0.293	2.090

Table 2. Viscoelastic Model constants.

Viscoelastic parameters	Bergstrom-Boyce (Chain B)	Bergstrom-Boyce (Chain C)
nkt	0.152	2.9077
N	3.559	27.262
C_1	1.1251	4.798
C_2	-0.1027	-0.0554
m	0.8924	0.961

The material model identified in this way was used for simulation of compression test and results were compared with experimental data. This comparison shows a very good correlation as it is seen in Fig. 4 (a), (b) and (c) where nonlinear force-displacement characteristics are displayed.

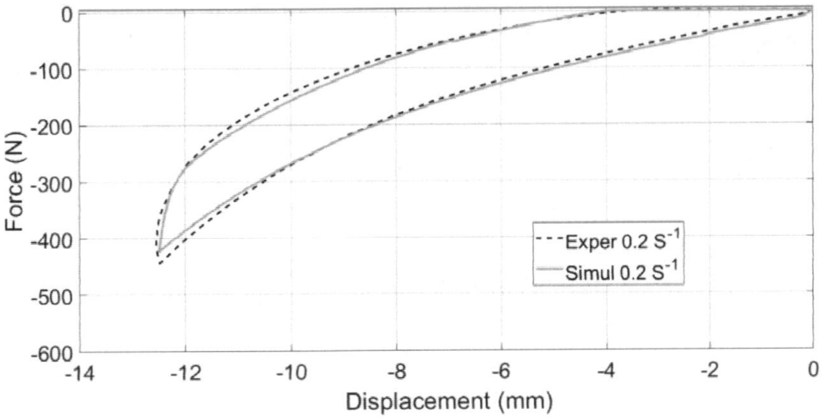

(a)

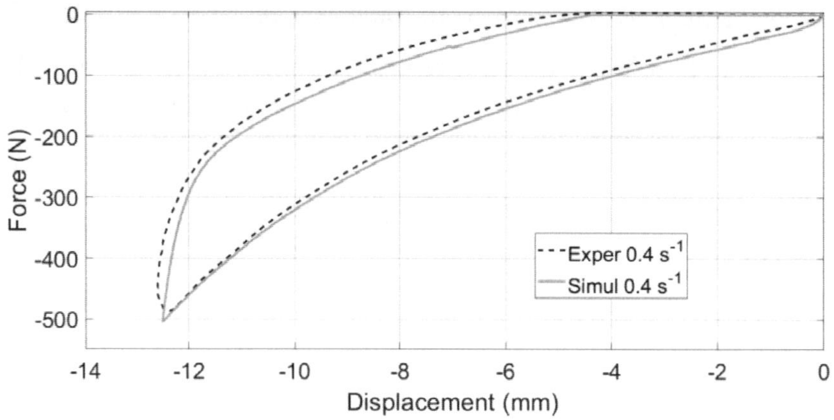

(b)

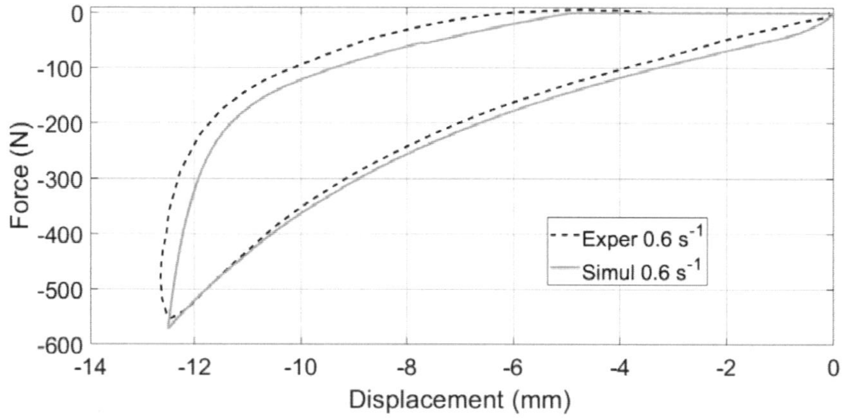

(c)

Fig.4 Experiment validation with FEA at different strain rates.

Conclusion

This study presents the mechanical response of new composite PolyJet material Agulus30 and VeroClear and the description of the hyperelastic and nonlinear viscoelastic properties. The cycle shear experiments were performed to determine the parameters of the hyperelastic and nonlinear viscoelastic model. The validation of the model was verified by a compression test with cylindrical sample. The comparison between experimental data and simulation results, in general, shows a good agreement, which is confirmed the accuracy and applicability of the PRF model.

Acknowledgment

This publication was written at the Technical University of Liberec as part of the project "Research of advanced materials, and application of machine learning in the area of control and modelling of mechanical systems" nr. SGS-2022-5072 with the support of the Specific University Research Grant, as provided by the Ministry of Education, Youth and Sports of the Czech Republic in the year 2022.

References

[1] Gaynor, A. T., Meisel, N.A., Williams, C.B. and Guest, J.K. “Multiple-Material Topology Optimization of Compliant Mechanisms Created Via PolyJet Three-Dimensional Printing,” *J Manuf Sci Eng*, **136**(6), pp. 1-11 (2014).

[2] Bass, L., Meisel, N.A. and Williams, C.B. "Exploring variability of orientation and aging effects in material properties of multi-material jetting parts," *Rapid Prototyp J*, **22**(5), pp. 826-834 (2016).

[3] Case, J.C., White, E.L. and Kramer, R.K. "Soft material characterization for robotic applications," *Soft Robot*, **2**(2), pp. 80-87 (2015).

[4] Hossain, M., Navaratne, R. and Perić, D. "3D printed elastomeric polyurethane: Viscoelastic experimental characterizations and constitutive modelling with nonlinear viscosity functions," *Int J Non Linear Mech*, **126**, 103546 (2020).

[5] Campoli, G., Borleffs, M.S., Amin Yavari, S., Wauthle, R., Weinans, H. and Zadpoor, A.A. "Mechanical properties of open-cell metallic biomaterials manufactured using additive manufacturing," *Mater Des*, **49**, pp. 957-965 (2013).

[6] Abayazid, F. F. and Ghajari, M. "Material characterisation of additively manufactured elastomers at different strain rates and build orientations," *Addit Manuf*, **33**, p. 101160 (2020).

[7] Szurgott, P. and Jarzębski, Ł. "Selection of a hyper-elastic material model – A case study for a polyurethane component," *Lat Am J Solids Struct*, **16**(5), pp. 1-16 (2019).

[8] Adams, R., Soe, S.P., Santiago, R., Robinson, M., Hanna, B., McShane, G., Alves, M., Burek, R. and Theobald, P. "A novel pathway for efficient characterisation of additively manufactured thermoplastic elastomers," *Mater Des*, **180**, p. 107917 (2019).

[9] Wei, W., Yuan, Y. and Gao, X. "Effects of Large Deformation and Velocity Impacts on the Mechanical Behavior of Filled Rubber: Microstructure-Based Constitutive Modeling and Mechanical Testing," *Polymers*, **12**(10), 2322 (2020).

[10] Bergström, J.S. and Boyce, M.C. "Constitutive modeling of the large strain time-dependent behavior of elastomers," *J Mech Phys Solids*, **46**(5), pp. 931-954 (1998).

[11] Arruda, E.M. and Boyce, M.C. "A three-dimensional constitutive model for the large stretch behavior of rubber elastic materials," *J Mech Phys Solids*, **41**(2), pp. 389-412 (1993).

Residual Stress Analysis of Additively Manufactured AlSi10Mg Alloy

Jiří Čapek[1,a], Karel Trojan[1,b], Radim Halama[2,c], Nikolaj Ganev[1,d], Jiří Hajnyš[2,e], Michal Kořínek[2,f] and Kamil Kolařík[1,g]

[1]*Czech Technical University in Prague; Trojanova 13, 120 00 Prague, Czech Republic;*
[2]*VSB - Technical University of Ostrava; 17. listopadu 2172/15, 708 00 Ostrava - Poruba, Czech Republic;*

[a]*jiri.capek@fjfi.cvut.cz*, [b]*karel.trojan@fjfi.cvut.cz*, [c]*radim.halama@vsb.cz*, [d]*nikolaj.ganev@fjfi.cvut.cz*, [e]*jiri.hajnys@vsb.cz*, [f]*michal.korinek@vsb.cz*, [g]*kamil.kolarik@fjfi.cvut.cz*

Abstract: The paper deals with the residual stress analysis of additively manufactured AlSi10Mg alloy. The influence of the printing strategy of SLM technology was found. The inhomogeneities of the residual stresses around the circumference of the test sample were analysed, moreover; only the tensile residual stresses were determined on the surface of as-built samples. Zones of steep changes in residual stresses (mainly tensile) can be a potential area for crack initialisation. Inherent anisotropy was studied by tracking the yield surface, from these the yield strength of 160 MPa was determined, moreover; the decrease in tensile residual stresses was estimated.

Keywords: Additive manufacturing; X-ray diffraction; Residual stresses; AlSi10Mg alloy; Inherent anisotropy

1 Introduction

Additive Manufacturing (AM), precisely Selective Laser Melting (SLM) technology, is a promising metal powder consolidation method and offers outstanding parts production opportunities. It is based on selectively melting parts of a thin flat powder bed in layers using a scanning energy source to produce 3D parts. The intricacy of the SLM process results in the magnitude and orientation of residual stresses (RS) being highly dependent on laser power, scanning speed, scanning strategy, and other processing parameters [1].

The effect of RS manifests itself in different ways. In particular, high tensile RS can lead to deformation and subsequent cracks in printed parts, which can disrupt the overall strength of the part. They can also affect the geometry of the printed object. Some shapes and structures are more sensitive to RS than others, especially in the horizontal printing strategy. Generally, tensile RS in the sub/surface layer reduce mechanical as well as corrosion properties. Furthermore, the notch toughness, fatigue resistance, and wear resistance properties are reduced; moreover, such factors as crack propagation rate, corrosion cracking, and intergranular corrosion are accelerated.

The initiation of fatigue cracks and their propagation play a significant role in fatigue properties with a strong dependence on surface roughness, microstructure parameters (dislocation density, crystallite size, microstrains), and RS [2]. In the study [3], the explicit correlation between the initialisation of surface fatigue cracks and its propagation with RS has been investigated using X-ray diffraction. It was found that not only RS but also the distribution of microstructure parameters play a crucial role in fatigue. In a previous study [4], a correlation was observed between the relaxation of structure-sensitive characteristics and the formation of surface fatigue cracks for welded materials. The basis of this investigation will be applied to this research, which deals with the study of additively manufactured AlSi10Mg.

2 Materials and methods

For the fabrication of AlSi10Mg, for chemical composition see Table 1, tubular samples with 12.5 mm outer diameter and 10 mm inner diameter from a recycled powder Renishaw AM400 machine was used with the printing parameters shown in Table 2. During fabrication, the chamber was filled with argon with purity 5.0 to prevent oxidation and the building platform was preheated to a temperature of 170 °C. Five samples were prepared according to the printing strategy (slope related to the Z axis); 0°, 10°, 20°, 30°, and 40°. No mechanical or heat treatment was done on the printed samples to maintain a fine-grained microstructure from SLM technology, therefore; as-built samples were analysed.

Table 1. Chemical composition of AlSi10Mg alloy.

Element	Composition [wt.%]
Si	9.0–11.0
Mg	0.20–0.45
Fe	≤0.55
Cu, Mn, Zn, Ni, Pn, Sn	each ≤0.1
Ti	≤0.15
Al	Bal.

Table 2. Printing parameters.

Parameter	
Laser power	350 W
Spot size	80 μm
Layer thickness	30 μm
Printing strategy	Meander
Laser speed	1150 mm/s

Macroscopic residual stresses (RS) were determined using an Empyrean diffractometer by X-ray diffraction (XRD) when the *{311}* diffraction line of the aluminium phase was measured using chromium radiation in the direction parallel to the rotation axis of the sample (axial direction). The diffraction angles $2\theta^{hkl}$ were determined from the diffraction lines $K\alpha_1$ of the planes *{311}* of Al phase. Diffraction lines $K\alpha_1$ were fitted using the *Pearson VII* function and Rachinger's method was used for the separation of the diffraction lines $K\alpha_1$ and $K\alpha_2$. For RS determination, the $\sin^2\psi$ method and X-ray elastic constants $\frac{1}{2}s_2 = 19.05$ TPa^{-1} and $s_1 = -4.89$ TPa^{-1} were used. The experimental errors given for each value are the standard deviations of the calculation of the RS [5]. The irradiated volume was defined by the experiment geometry, the effective penetration depth of the X-ray radiation (approx. 6–11 and 5–20 μm), and the pinhole size (1×0.5 and 3×0.5 mm^2) for chromium and cobalt radiation, respectively.

The tracking of the yield surface is a semi-destructive and online method of determining the specific yield stress. Using this method, we can determine the extent of the elastic region under any combination of tensile and torsional loads on a single sample [6]. This method is based on the deviation from linearity in the form of the effective increase in plastic deformation, which is given by Eq. (1)

$$\Delta\varepsilon_{eff} = \sqrt{\Delta\varepsilon_{pl}^2 + \frac{1}{3}\Delta\gamma_{pl}^2}, \tag{1}$$

where $\Delta\varepsilon_{pl}$ and $\Delta\gamma_{pl}$ are increments of plastic deformation in the axial and torsional directions. The plasticity criterion is met as soon as $\Delta\varepsilon_{eff}$ reaches the prescribed value, which was 50 με in the following experiments. A total of 16 combinations of loading paths were used to determine the elastic area, as in [7]. Thin-walled samples with geometry for low-cycle fatigue [8] are used for these tests. The testing equipment for the tracking tests was the LabControl biaxial hydraulic machine with a maximum load of 100 kN/1000 Nm, and the test was stress controlled at a constant rate of 1 MPa/s. The sample that was tilted by 40° from the vertical axis during printing was tracked. Tracking was repeated a total of three times to ensure statistical significance.

3 Results and discussions

Using the X-ray diffraction phase analysis, three phases were identified in the irradiated volume: approx. 85 wt.% of Al, 15 wt.% of Si, and 0.1 wt.% of intermetallic phase Mg_2Si (according to Rietveld refinement method). It is necessary to mention that, according to the chemical composition, the microstructure should be hypoeutectic, i.e., α(Al)+Si+Mg_2Si, in other words: solid solution of Al, eutectic Al+Si, and intermetallic Mg_2Si. The average crystallite size of Al and Si is 122 and 16 nm, respectively, which correlates with the fact that the matrix is coarser-grained and eutectic fine-grained in hypoeutectic alloy. Moreover, the results from Al phase are influenced by both Al, in solid solution and eutectoid too.

3.1 Residual stresses around the circumference

The distribution of the RS (Al phase) was analysed around the outer circumference of the tubular in the axial direction; see Figs. 1–2. The 0° position within the circumference was pointing down to the building platform, i.e. the printing supports side.

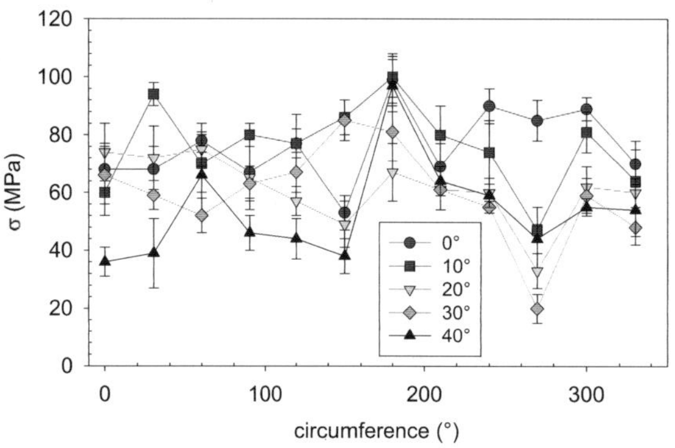

Fig. 1 Distribution of residual stresses around the circumference depending on the printing slope (related to Z axis).

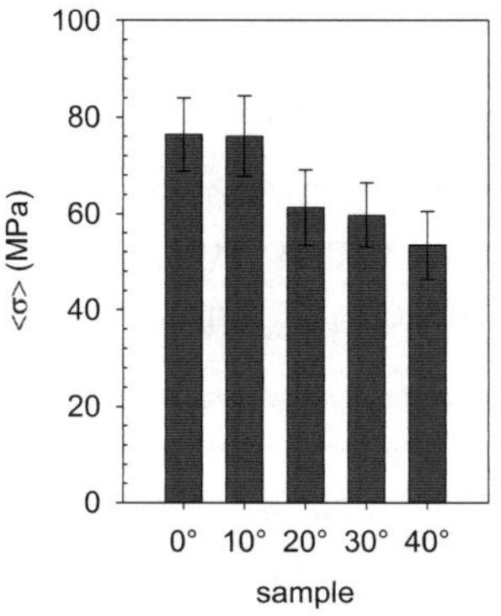

Fig. 2 Average residual stresses around the sample circumference depending on the printing slope.

The circuit distribution of RS of Al phase of all samples is inhomogeneous, moreover; all RS are tensile; see Fig. 1. The RS maxima were systematically detected at 180°, which is opposite to the printing support side. On the other hand, the minima of tensile RS were found at 270°. This effect is clearly related to the printing strategy, i.e., the inclination of the sample axis to the building platform. According to Fig. 2, the average RS around the sample circumference have a decreasing character with increasing printing slope. However, samples with higher slope have higher inhomogeneities around the sample circumference.

The circuit inhomogeneities should have a negative influence on the mechanical properties such as fatigue life, where significant changes in RS create so-called surface microstructure notches which are areas prone to the formation and propagation of fatigue cracks.

3.2 Tracking of the yield surface

The size of the elastic region is shown in Fig. 3. The effect of tracking on the distribution of RS is depicted in Fig. 4.

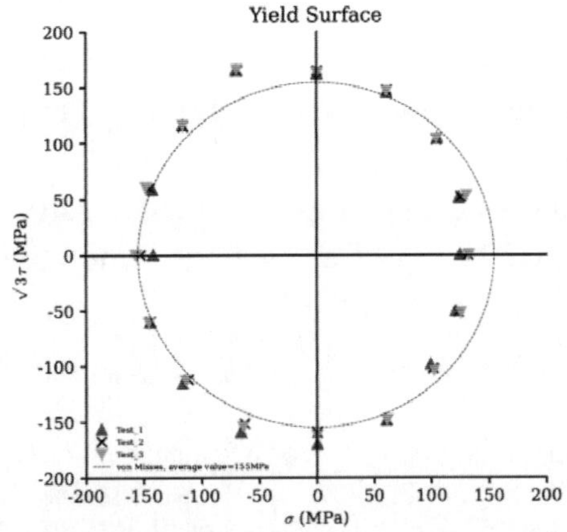

Fig. 3 The size of the elastic region for the sample printed with slope of 40°.

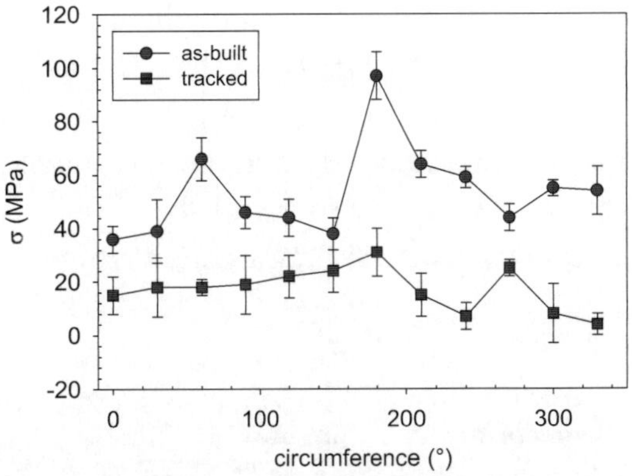

Fig. 4 Comparison of residual stresses of tracked and as-built 40° sample.

The measured data were fitted with the von Mises condition with a yield strength of an average value of 160 MPa (corresponding to 0.005% of plastic strain). The same value was evaluated from a tensile test. The yield strength (corresponding to 0.2% of plastic strain) is given by around 255 MPa. Compared to the maximum value of RS (Fig. 1), which is 100 MPa for the aluminium phase, it is possible to state that in some surface RS reach 63% of the yield strength and, therefore; 60 MPa of external tensile stress will result in plastic deformation. The higher yield strength in compression, which is obvious from Fig. 3, is the confirmation of the tensile residual stress observed by XRD analyses. Nevertheless, the results from Al phase are influenced by both Al, in solid solution and eutectoid too, where one of the Al phase should be harder and, therefore; one Al phase should be in elastic and the second in the plastic region.

Tracking of the yield surface leads to cyclic plastic deformation, which reduces the average crystallite size from values of 122 and 16 nm to 116 and 14 nm for Al and Si phases, respectively. The RS of Al phase were redistributed/reduced and homogenized their values; see Fig. 4.

Conclusion

The slope of the printing (printing strategy) does not affect the phase composition, nevertheless; it affects the values of RS and their inhomogeneities around the circumference of tubular samples. Moreover, all analysed RS were tensile. The difference between the maximum and minimum values around the circumference reaches up to 60 MPa. These inhomogeneities, the so-called surface microstructure notches (step changes in RS and/or microstructure parameters) are prone to fatigue crack formation, which has a major influence on fatigue life.

The worst low-cycle fatigue life was found for the 0° sample (results will be the subject of another article), probably due to the presence of pores in the critical layer parallel to the printing platform. Furthermore, the decrease in fatigue life can be attributed to the stress inhomogeneities of the RS of the 40° sample.

The yield strength of 160 MPa of the 40° sample was determined by tracking. It can be assumed that this value is lower than for conventional-made samples because it is related to the real structure of the SLM printed samples, i.e., the presence of pores, printing layers, etc.

Acknowledgements

Measurements were supported by the project 23-05338S of the Czech Science Foundation. The work of CTU staff was supported by the grant number CZ.02.1.01/0.0/0.0/16_019/0000778 "Center for Advanced Applied Science" within the Operational Program Research, Development and Education supervised by the Ministry of Education, Youth and Sports of the Czech Republic and the Grant Agency of the Czech Technical University in Prague, grant No. SGS22/183/OHK4/3T/14.

References

[1] Bartlett, J.L. and Li, X. "An overview of residual stresses in metal powder bed fusion," *Additive Manufacturing*, **27**, pp. 131-149 (2019).

[2] Schneller, W., Leitner, M., Pomberger, S., Springer, S., Beter, F. and Grün, F. "Effect of post treatment on the microstructure, surface roughness and residual stress regarding the fatigue strength of selectively laser melted AlSi10Mg structures," *Journal of Manufacturing and Materials Processing*, **3**(4), pp. 89 (2019).

[3] Čapek, J., Trojan, K., Kec, J., Černý, I., Ganev, N. and Němeček, S. "On the Weldability of Thick P355NL1 Pressure Vessel Steel Plates Using Laser Welding," *Materials*, **14**(1), pp. 131 (2020).

[4] Černý, I., Čapek, J., Kec, J., Trojan, K., Ganev, N. and Němeček, S. "Fatigue of Laser-Welded Structures: Role of Residual Stress," in *Comprehensive Structural Integrity*, Aliabadi F. and Soboyejo W., ed. (Elsevier Science, 2023), ISBN 9780128229446.

[5] Kraus, I. and Ganev, N. Technické aplikace difrakční analýzy (ČVUT, Praha, 2004).

[6] Štefan, J., Parma, S., Marek, R., Plešek, J., Ciocanel, C. and Feigenbaum, H. "Overview of an Experimental Program for Development of Yield Surfaces Tracing Method," *Applied Sciences*, **11**(16), 7606 (2021).

[7] Gil, C.M., Lissenden, C.J. and Lerch, B. A. "Determination of Yield in Inconel 718 for Axial-Torsional Loading at Temperatures up to 649 C," *NASA/TM*, pp. 1-33 (1998).

[8] ASTM E2207 - 15 Standard Practice for Strain-Controlled Axial-Torsional Fatigue Testing with Thin-Walled Tubular Specimens. West Conshoshocken (USA: ASTM International, 2015).

Bird Strike Tests Verification of Simulation for Development of Composite Inlet for Tilt-rotor Aircraft

Radek Doubrava[1,a], Martin Oberthor[1,b] and Petr Bělský[1,c]

[1] *VZLÚ – Czech Aerospace Research Centre, Beranových 130, Prague 9, 199 05, Czech Republic;*

[a] *doubrava@vzlu.cz*, [b] *oberthor@vzlu.cz*, [c] *belsky@vzlu.cz*

Abstract: Bird strikes are an important phenomena that must be considered when designing aircraft. Most major bird strike incidents result in engine damage. Because an engine is the sole thrust-providing system of an aircraft, the effect of bird strikes on engine inlets and systems must be investigated and mitigated to the maximum extent. Especially in the case of convertible aircraft, such as an aircraft with tilting rotors, this effect is critical from the point of view of the operation, from the point of view of flight mechanics, and the overall control of the aircraft. International certification regulations require that all forward-facing aircraft components be proven to withstand bird strikes to a certain level before they can be used in an aircraft. A bird impact test provides a direct method for determining bird strike resistance; however, the design of aircraft structures typically involves numerous iterations, from design to manufacturing to testing and back, requiring that many bird impact tests be conducted. This is not only time-consuming but also costly. Furthermore, the experimental data from these tests are frequently narrowly focused, acting as a barrier to their direct use in refining structural designs. Owing to these shortcomings, several numerical methods have been developed to simulate bird or hailstone strikes to reduce the number of intermediate tests required and subsequently shorten the duration of the component design phase. The paper aims to experimentally validate numerical models for verifying composite air duct design for new tilt-rotor aircraft using flat and simple curved test panels.

Keywords: Bird strike; Impact; Composite

1 Introduction

Foreign object impacts, such as birds, hail, debris, etc., are important phenomena that must be taken into consideration when designing aircraft. The critical parts of planes or helicopters are windshield, nose, fuselage panels, wing and empennage leading edges, rotor blades, fan blades, engine cowlings, and inlets [1].

International certification regulations require that all forward-facing aircraft components should be proven to withstand bird strikes to a certain level before they can be employed in an aircraft [2]. A bird impact test provides a direct method for determining bird strike resistance; however, the design of aircraft structures typically involves many iterations from design to manufacturing to testing and back, requiring that many bird impact tests be conducted. These empirical verifications, which cause damage to expensive prototypes or part of structure [3-5], and the biological hazard of using real birds, can be costly and time-consuming. Furthermore, experimental data from these tests are often narrowly focused, constituting a barrier to their direct use in refining structural design. Owing to these shortcomings, several numerical methods based on CEL (Coupled Eulerian-Lagrangian) [6] or SPH (Smoothed Particle Hydrodynamics) [7-9, 15] have been developed to simulate bird strikes to reduce the number of intermediate tests required and subsequently shorten the duration of the component design phase. More bird models are verified only on isotropic material or impact on the rigid wall [10,11].

The principal objective of the present work is to provide a more universal experimental procedure, applicable e.g., for a different type of material or technology production, for verification of a numerical model for bird strike analysis on the real composite structure. The verified numerical simulation will allow the design to be optimized for the next to satisfy certification on final design respectively expensive prototypes.

The testing program was established to assist in the selection of composite material and optimised lay-up from the point of view of energy absorption from bird impacts on the composite inlet for a new generation of tilt-rotor aircraft. The flat and simple curved test specimen used in the high-speed impact resistance verification tests was designed to provide confirmation of the performance of the selected composite material and to assist in the finite element modelling of the global structure [13]. Fig. 1 shows the building block diagram for the proposed approach.

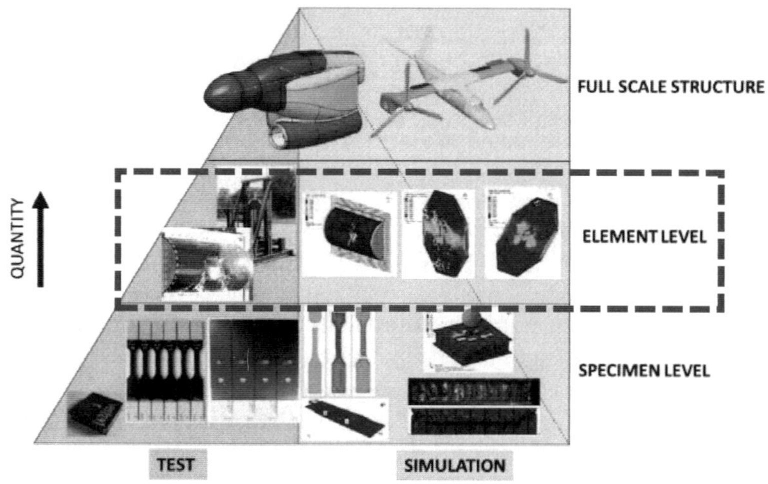

Fig. 1 Building block diagram (BBD) for bird strike resistance analysis of composite inlet.

2 Materials and methods

The physical bird impact tests were performed at the Czech Aerospace Research Centre (VZLÚ) according to airworthiness requirements [1]. The real bird projectiles (chicken) were used during the experimental verification according to airworthiness requirements [14]. The weight of the bird projectile was 1 kg [2]. The projectiles were accelerated by compressed air through the smooth borehole of a gun barrel up to the required velocity according to specifications. Fig. 2 shows the air gun test facilities.

Fig. 2 VZLU air gun for bird strike tests.

The flat and curved panels were defined in dependency on the preliminary design of the final part, bird mass, and impact speed range. The flat test specimens represented monolithic and sandwich structure designs (see Fig. 3). The simple curved test specimens represented monolithic structure design (see Fig. 3). The partner of TRIcEPS project Composite Impulse GmbH & Co manufactured the flat and curved test specimens.

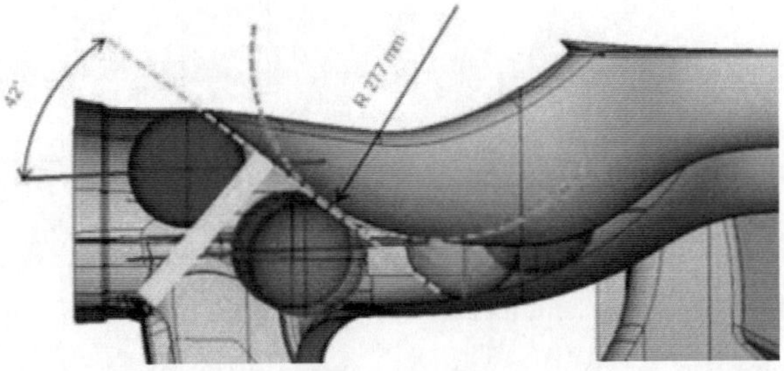

Fig. 3 Analysis of bird strike on the critical parts of the inlet for design of replacement test panels.

The VZLU test rigs designed from the point of view of numerical boundary conditions were used for the attachment and instrumentation of the tests (Fig. 4) [13].

Fig. 4 VZLU test rigs to attached of curved (left) and flat (right) test specimen for bird strike tests.

For the flat test specimen, it was realised three 42° and three 52° declined impacts with a range of impact speed from 307 to 428 km/h. For simple curved test specimens, it was realised 3 bird strike tests with a range of impact speeds from 100 to 204 km/h. The speed of impact was defined on the result of the previous test from the point of view of panel damage or failure. The main criterion was barely visible damage.

Standard qualitative techniques for high-speed impact analysis, such as high-speed cameras and post-test non-destructive techniques (NDT), were used for the analysis of the test results and verification of the numerical model. A technique of dynamic displacement measurement by using a laser triangulation sensor during the impact process was used for quantitative analysis and a more precise verification of the data obtained from the numerical model (Fig. 5) [13]. The measurement range of the laser sensors was 300 mm (centre of test specimen) and 40 mm (test rig frame). The actual bird impact point was then determined via the high-speed camera record and visual inspection. The measurement data acquisition rate is adjustable over a range from 1.5 to 49.1 kHz and was set at 20 kHz during testing. The resolution of the sensor was 0.3 μm.

Fig. 5 Displacement measurement during the bird strike test.

The video images were recorded by three high-speed cameras; the two cameras, recorded the course of the impact from the front side and backside, and one camera was used to measure the projectile speed.

The reaction force during the bird strike of curved panels was measured by load cells (see Fig. 6). Three load cells of KISTLER 9105A type were used in the test rig. The signal from load cells was stored using BMC Messsysteme GmbH data acquisition system completed with Vishay 2230 type amplifier. The measurement frequency was 10 kHz with 50,000 sample/s for load cells.

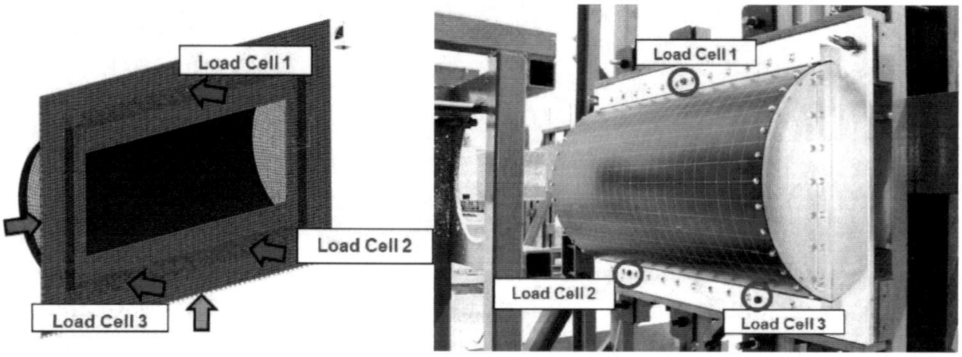

Fig. 6 Measurement of reaction force during bird strike on curved panels by load cells.

NDT of bird-strike test panels was performed before and after bird-strike tests. Monolithic test panels were inspected by visual and ultrasonic methods by immersion C-scan (UT-PE method) (see Fig. 7). Visual and mechanical impedance inspection were used for the NDT of the sandwich test sample.

Fig. 7 Example of C-scan inspection of flat test panel in VZLU immersion tank.

3 Test result analysis

For the verification of numerical analyses, the experimental results were evaluated in terms of qualitative and quantitative parameters:

- images from high-speed cameras,
- visual and NDT inspection before and after the tests,
- displacement measurements using the laser triangulation sensor (flat panels),
- reaction force measurement (curved panels).

3.1 Flat test specimen

The flat test panels represented the real structure in the area of oblique impact on the part of the air duct and the filter door. The experiments confirmed the higher impact resistance of the monolithic construction compared to the sandwich construction of the filter door.

Fig. 8 - Fig. 10 show examples of high-speed camera picture analysis from the point of view of test specimen displacement and failure initialization.

Fig. 8 Example of high-speed camera pictures during the bird-strike test on the monolithic composite test panel for impact speed 403 km/h.

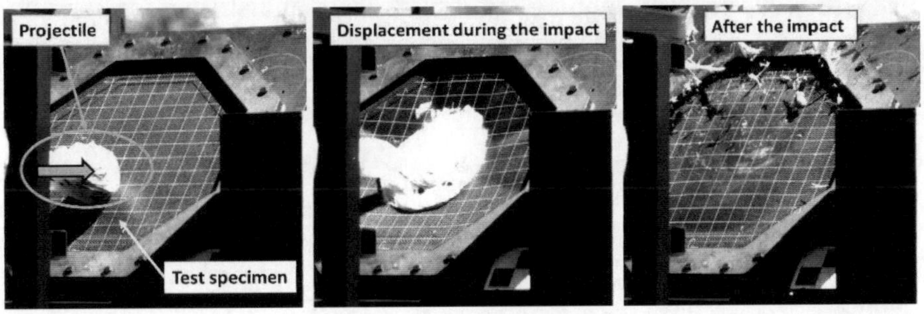

Fig. 9 Example of high-speed camera pictures during the bird-strike test on the sandwich composite test panel for impact speed 332 km/h.

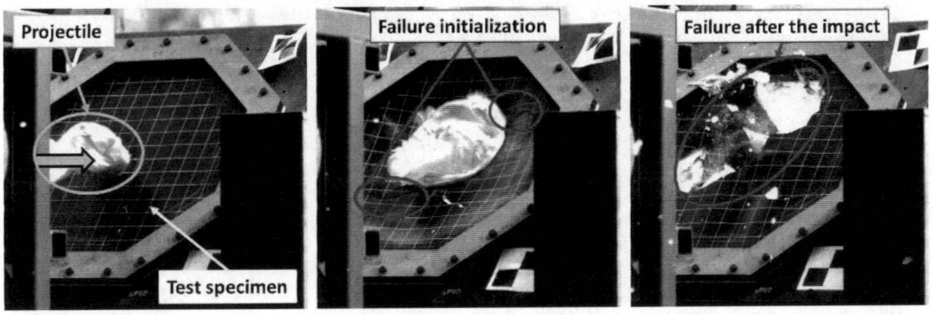

Fig. 10 Example of high-speed camera pictures during the bird-strike test on the sandwich composite test panel for impact speed 392 km/h.

Fig. 11 shows the result of the experimental verification of bird impact on monolithic (42°, 403 km/h) and sandwich (52°, 392 km/h) composite panels at the same impact speed.

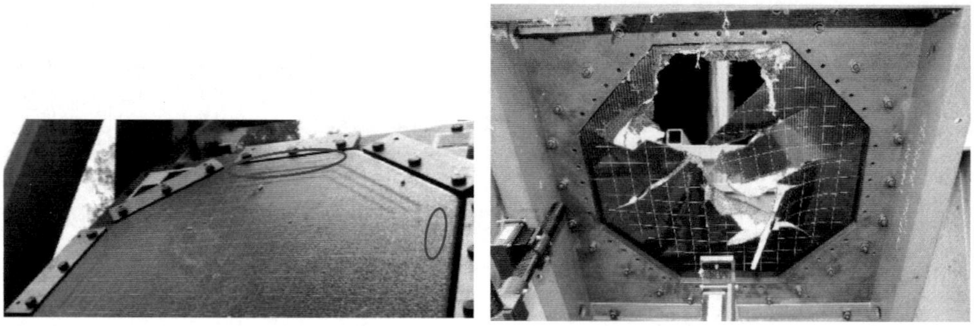

Fig. 11 Visible damage and failure after the test on the monolithic (left) and sandwich (right) composite test panel.

The visual results were simultaneously compared with the NDT inspection to obtain information on the size and character of damage and compare them with numerical models (see Fig. 12).

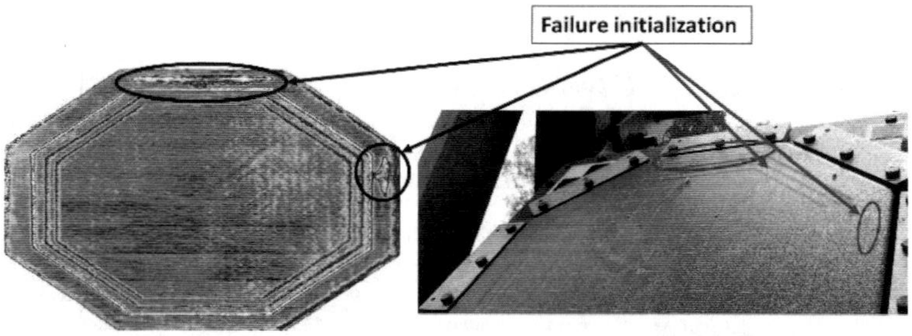

Fig. 12 Comparison of NDT and visual inspection of flat monolithic composite panel after the test.

Fig. 13 shows the results of measuring the deformation in the centre of the panel using laser triangulation sensors. These results confirm the greater effectiveness of the sandwich composite structure in terms of damping but less strength and stiffness in terms of impact energy absorption.

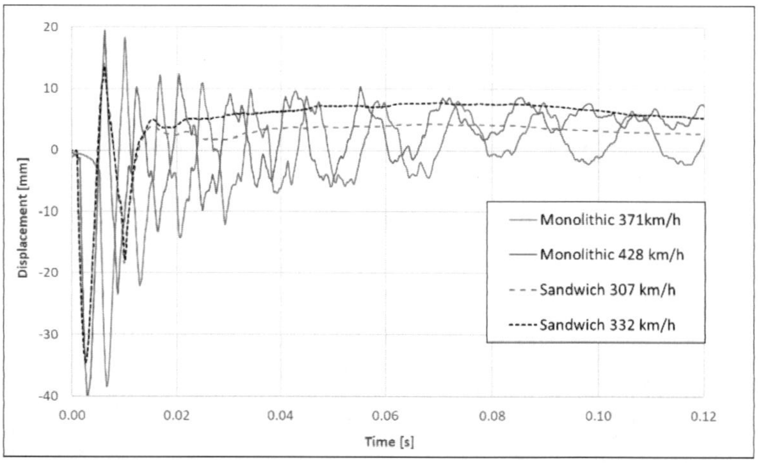

Fig. 13 Results of measuring the displacement in the centre of the flat test panels during impact.

3.2 Curved test specimen

The curved test panels represented the real structure in the area for the gear box. The experiment made it possible to assess the effect of curvature on the stiffness and strength of a thin-walled composite structure close to the real shape of a real air duct.

Fig. 14 shows an example of high-speed camera picture analysis from the point of view of test specimen displacement and failure initialization.

Fig. 14 Example of high-speed camera pictures during the bird-strike test on the curved test panel for impact speed 185 km/h.

An important result from the point of view of numerical model verification is the size and extent of damage to the composite structure after a defined projectile impact. Fig. 15 shows the influences of speed impact on the size of damage on a simple curved composite panel.

Fig. 15 The influences of speed impact on the size of damage on simple curved composite panel.

Fig. 16 shows the results of measuring reaction forces during the impact. These results confirm the influences of impact speed on the size and course of the reaction force.

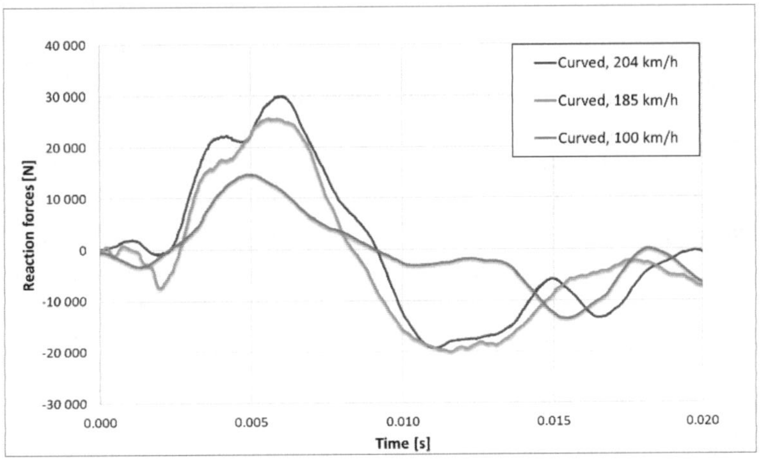

Fig. 16 Reaction forces analysis from bird-strike test on curved composite panels.

Conclusion

The obtained results from the experiments confirmed, in most cases, the expected resistance values of the test specimens determined based on preliminary numerical simulations.

The application of measuring systems for high-speed loading, such as laser distance meters or force meters for measuring reaction forces, will enable quantitative assessment and verification of numerical models.

The obtained results from the experimental measurement and the subsequent NDT analysis will enable the refinement of the material model for simulations on a real structure.

Acknowledgements

The authors are grateful to the European Commission for the support to the present work, performed within the EU CleanSky 2 project TRIcEPS (Grant Agreement No.831879).

References

[1] Hedayati, R. and Sadighi, M. *Bird Strike: An Experimental, Theoretical and Numerical Investigation* (Woodhead Publishing, Elsevier, 2015). ISBN: 978-0-08-100093-9.

[2] European Safety Agency (EASA), Certification Specification for Large Rotorcraft CS-29, Amendment 3, Cologne (2012).

[3] Georgiadis, S., Gunnion, A.J., Thomson, R.S. and Cartwright, B.K. "Bird-strike simulation for certification of the Boeing 787 composite moveable trailing edge," *Compos Struct*, **86**(1-3), pp. 258-268 (2008).

[4] Guida, M., Marulo, F., Meo, M. and Russo, S. "Certification by birdstrike analysis on C27J fullscale ribless composite leading edge," *Int J Impact Eng*, **54**, pp. 105-113 (2013).

[5] Poola, R., "Bird Strike Impact Analysis of Vertical Stabilizer Structure Using Abaqus/Explicit," SIMULIA India Regional Users Meeting 11 (2011).

[6] Smojver, I., Ivancevic, D. "Coupled Euler Lagrangian Approach Using Abaqus/Explicit in the Bird Strike Aircraft Damage Analysis," in *Proceedings of the 2010 Simulia Customer Conference*, Providence, Rhode Island, 2010.

[7] Allaeys, F., Luyckx, G., Van Paepegem, W. and Degrieck, J. "Numerical and experimental investigation of the shock and steady state pressures in the bird material during bird strike," *Int J Impact Eng*, **107**, pp 12-22 (2017).

[8] Grimaldi, A., Sollo, A., Guida, M. and Marulo, F. "Parametric study of a SPH high velocity impact analysis – A birdstrike windshield application," *Compos Struct*, **96**, pp. 616-630 (2013).

[9] Cervi, J.P. "Aircraft Birdstrike Simulation Using Smoothed Particle Hydrodynamics," in *Proceedings of the XXIV Congreso sobre Métodos Numéricos y sus Aplicaciones*, Santa Fe, Argentina, 2019, pp. 1479-1488.

[10] Wilbeck, J.S. Air Force Materials Laboratory, Technical Report AFML-TR-77-134 (1977)

[11] Hedayati, R. and Ziaei-Rad, S. "A new bird model and the effect of bird geometry in impacts from various orientations," *Aerosp Sci Technol*, **28**(1), pp. 9-20 (2013).

[12] ASTM D7136M:07 Standard Test Method for Measuring the Damage Resistance of a Fiber-Reinforced Polymer Matrix Composite to a Drop-Weight Impact Event.

[13] Doubrava, R., Oberthor, M., Bělský, P. and Raška, J. "High-speed impact assessment for composite air inlet," *International Journal of Structural Integrity*, **11**(5), pp. 723-736 (2019).

[14] ASTM F330:2010 Standard Test Method for Bird Impact Testing of Aerospace Transparent Enclosures.

[15] Abdullah, N.A., Yusoff, M.D., Shahimi, S.S. and Ahmad, M.I.M. "Numerical modelling of bird strike on aerospace structures by means of coupling FE-SPH," *International Journal of Integrated Engineering*, **13**(5), pp. 185-193 (2021).

Simulation of Micromechanics of a Unidirectional Laminated Composite

Johnson Ezenwankwo[1,a], Yordan Kyosev[2,b], Iva Petríková[1,c] and Josef Žák[1,d]

[1]*Technical University of Liberec, 46117, Czech Republic;*
[2]*The Leibnitz Institute of Polymer Research, TU-Dresden, Dresden, Germany;*

[a]*johnson.ezenwankwo@tul.cz,* [b]*yordan.kyosev@tu-dresden.de,* [c]*iva.petrikova@tul.cz,* [d]*josef.zak@tul.cz*

Abstract: Composite materials are among the most sought-after materials in major aerospace, automotive, naval applications of the 21st century. It is therefore paramount to conduct detailed materials analysis to obtain the elastic properties of attendant materials with respect to their dynamics in operation. This is technically cost-effective in eliminating repeated laboratory testing for very minute changes in input parameters. The model is a square-type unidirectional (UD) representative volume element (RVE) modelled in Abaqus CAE 6.14. An infinitesimal remote strain induced by means of a small displacement (15 - 20% of the boundary box length) from a reference/remote point tied kinematically and periodically to the face in question. The mesh sensitivity study was carried out in Ansys R22 because of its practical parametric sweep function, although mesh size was limited to 3 μm below which the number of nodes is not acceptable for the sake of license. The respective plots and attendant discussions are presented in the result section.

Keywords: Function; Unidirectional; Composites

1 Introduction

Composite materials have found widening applications in science and engineering. They present such technically desirable properties as good strength, low weight, high machinability, paths for innovative design changes, and corrosion resistance. They are not entirely devoid of challenges; their hydrophilic nature, high production cost and disposal problems are some areas that have bedevilled their usage [1]. Composites and their reinforcement textiles have vast application and scope; for example the weaving structure of warp knotted fabrics was extensively described by Kyosev Y. [2,3]. Reuss and Voigt proposed one of the earliest means of finding the mechanical properties of a composite in the rule of mixtures (ROM) [4,5]. In micromechanical analysis, homogenization is used to determine the engineering constants of a representative volume element (RVE). This is a fundamental step towards the practical design of composite structures and laminates [6]. The modified ROM was also developed to address the disagreement with experimental results. Other models for estimating the homogenized mechanical properties of an RVE exist. For the Young's modulus, shear modulus and Poisson's ratio, the homogenized orthotropic material properties are taken from literature [7]. The homogenized properties of an RVE are evaluated. To simplify the solutions, only the square-type model will be used; it is one of the simplest types of RVE when compared to randomly-distributed short fibres, particle fibres, hexagonal and diamond type. A mesh sensitivity study will be done using Ansys 2022 R2. This offers relative ease of manipulating the parametric sweep function of generating data of independent variables. Further future work will be on using the result obtained to perform a macromechanical analysis of a composite laminate plate of n-plies using safety factor (SF) and failure index (FI) as key performance indices for composite integrity. Input parameters are created as a text file and imported into the FEM platform at each required step. Several authors have expounded on laminated composites' micro and macromechnical analysis and investigated their various failure models [8,9]. The very popular Chamis micromechanical model equations are widely employed to evaluate the RVE composite elastic properties. The idea being that both matrix and fibres are linear and elastic, and that the fibres are spacially distributed in a periodic fashion in either square or hexagonal type array [10,11]. In this study the Abaqus CAE tool is used to investigate the engineering constants of an RVE under the six standard load cases of tension in x, y and y – pure shear in xy, xz, xw. The Dirichlet boundary condition is imposed as against the periodic and Neumann boundary condition. The results obtained showed good agreement with the verified results from Python script developed by similar model [12].

2 FEM setup

2.1 Micromechanical analysis

The micromechanical analysis is performed using Abaqus CAE and the classic method of manually imposing periodic boundary condition (PBC) on a square type RVE in Abaqus and using a pre-existing plug-in [12] to compare the results obtained. Two models of same geometrics are generated using TexGen standalone open software and CMAS plug-in with Ansys r22. The latter showed better compatibility with other FEM tools, both for manually induced PBC and automatically-implemented ones through other plug-ins. The geometric properties are given in Table 1 below:

Table 1. RVE geometric properties.

	Domain		Fibre		$V_f \approx 0.448$
	(µm)				
	start	*end*	*start*	*end*	
X	0.0	100	0.0	100	D_f (µm) ≈ 75.7
Y	-50	50	0.0	0.0	
Z	-50	50	0.0	0.0	

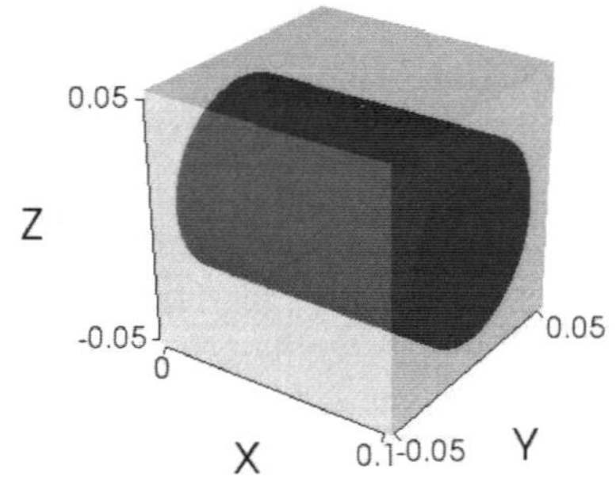

Fig. 1 Square-type RVE model (mm).

For all cases, the model is imported while retaining internal boundaries so that fibre and matrix can be assigned different material properties and selected as separate entities during the mesh creation process. The model from TexGen and Ansys r22 CMAS plug-in are exported as faceted models, as against smooth, since this improves the possibility of convergence during the solution. The following section will describe the implementation of periodic boundary condition (PBC) and sweep meshing sequences to ensure a conformal mesh on opposing sides of the model [13].

2.2 Periodic boundary condition and meshing

In mechanics of solids, the rule of thumb holds that the displacement vector component orthogonal to the plane of reference is zero. The rotational vector components parallel to the plane are also zero. But in solid models, rotational vector is not active; only the translational vector is considered. Periodic boundary condition (PBC) can be stated mathematically as follows:

$$r_i^a - r_i^b = 0, u_i^a = u_i^b + \bar{\varepsilon}_{ij}(x_j^a - x_j^b), \tag{1}$$

where a and b represent the corresponding surface pairs that describe the RVE, x_j is the distance between the opposite sides, u_i represents the displacement, and ε is the strain homogenous strain across the surface of the RVE. A critical part of the boundary condition is that the nodes along the symmetry plane for each case remained along the symmetry plane both before and after loading. For each simulation case, a total node-node constraint is applied to the symmetry plane, so there is no displacement in the direction perpendicular to the symmetry plane. An arbitrary plane is chosen as the symmetry plane and is retained for all conditions. In applying PBC, it is ensured that there exists equality in values of parameters on opposing edges that are *kinematically* linked to conform to the canonical equation in (1). This is typified by the idea that the meshing on each side of the geometry should be identical and conformal [14].

2.3 Material properties

The material properties of the matrix (modelled as Epoxy) and fibre (E-Glass) are given in the table below obtained from [8]. In a further work, the material property of the interface region will be modelled as data points of stress and their corresponding strain.

Table 2. Material properties.

	Matrix (Epoxy Resin)	Fibre (E-Glass)
Young's Modulus	3790 MPa	73000 MPa
Poisson's Ratio	0.35	0.22
Shear Modulus	4200 MPa	4345 MPa
Yield Strength	54.6 MPa	

3 Model setup

3.1 Using Abaqus CAE

In this method, the edges are tied all round with a representative loading condition whilst systematically excluding the three faces, XF, YT and ZF. Six sets capturing the faces of the RVE are creating at each loading front: X_{Front} (XF), X_{Back} (XB), Y_{Top} (YT), Y_{Base} (YB), Z_{Front} (ZF) and Z_{Back} (ZB).

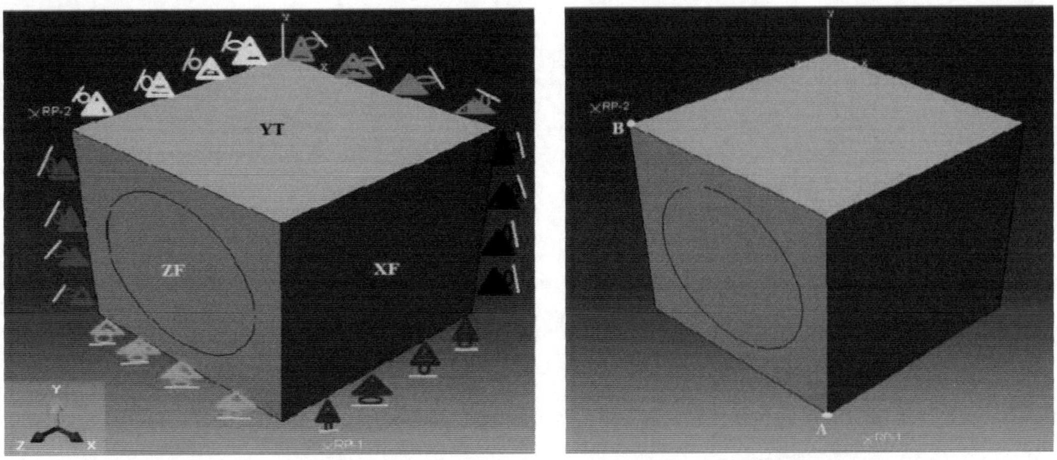

(a) (b)

Fig. 2 (a) Schematic of representative loading condition, (b) reference points (RF) A and B.

3.2 *Meshing the part*

Abaqus is equipped with one of the most efficient and accurate inbuilt meshing functions. Then a suitable element type is chosen. The number of divisions on each side of the cube representing the matrix can be seeded by parts or by edge length. If there is a remarkable size difference between constituent parts or deep contours which can present messing problems in some areas, the seeding could be done separately in parts, and *by parts* or by *edge length*. The mesh control option is used to determine the *element shape* and the *meshing technique* employed. The sweep technique is particular important for the periodic boundary condition so that there are identical meshing sequence on opposite face. The basic meshing metrics are shown below. Fig. 3 shows the periodic mesh of equal number of nodes on each pair opposite side.

Table 3. Meshing details.

Technique	Element Shape	Edge Length Division (mm)	Number of Divisions on Edge	Total Number of Elements:Nodes
Sweep	Hex/Hex dominated	0.005	= 0.1/0.005 = **20**	9560:10899

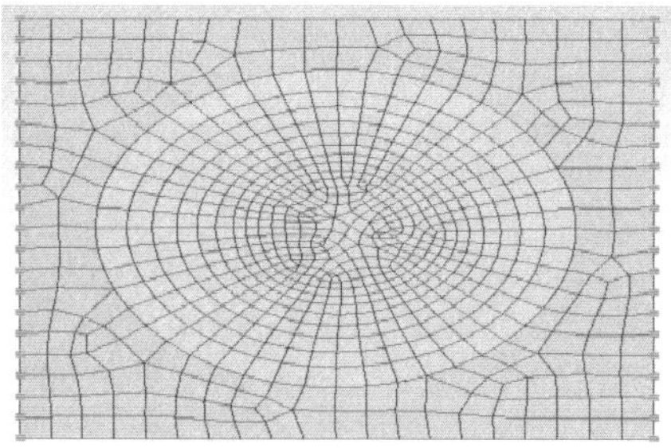

Fig. 3 Conformal periodic sweep mesh.

3.3 *Reference loading*

The reference points X^{RP-1} and X^{RP-2} are positioned at an arbitrarily small distance away (in this case 20 µm = 0.02 mm) from two chosen vertices whose coordinates are given in the *Table 2* below.

Table 4. Reference load points of Fig. 2b.

Vertex ID	Coordinates					
	x		y		z	
	Node	Load Point	Node	Load Point	Node	Load Point
A	0.1	0.12	-0.1	-0.1	0.1	0.1
B	0	-0.02	0	0	0.1	0.1
			Load Size			
Direction	(δ_x)	15 µm	(δ_y)	15 µm	(δ_y)	20 µm

Transverse Loading in x-direction

An infinitesimal displacement load of 12 µm is applied on the X^{RP-1} (node A, Fig. 2b) while maintaining a linear constraint equation linking X^{RP-1} with X_{Front} so that,

$$U(X)^{X_{Front}} = U(X)^{\mathrm{XRP\text{-}1}} \tag{2}$$

The same is applied at X^{RP-2} and a third one at X^{RP-1} again for ZF at magnitudes of 15 μm and 20 μm respectively.

Transverse Loading in y-direction

Similarly, a reference point 2 (X^{RP-2}) is used to create a load location for the Y_{Top}. So that what is taking place at X^{RP-2} is transferred to the the Y_{Top} in that same direction.

$$U(Y)^{Y_{Top}} = U(Y)^{\mathrm{XRP\text{-}2}} \tag{3}$$

Longitudinal Loading in z-direction

As in the cases of the X and Y fronts, a load which is tied to point 1(X^{RP-1}) is specifically oriented kinemtically so that its attributes are presented as loading in the z-direction as below. The load is upped by about 33%

$$U(Z)^{Z_{Front}} = U(Z)^{\mathrm{XRP\text{-}1}} \tag{4}$$

4 Boundary condition

Restriction is needed to so that during loading only the side in question engages in displacement. Using the initial boundary condition for the uniaxial case, we fix XB in the x-direction, YB in the y-direction and ZB in the z-direction –all using roller boundary condition.

4.1 Constraint equations

The constraint equations are in agreement with those used to develop the Abaqus plug-in for RVE homogenization developed by Omairey S. And the periodicity condition in Digimat [12,15,16]. The constraint equations will involve actually implementing the relationships described in (1) – (4). A set of constraints is culled up using *equation* function in Abaqus. The first is to constrain the behaviour of XF WITH X^{RP-1}.

Table 5. Tensile load constrain equation variables on load points.

	x	**y**	**z**
Coefficient	1	1	1
	-1	-1	-1
Set Name	XF	YT	ZF
	X^{RP-1}	X^{RP-2}	X^{RP-1}
DOF	1	2	3
	1	2	3

Table 5 above captures the constraint equation variables for the infinitesimal representative loading in the x, y and z directions of the RVE. The negatives in the second terms of the coefficients come from moving the non-zero parameters to the left-hand side of the kinematic equation. The loading schematics and the boundary conditions are shown below in Fig. 4 where the reference points 1 and 2 are used to apply loading in the *x* and *z*, and then *y* directions, respectively.

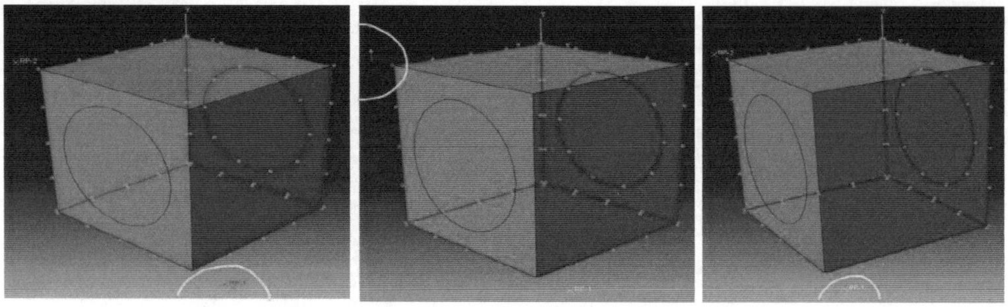

Fig. 4 Boundary condition and x, y and z loading.

4.2 Pure shear loading

Before discussing the result from the tension loading, it's worthy to expound on the concept of adding pure shear load to the RVE. For this, loads are simultaneously applied to the YZ-plane and XY-plane with respect to the $X^{RP\text{-}1}$ and $X^{RP\text{-}2}$ respectively. It is also pertinent to observe the care in positioning and directing the constituent loads that make up the shear components to accommodate for the y-bound force in the positive y direction and the x-bound force.

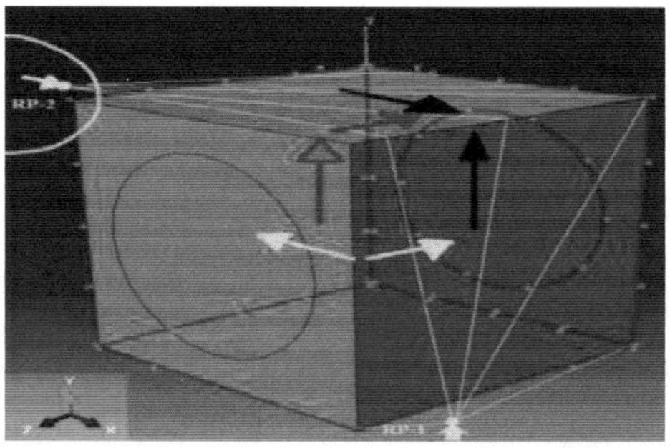

Fig. 5 Pure shear: τ_{xy}, τ_{xz} and τ_{yz} planes

The boundary condition involves imposing fixed constraints according to Table 6 below and then running separate simulations to obtain results for each case.

Table 6. Tensile load **c**onstrain equation variables on load points.

Pure Shear Direction	τ_{xy}		τ_{xz}		τ_{yz}	
Reference Side	YB	XB	*ZB*	*XB*	YB	ZB
Constrained Axis	*x, z*	*y, z*	*x, y*	*y, z*	*x, z*	*x, y*

As before, creating the correct boundary conditions plays a key role in ensuring accurate representation of a realistic model behaviour on which the results are hinged. For the first shear case (*XY* and *YX shear*), the boundary condition on the Y_{Base} and X_{Back} are that there are fixed constraints in the *x* and *z* directions, and then the y and z directions respectively. For the second shear case (*XZ* and *ZX shear*), the Z_{Back} and X_{Back} are fixed in the *x* and *z* directions, and then the y and z directions respectively, and finally, for the third case (*YZ* and *ZY shear*), the Y_{Base} and Z_{Back} are fixed in the *x* and *z* directions, and then the x and y directions respectively. All other variables not mentioned are not needed, and thus, are suppressed for the specific case. Tables 7 and 8 below show the shear load constrain dynamics and the load application schedule.

Table 7. Shear load constrain equation variables on load points.

Shear Equation Axes	XY	YX	XZ	ZX	YZ	ZY
Coefficient	1	1	1	1	1	1
	-1	-1	-1	-1	-1	-1
Set Name	X_{Front}	Y_{Top}	X_{Front}	Z_{Front}	Y_{Top}	Z_{Front}
	X^{RP-1}	X^{RP-2}	X^{RP-1}	X^{RP-2}	X^{RP-1}	X^{RP-2}
DOF	2	1	3	1	3	2
	2	1	3	1	3	2

Table 8. Reference shear load information.

Shear Load ID	XY	YX	XZ	ZX	YZ	ZY
Reference Point	X^{RP-1}	X^{RP-2}	X^{RP-1}	X^{RP-2}	X^{RP-1}	X^{RP-2}
Axis	+y	+x	+z	+x	+z	+y
Load Size	15 μm	15 μm	20 μm	15 μm	20 μm	15 μm

4.3 Mesh sensitivity study

Meshing a part is a vital part of FEM since it stems from the discretization of the geometry into smaller parts (elements) which are solved and generalized to give the result of the whole system. For this section, the Material Modeller built into newer versions Ansys is used. This is chosen because it offers great convenience in performing the needed parametric sweep of independent variables. Originally the model was meshed with a maximum mesh size of 5 μm. The sensitivity involved performing a sweep of parameters from 3 to 10 μm. The values remained remain tightly close with very infinitesimal margins.

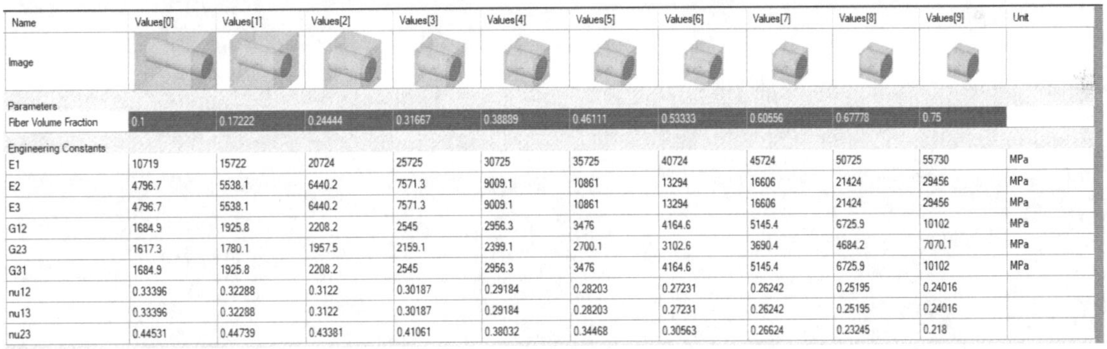

Name	Values[0]	Values[1]	Values[2]	Values[3]	Values[4]	Values[5]	Values[6]	Values[7]	Values[8]	Values[9]	Unit
Image											
Parameters											
Fiber Volume Fraction	0.1	0.17222	0.24444	0.31667	0.38889	0.46111	0.53333	0.60556	0.67778	0.75	
Engineering Constants											
E1	10719	15722	20724	25725	30725	35725	40724	45724	50725	55730	MPa
E2	4796.7	5538.1	6440.2	7571.3	9009.1	10861	13294	16606	21424	29456	MPa
E3	4796.7	5538.1	6440.2	7571.3	9009.1	10861	13294	16606	21424	29456	MPa
G12	1684.9	1925.8	2208.2	2545	2956.3	3476	4164.6	5145.4	6725.9	10102	MPa
G23	1617.3	1780.1	1957.5	2159.1	2399.1	2700.1	3102.6	3690.4	4684.2	7070.1	MPa
G31	1684.9	1925.8	2208.2	2545	2956.3	3476	4164.6	5145.4	6725.9	10102	MPa
nu12	0.33396	0.32288	0.3122	0.30187	0.29184	0.28203	0.27231	0.26242	0.25195	0.24016	
nu13	0.33396	0.32288	0.3122	0.30187	0.29184	0.28203	0.27231	0.26242	0.25195	0.24016	
nu23	0.44531	0.44739	0.43381	0.41061	0.38032	0.34468	0.30563	0.26624	0.23245	0.218	

Fig. 6 Parametric sweet of the 5 μm mesh size.

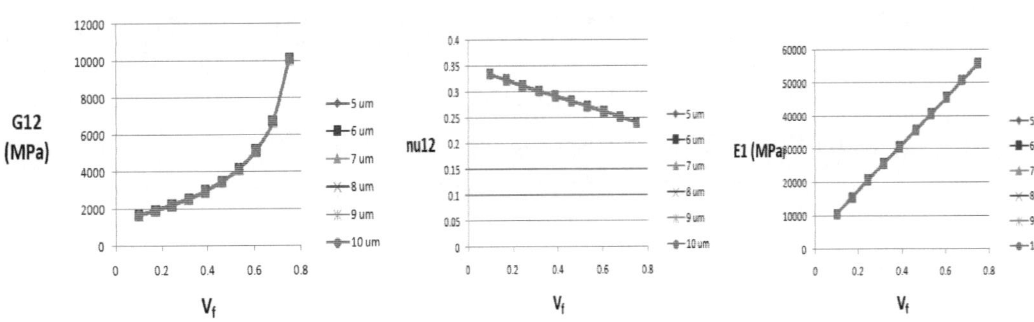

Fig. 7 Mesh sensitivity plot for G_{12}, nu_{12} and E_1.

5 Results and discussion

To present the results, the first idea is to plot the stress-strain curve for the three tensile load cases in X, Y and Z directions. Then a comparison is made of the plots of the shear cases considering the shear cases for each pair of planes (XY, YZ, and ZX planes). Stress-strain curve is obtained from the Abaqus post processor by invoking the excel plugin that exports the ramped-to-target-value data for the force and displacement. The stress and strain values are calculated from tables using the value of the measurement of the side of the bounding box (matrix) and consequently its cross-sectional area. The values of the Young's modulus are given in the table below and compared with results from a python-based Abaqus Plugin developed by Sadiq O. et al [12].

Table 9. Young's Modulus.

E_1		E_2		E_3	
Abaqus Simulation	Plugin (Abaqus)	Abaqus Simulation	Plugin (Abaqus)	Abaqus Simulation	Plugin (Abaqus)
7.0372GPa	10.10GPa	7.0049GPa	10.12GPa	32.926 GPa	32.915 GPa

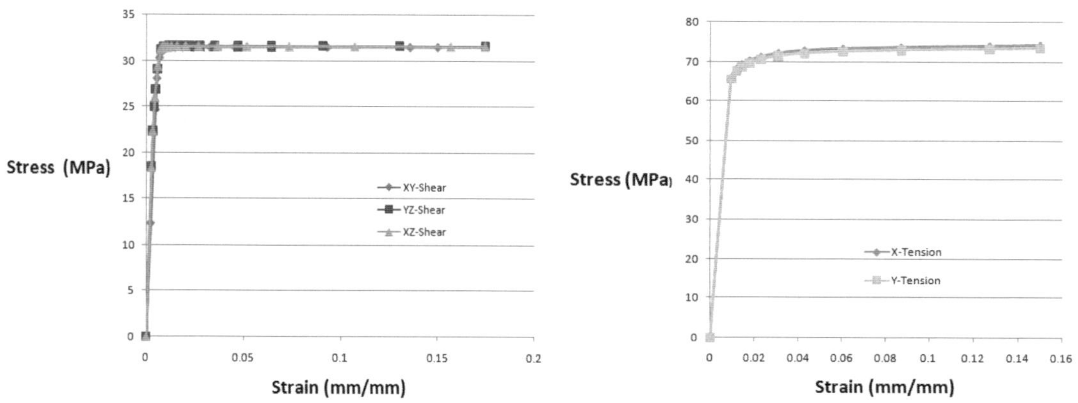

Fig. 8 Stress-strain result plots.

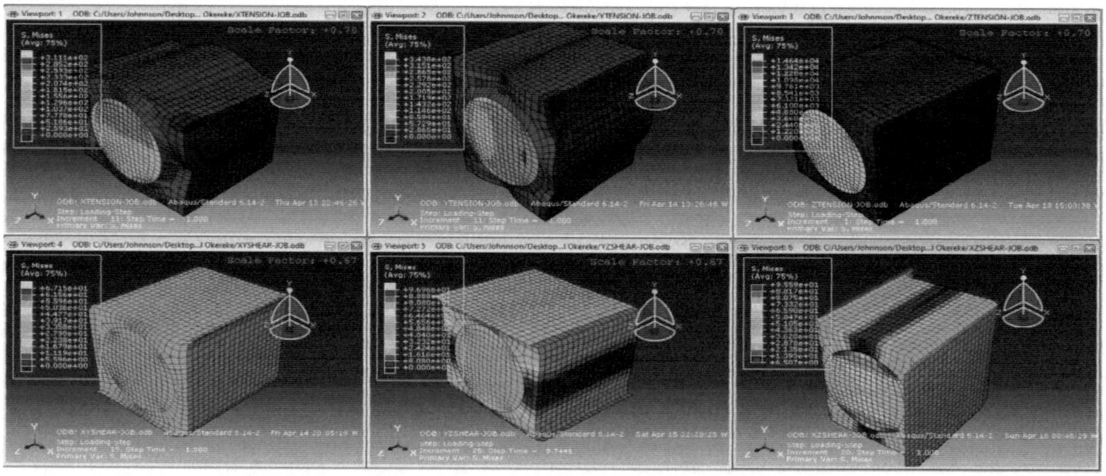

Fig. 9 Contour plots of the 6 load cases (*top row*: tension, *bottom row*: shear).

5.1 Stress and stain deliverables

Fig. 9 above shows the linked viewport of the 6 load cases. All components of stress *(s)*, strain *(E)*, reaction force *(RF)*, displacement *(u)* and plastic strain types *(PE, PEEQ and*

PEMAG) can be deduced from the virtualization tab. Loaded to the x-y plot function and exported as an excel file. The equivalent plastic strain is recorded as 1.5% while the von Mises stress is 311.1MPa for X-tension load case. The same is done for the remaining 5 load cases.

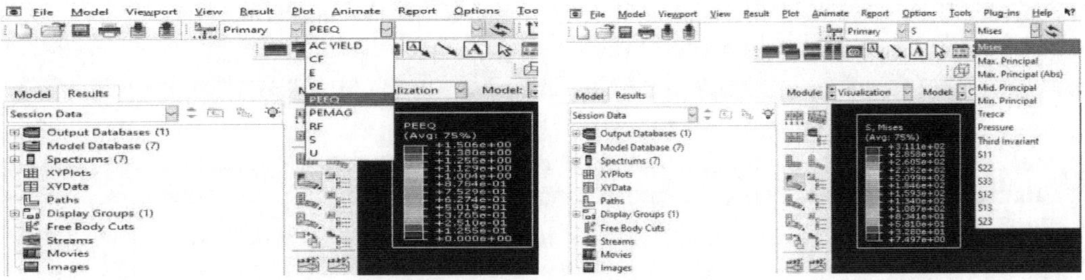

Fig. 10 Virtualization and exporting data points.

Conclusion

In general, Abaqus CAE among other tools has proven to be one of the best and user-friendly tools used by researchers and engineers alike in solving complex and delicate technical problems that require modelling and virtual solutions. Further work will involve incorporating the thermal aspect of response of UD RVE in Comsol Multiphysics and Digimat homogenization kit. As discussed earlier, modelling the interface and its plastic behaviours is one key aspect that further improves on the realistic approach of describing a UD-RVE.

Acknowledgement

This publication was written at the Technical University of Liberec as part of the project "Research of advanced materials, and application of machine learning in the area of control and modeling of mechanical systems" nr. SGS-2022-5072 with the support of the Specific University Research Grant, as provided by the Ministry of Education, Youth and Sports of the Czech Republic in the year 2022.

References

[1] Sathishkumar, T. P., Naveen, J. and Satheeskumar, S. "Hybrid fibre reinforced composites – a review," *Journal of Reinforced Plastics and Composites*, **33**(5), pp. 454-471 (2014).

[2] Kyosev, K.K. *Warp Knitted Fabrics Construction,* (CRC Press, 1st Edition, 2019).

[3] Kyosev, K.K. "Process Based Method for Pattern Development of Narrow Woven Complex Profiles," in *Advanced Weawing Technology*, Kyosev, K.K. and Bossou F. *eds.* (Springer Cham., Switzerland, 2022).

[4] Reuss, A. "Berechnung der Fließgrenze von Mischkristallen auf Grund der Plastizitätsbedingung für Einkristalle," *Z angew Math Mech*, **9**(1), pp. 49-58 (1929). ISSN 1521-4001.

[5] Voigt, W. "Ueber die Beziehung zwischen den beiden Elasticitätsconstantenisotroper Körper," *Annalen der physik*, **274**(12), pp. 573-587 (1889). ISSN 0003-4916.

[6] Yan C. "On the homogenization and de-homogenization of composite materials," PhD Thesis, Drexel University, PA, USA, 2019.

[7] Tsai, S.W. and Hahn, H.T. *Introduction to composite materials* (CRC Press, 2018).

[8] Kaw, A.K. *Mechanics of composite materials* (Taylor & Francis Group, LLC, 2006).

[9] Chamis, C.C. "Simplified Composite Micromechanics Equations for Hygral, Thermal and Mechanical Properties," in *Proceedings of the Anonymous Annual Conference of the Society of the Plastics Industry (SPI),* Houston, USA, 1983, pp. 1-10.

[10] Voyiadjis, G.Z. and Kattan, P.I. *Mechanics of Composite Materials with MATLAB* (Springer, Germany, 1st Edition, 2005).

[11] Gay, D. Composite Materials: Design and Application (CRC Press; 3rd Edition, 2014).

[12] Omairey, S.L., Dunning, P.D. and Sriramula, S. “Development of an ABAQUS plugin tool for periodic RVE homogenization,” *Eng Comput*, **35**, pp. 567-577 (2019).

[13] Michael, O.I. and Akoyomare, A.I. "Virtual Framework for Prediction of the Full-Field Elastic Response of Unidirectional Composites," *Comp Mater Sci,* **70**, pp. 82-99 (2013).

[14] Nguyen, V.D., Bechet, E., Geuzaine, C. and Noels, L. "Imposing Periodic Boundary Condition on Arbitrary Meshes by Polynomial Interpolation", *Comp Mater Sci*, **55**, pp. 390-406 (2012).

[15] e-Xstream Engineering - Digimat, The Material Modeling Tool, May, *Available from* *http://www.e-xstream.com*., *Assessed* 2018-05-02.

[16] Xia, Z, Zhang, Y and Ellyin, F. “A Unified Periodical Boundary Conditions for Representative Volume Elements of Composites and Applications,” *Int J Solids Struct,* **40**(8), pp. 1907-1921 (2003).

Analysing the Influence of the Inner Pressure on the Modal Parameters of the Vessel

Martin Hagara[1,a], Pavol Lengvarský[1,b], Róbert Huňady[1,c], Miroslav Pástor[1,d] and Peter Palička[1,e]

[1]*Technical University of Košice, Faculty of Mechanical Engineering, Department of Applied Mechanics and Mechanical Engineering, Letná 1/9 Košice-Sever 042 00, Slovakia;*

[a]*martin.hagara@tuke.sk*, [b]*pavol.lengvarsky@tuke.sk*, [c]*robert.hunady@tuke.sk*, [d]*miroslav.pastor@tuke.sk*, [e]*peter.palicka@tuke.sk*

Abstract: In dynamic analysis of structures, modal parameters estimation plays an important role. The paper describes the experimental assessment of the modal parameters, i.e. mode shapes, natural frequencies and damping ratios, of the pressure vessel filled with helium under two states, i.e. by its inner pressure equal to 80 bar and the analysis of the empty vessel, whereby the inner pressure is expected to be 0 bar. Moreover, it provides a description and application of the simple and effective method for the determination of Rayleigh coefficients α and β, which are commonly used in the finite element analysis as an approximation to viscous damping. The obtained results prove that the inner pressure has significant influence on the analysed modal parameters. As the analyses of the vessel were realised under free-free boundary conditions, the results obtained can be simply used to correlate finite element analytical model.

Keywords: Experimental Modal Analysis; Modal parameters; Rayleigh damping model; Inner pressure; Vessel

1 Introduction

Damping plays an important role when analysing the dynamic behaviour of structures. Viscous and hysteretic damping models are the most common damping models used in civil, mechanical, material, and automotive engineering, respectively. The viscous damping model assumes that the damping force is proportional to the velocity, and thus the damping ratio and natural frequency are independent of the excitation level. On the other hand, the hysteretic damping model refers to a situation where the damping force or coefficient depends on the displacement or strain history of the system [1].

One of the most common damping specified in time history response analysis is Rayleigh damping, a form of viscous damping proportional to a linear combination of mass **M** and stiffness **K** matrices according to

$$\mathbf{C} = \alpha\mathbf{M} + \beta\mathbf{K}, \tag{1}$$

where coefficients α and β are not generally known, but calculated from the modal damping ratios [2,3]. Traditionally, two modes of vibration (i^{th}- and j^{th}-order) are selected and their damping ratios ζ_i, ζ_j and their natural damped frequencies ω_i, ω_j obtained through experimental modal analysis (EMA) are used to determine α and β [4,5] using the following relation

$$\begin{Bmatrix} \alpha \\ \beta \end{Bmatrix} = \frac{2\omega_i\omega_j}{\omega_j^2 - \omega_i^2} \begin{pmatrix} \omega_j & -\omega_i \\ -\dfrac{1}{\omega_j} & \dfrac{1}{\omega_i} \end{pmatrix} \begin{Bmatrix} \zeta_i \\ \zeta_j \end{Bmatrix}. \tag{2}$$

The determination of the damping ratio ζ_j as a function of frequency is then realized according to

$$\zeta_j = \frac{\alpha}{2\omega_j} + \frac{\beta\omega_j}{2}. \tag{3}$$

Typical dependence of damping ratio on the frequency by considering the Rayleigh damping model is shown in Fig. 1.

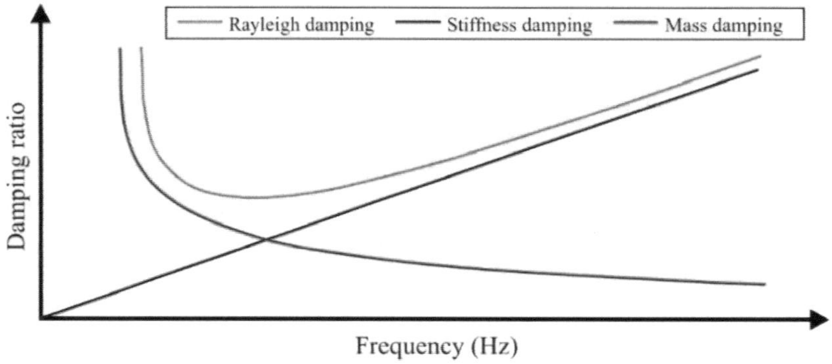

Fig. 1 Typical dependence of damping ratio on the frequency.

2 Measurement and methods

The estimation of the modal parameters, including Rayleigh damping coefficients, was carried out at two states of a pressure vessel (Fig. 2). Firstly, when it was filled with helium at 80 bar and secondly, when the vessel was empty, and thus the inner pressure was 0 bar (or low value approaching 0 bar caused by the remaining unreleased helium). The basic dimensions of the pressure vessel are the length 260 mm and outer diameter 103 mm. The mass of the filled vessel is approximately 1770 g. According to [6], the vessel is formed by welding the central cylindrical part to the bottom or cap of the vessel. A glycerine manometer GS04/20 was used to determine the values of the inner pressure.

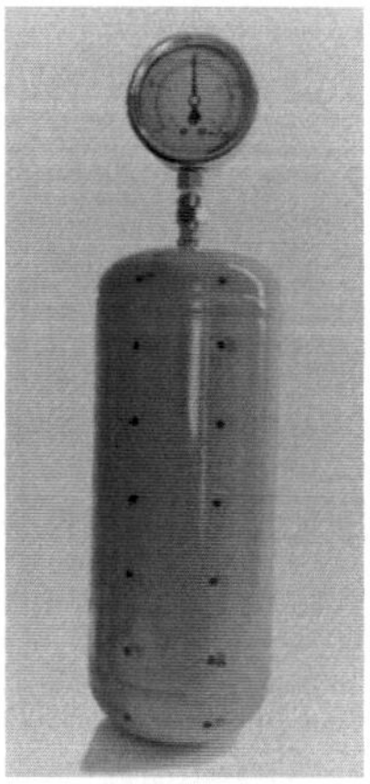

Fig. 2 Analysed pressure vessel.

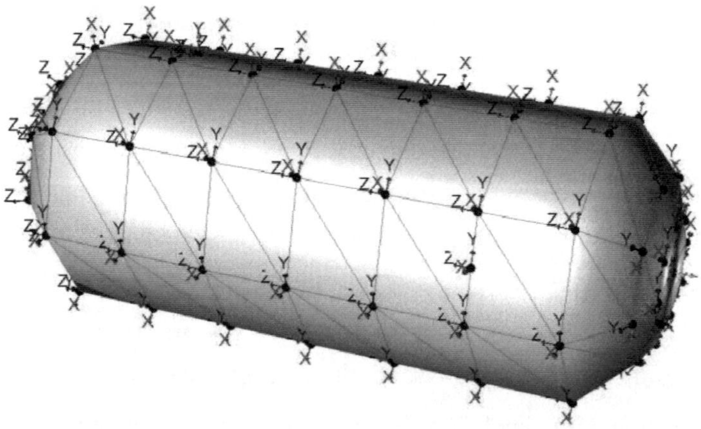

Fig. 3 The geometry of the pressure vessel, created in MTC hammer software, used to perform the experimental modal analysis.

2.1 Experimental modal analysis of the pressure vessel

Experimental data of frequency response functions (FRFs) form was acquired using the Brüel&Kjær Pulse® system with the LAN-XI 3050 module. The Pulse MTC Hammer software was used to perform the measurements, in which a simplified pressure vessel geometry was created (Fig. 3). Soft rubber cords were used to suspend the vessel from a rigid frame (Fig. 4).

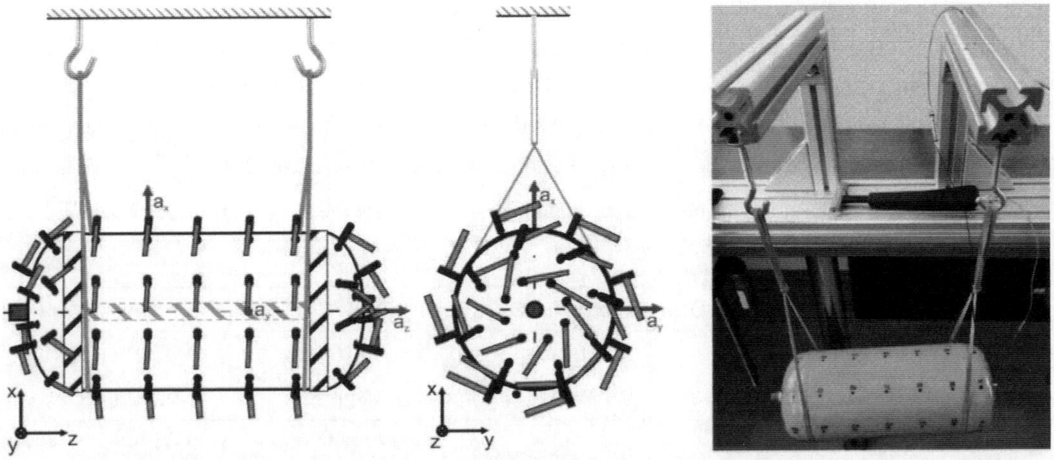

Fig. 4 Representation of reference points, accelerometer locations, and free-free boundary conditions.

Measurement in the form of the fixed embedded accelerometer and roving hammer-impact method was carried out using a Brüel&Kjær 8206 modal hammer with a plastic tip. At the same time, the structure's response in the three mutually perpendicular X, Y, and Z directions was recorded successively at three measurement points by a Brüel&Kjær 4374 single-axis miniature acceleration sensor. The resulting signals were obtained by averaging using the linear peak hold average method of the fast Fourier transform (FFT) signals, which increased the accuracy of the FRFs. The complex mode indicator functions (CMIFs) obtained from the measurement are depicted in Fig. 5.

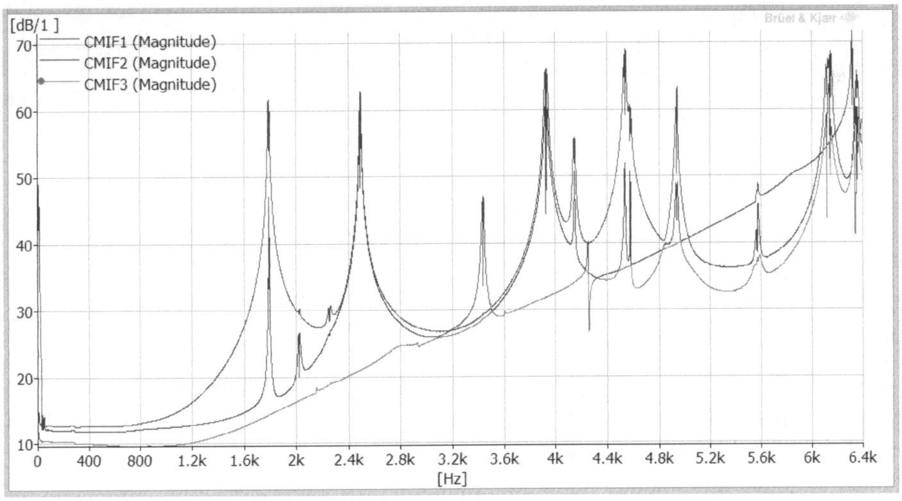

Fig. 5 Complex modes indicator functions (CMIFs) obtained by experimental modal analysis of the filled pressure vessel.

The obtained frequency response functions were evaluated in Pulse Reflex®, providing information on the natural frequencies, damping ratios, and individual mode shapes of the vessel vibration in the frequency range from 0 Hz to 6400 Hz. Stability diagrams resulting from the Complex Mode Indicator Function (CMIF) were used to estimate the vibration modes. The stability diagrams were plotted sequentially for the selected frequency bands, while the Rational Fraction Polynomial-Z method was used to identify the stability poles. Based on the most stable poles, the possible modes of vibration of the vessel were determined over the entire analysed frequency range. Since, in some cases, modes were estimated that were not directly related to the vibration of the pressure vessel but only of its valve, these modes were removed from the estimated modes of vibration. In the case of a vessel loaded with an internal pressure of 80 bar, these were the modes with the following frequencies 2021 Hz, 3436 Hz, and 3440 Hz. The selected possible modes were analysed using Auto Modal Assurance Criterion (Auto MAC) method based on orthogonality checking. The natural frequencies and damping ratios of the vessel vibration were obtained from the experimental modal analyses of the filled and empty pressure vessel, as shown in Table 1.

Table 1. The natural frequencies and damping ratios obtained for the filled and empty vessel.

	80 bar		**0 bar**	
	Frequency (Hz)	**Damping ratio (%)**	**Frequency (Hz)**	**Damping ratio (%)**
Mode 1	1788	0.1161	1665 (↓6.9%)	0.0579 (↓50.1%)
	1794	0.1201	1673 (↓6.8%)	0.0593 (↓50.6%)
Mode 2	2489	0.0941	2209 (↓11.3%)	0.0499 (↓47.0%)
	2490	0.0948	2211 (↓11.3%)	0.0503 (↓47.0%)
Mode 3	3923	0.0877	3754 (↓4.3%)	0.0453 (↓48.3%)
	3935	0.0881	3765 (↓4.3%)	0.0469 (↓46.8%)
Mode 4	4146	0.0847	3830 (↓7.6%)	0.0420 (↓50.4%)
	not found		not found	
Mode 5	4533	0.0773	4501 (↓0.7%)	0.0458 (↓40.8%)
	4536	0.0785	4504 (↓0.7%)	0.0436 (↓44.5%)
Mode 6	4945	0.0759	4677 (↓5.4%)	0.0442 (↓41.8%)
	4948	0.0757	4685 (↓5.3%)	0.0448 (↓40.8%)
Mode 7	5560	0.0622	5593 (↑0.6%)	0.0410 (↓34.1%)
	5571	0.0609	5617 (↑0.8%)	0.0403 (↓33.8%)
Mode 8	6113	0.0872	6009 (↓1.7%)	0.0465 (↓46.7%)
	6141	0.0877	6028 (↓1.9%)	0.0474 (↓45.9%)
Mode 9	6347	0.0771	6033 (↓4.9%)	0.0466 (↓39.5%)
	6356	0.0792	6044 (↓4.9%)	0.0469 (↓40.8%)
Mode 10	not found		6222	0.0439
			6248	0.0446

In both analysed cases, the so-called multiple modes of vibration typical for axisymmetric structures were found. In the case of an empty pressure vessel, the first ten vibration modes were estimated in the frequency range from 0 Hz to 6400 Hz. The tenth mode was not identified for the filled vessel because its corresponding frequency exceeds the analyzed frequency range. A decrease of up to 11.3% can be seen for almost all natural frequencies. Only in the case of mode seven, this change did not occur, even though the natural frequency increased by 0.6-0.8%. However, this mode differed in character from the others (significant vibration in the valve region visible from the FEM analysis not included in this paper). The estimated mode shapes of the vessel obtained from the experimental modal analysis are shown in Fig. 6. The individual vessel mode shapes were identical for both analysed cases.

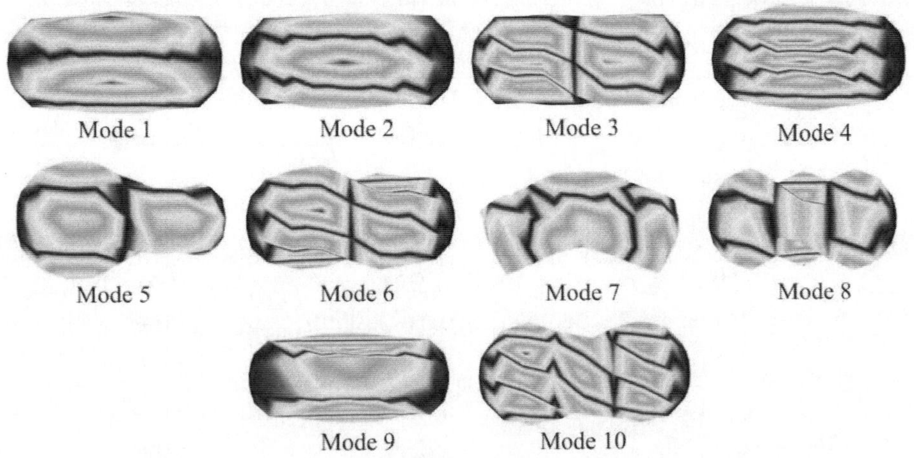

Fig. 6 Mode shapes of the analysed vessel obtained by experimental modal analysis.

2.2 Estimation of Rayleigh damping coefficients

The coefficients α and β for estimating the Rayleigh damping according to relation (1) were calculated from the experimentally determined values of the natural frequencies and damping ratios of the individual modes for both states of the pressure vessel. Estimates of the above coefficients were obtained sequentially for the individual frequency bands defined by the nth and first vibration modes but only for the first from the multiple modes. The obtained results are presented in Table 2 and Table 3.

In both analysed states of the vessel, a very low value of the β coefficient approaching zero can be observed, i.e., the damping of the vessel is proportional to its mass. This case is characterized by very low damping at higher structure modes. Comparing the coefficient α, it is possible to observe a decrease of more than 50% in the case of an empty vessel.

Table 2. Rayleigh damping coefficients for the analysed vessel filled with helium at 80 bar.

	Frequency (Hz)	**Damping ratio (%)**	**β**	**α**
Mode 1	1788	0.1161		
Mode 2	2489	0.0941	2.836E-08	22.509
Mode 3	3923	0.0877	3.560E-08	21.595
Mode 4	4146	0.0847	3.269E-08	21.963
Mode 5	4533	0.0773	2.617E-08	22.785
Mode 6	4945	0.0759	2.509E-08	22.922
Mode 7	5560	0.0622	1.588E-08	24.084
Mode 8	6113	0.0872	3.031E-08	22.263
Mode 9	6347	0.0771	2.416E-08	23.038

Table 3. Rayleigh damping coefficients for the empty analysed vessel (0 bar pressure).

	Frequency (Hz)	Damping ratio (%)	β	α
Mode 1	1665	0.0579		
Mode 2	2209	0.0499	2.076E-08	9.845
Mode 3	3754	0.0453	2.067E-08	9.854
Mode 4	3830	0.042	1.720E-08	10.234
Mode 5	4501	0.0458	1.998E-08	9.929
Mode 6	4677	0.0442	1.836E-08	10.107
Mode 7	5593	0.041	1.482E-08	10.495
Mode 8	6009	0.0465	1.747E-08	10.205
Mode 9	6033	0.0466	1.750E-08	10.201
Mode 10	6222	0.0439	1.565E-08	10.404

The minor differences in the values of the parameter α obtained in both measurements for different defined frequency bands declare that the Rayleigh damping model estimated for the whole analysed frequency range (between modes 1 and 9 for a vessel with 80 bar gas pressure, respectively 1 and 10 for an empty vessel) should be sufficiently accurate also in the case of the vibration analysis of the lower modes.

A comparison of the dependence of the damping ratio on frequency obtained for different defined frequency bands in the case of the two analysed vessel states can be seen in Fig. 7-9.

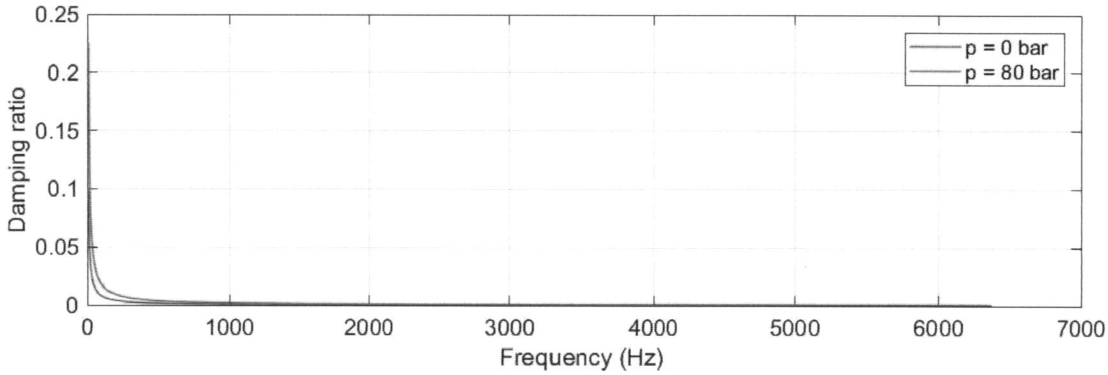

Fig. 7 Frequency dependence of Rayleigh damping ratio obtained for the frequency band ω_2-ω_1.

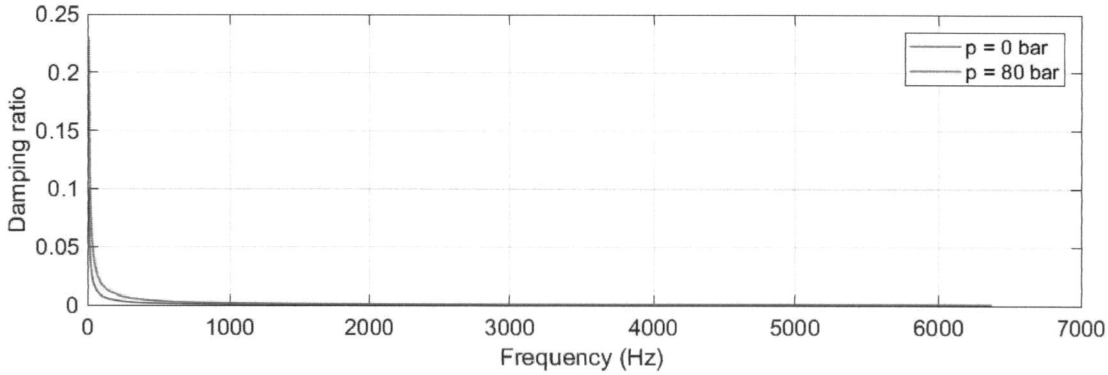

Fig. 8 Frequency dependence of Rayleigh damping ratio obtained for the frequency band ω_6-ω_1.

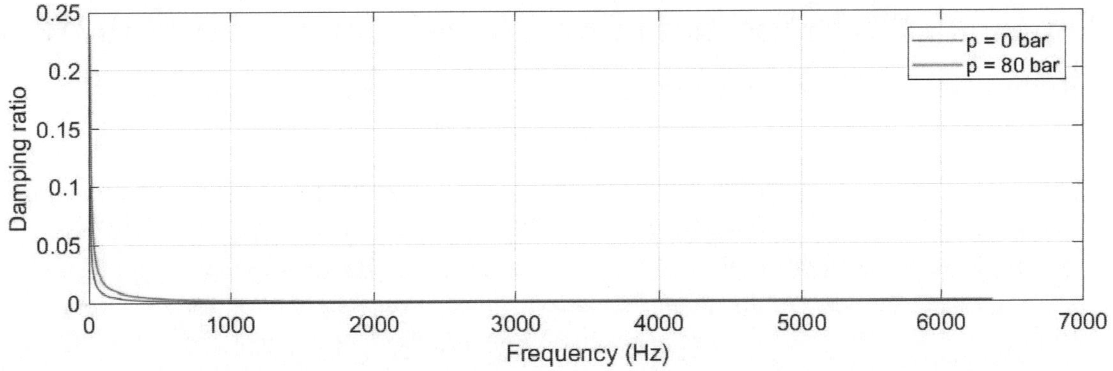

Fig. 9 Frequency dependence of Rayleigh damping ratio obtained for the frequency band ω_9-ω_1.

Conclusion

The paper describes the methodology for the assessment of the modal parameters of the pressure vessel. The analysis of the mode shapes, natural frequencies, and damping ratios was realized for two different states of the vessel, i.e., the vessel filled with helium at a pressure of 80 bar and an empty vessel, whereby the inner pressure of the gas was expected to be 0 bar. The obtained results confirm that the inner pressure influences the analysed quantities, i.e., the higher the pressure, the higher the natural frequencies and damping ratios. On the other hand, there is no difference in the vibration mode shapes.

The second part of the paper deals with estimating the vessel Rayleigh damping model, a form of viscous damping proportional to the linear combination of mass and stiffness matrices often used in finite element analysis. According to Fig. 1, it was obtained in both cases that the damping of the vessel is proportional to its mass. Moreover, it can be concluded that the inner pressure also influences the Rayleigh damping coefficients, i.e., α and β, which decrease with decreasing inner pressure. As the experimental modal analyses of the vessel were realized using soft rubber cords to minimize the effect of bond stiffness, the authors aim to use the obtained results in the vessel numerical model verification process.

Acknowledgements

This work was supported by the Ministry of Education, Science, Research and Sport of the Slovak Republic under the project VEGA 1/0516/22.

References

[1] Liu, M. and Corman, D.G. "Formulation of Rayleighdamping and its extensions," *Computers and Structures*, **57**(2), pp. 277-285 (1995). ISSN 1879-2243.

[2] Geraschenko, V.S., Grishin, A. S. and Gartung, N. I. "Approaches for the calculation of Rayleigh damping Coefficients for a time-history analysis," *Structures under Shock and Impact*, **XV**, pp. 227-237 (2018). ISSN 1743-3509.

[3] Xu, J., Huang, Y and Qu, Z. "The high-frequency oscillation in sstems with Rayleigh damping model," in *Proceedings of the 12th Canadian Conference on Earthquake Engineering*, Quebec City, 2019, ed. (QC, Canada, June 17-20), pp. 1-8.

[4] Song, Z. and Su, C. "Computation of Rayleigh Damping Coefficients for the Seismic Analysis of a Hydro-Powerhouse," *Shock and Vibration*, **2017**, pp. 1-11 (2017). ISSN 1875-9203.

[5] Rahul, B., Dharani, J. and Balaji, R. "Optimal Method for Determination of Rayleigh Damping Coefficients for Different Materials using Modal Analysis," *International Journal of Vehicle Structures and Systems*, **13**(1), pp. 102-111 (2021). ISSN 0975-3540.

[6] EN 12205:2001 Transportable gas cylinders - Non-refillable metallic gas cylinders.

Uniaxial LCF Properties of Additively Manufactured In718

Radim Halama[1,a], Jakub Cienciala[1,b], Ajay Vignesh Natarajan[1,c] and Jiří Hajnyš[2,d]

[1] *VŠB - Technical University of Ostrava, Faculty of Mechanical Engineering, Department of Applied Mechanics; 17. listopadu 2172/15, 708 00 Ostrava - Poruba, Czech Republic;*

[2] *VŠB - Technical University of Ostrava, Faculty of Mechanical Engineering, Department of Machining, Assembly and Engineering Metrology; 17. listopadu 2172/15, 708 00 Ostrava - Poruba, Czech Republic;*

[a] *radim.halama@vsb.cz,* [b] *jakub.cienciala@vsb.cz,* [c] *ajay.vignesh.natarajan@vsb.cz,* [d] *jiri.hajnys@vsb.cz*

Abstract: The paper focusses on the stress-strain behaviour of Inconel 718 prepared by the selective laser melting technology. Uniaxial cyclic plastic properties are reported together with lifetimes of particular specimens. The transient behaviour of the material is a challenge for advanced constitutive modelling. The cyclic plasticity model considering a virtual backstress for memory surface evolution is used in the description of the strain-range-dependent transient behaviour. Strain-controlled cases considered in the study are predicted very well for strain amplitudes higher than 0.9%.

Keywords: Low cycle fatigue; Cyclic plasticity; Inconel 718; Additive manufacturing

1 Introduction

Recently, additive manufacturing has played an important role in technical practice not only in prototyping but also in the production of functional parts. This article focusses on the investigation of low-cycle fatigue (LCF) properties of the Ni-based superalloy Inconel 718. It is well known that the material is designed to maintain high strength and fatigue performance at high operating temperatures (650°C). There are many studies available in the literature that focus on the heat treatment of the additively prepared variant of Inconel 718 [1-3]. However, cyclic plastic properties and low cycle fatigue data are reported in the literature very rarely. This contribution brings new results of the stress-strain behaviour of Inconel 718 manufactured by selective laser melting (SLM), as well as further numerical modelling of its complexity under uniaxial loading.

2 Experiments

2.1 Specimens manufacturing

All specimens prepared for low-cycle fatigue testing were printed from a virgin powder with selective laser melting (SLM) technology using Renishaw AM500E. Specimens were printed in the Z direction (vertical orientation of the specimen axis) and were tested in as-built state. The main printing parameters used during the AM process are listed in Table 1. The chamber was filled with argon with purity 5.0 during fabrication to prevent oxidation. The building platform was preheated to 170 °C to reduce residual stresses during the AM process.

Table 1. Printing parameters used in AM process.

Specimens orientation	3D printing strategy	Laser power (W)	Laser rate (mm · s^{-1})	Layer thickness (µm)	Preheating of building platform (°C)
90°	"meander"	400	1150	60	170

2.2 Fatigue testing

Uniaxial fatigue tests were performed in tension-compression on the LabControl 100kN/120Nm biaxial electro-servohydraulic test machine. Strain-controlled tests were performed on a solid specimen with 6.3 mm diameter (up to 1% of the strain amplitude) and on a thin-walled tubular specimen with 11.3 mm inner diameter and 13.9 mm outer diameter.

2.3 Results

Cyclic stress-strain behaviour and lifetimes have been evaluated in constant strain amplitude tests using own Python scripts. The most important results are: lifetimes N_f, stress amplitude σ_a, plastic strain amplitude ε_{ap}, dynamic modulus E and cyclic yield stress Y (see Fig. 1). In general, dynamic modulus has different values in the tensile and compressive branches of the hysteresis loop (an aluminium alloy is considered in the scheme shown in Fig.1).

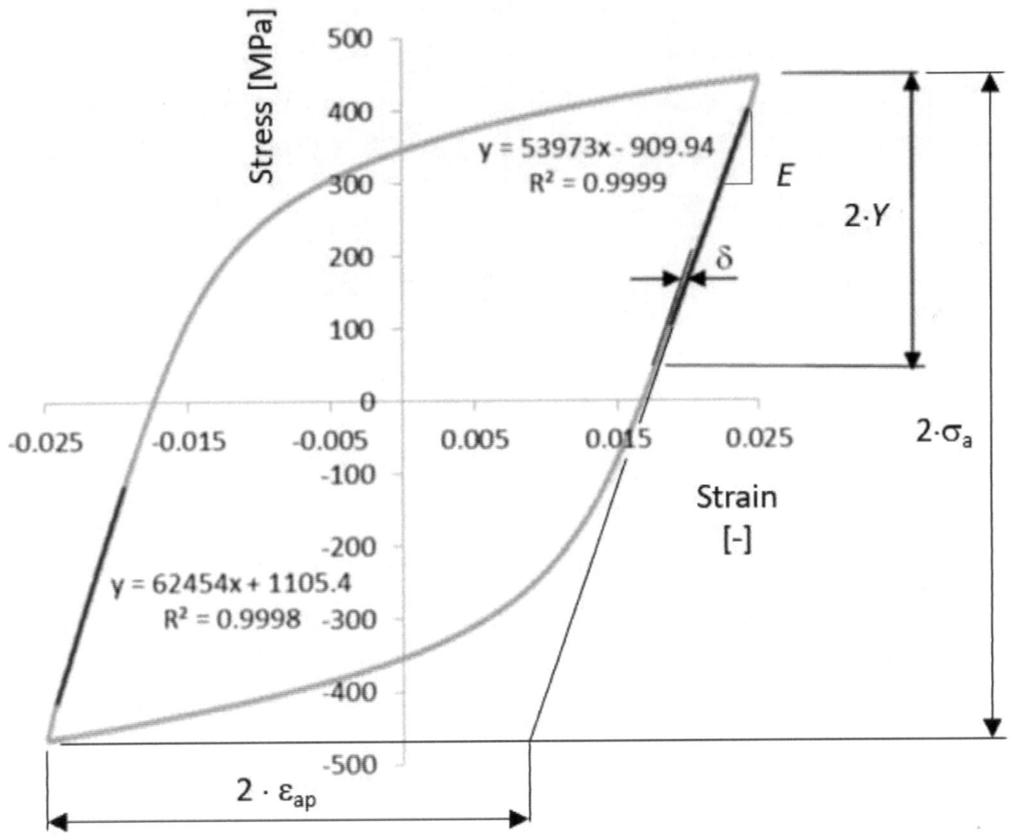

Fig. 1 Scheme of a hysteresis loop.

The results show a significant cyclic hardening in the initial cycles followed by a cyclic softening under strain amplitudes greater than 0.4% (Fig. 2). Saturated hysteresis loops at half-life are shown in Fig.3. The material also shows a non-Masing behaviour (Fig.4).

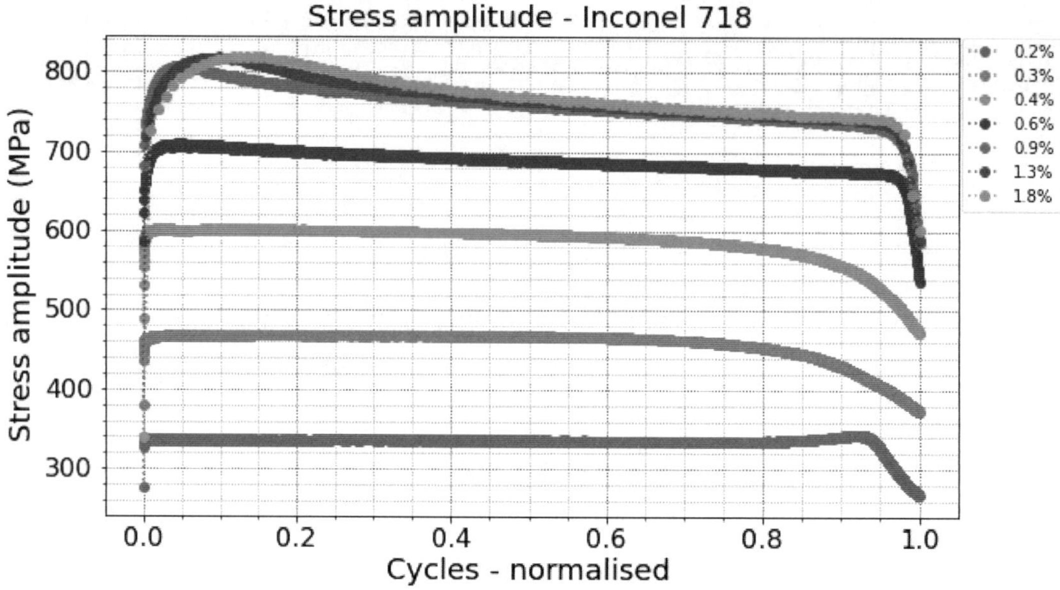

Fig. 2 Stress amplitude evolution of Inconel 718 manufactured by SLM technology.

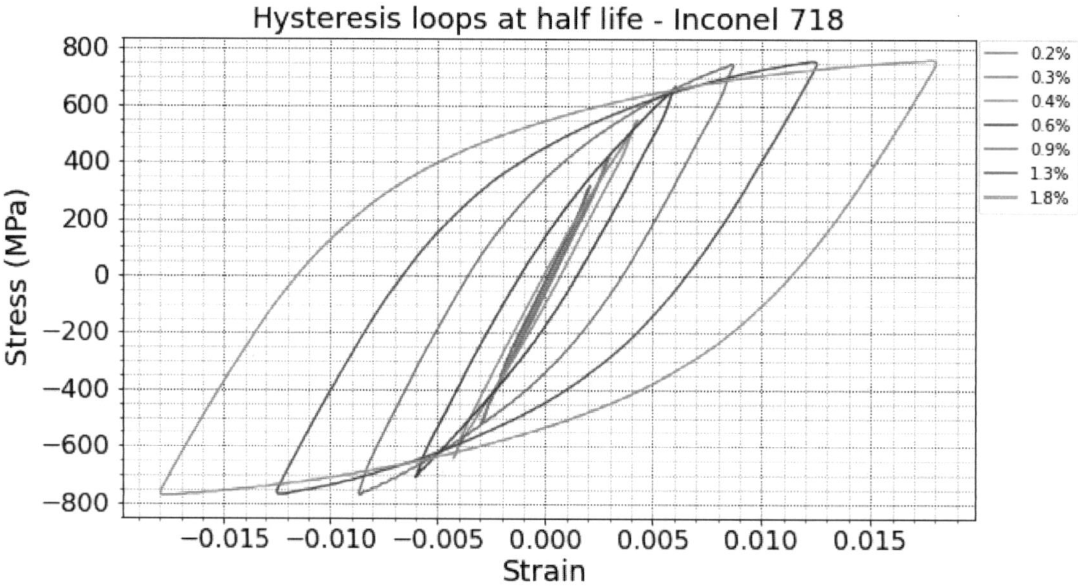

Fig. 3 Uniaxial hysteresis loops of Inconel 718 manufactured by SLM technology.

The most important results of the low cycle fatigue tests are reported in Table 2. The stress amplitudes correspond to the half-life. The number of cycles to failure was evaluated with a criterion of 25% decrease from the asymptotic line created in the diagram stress amplitude vs number of cycles.

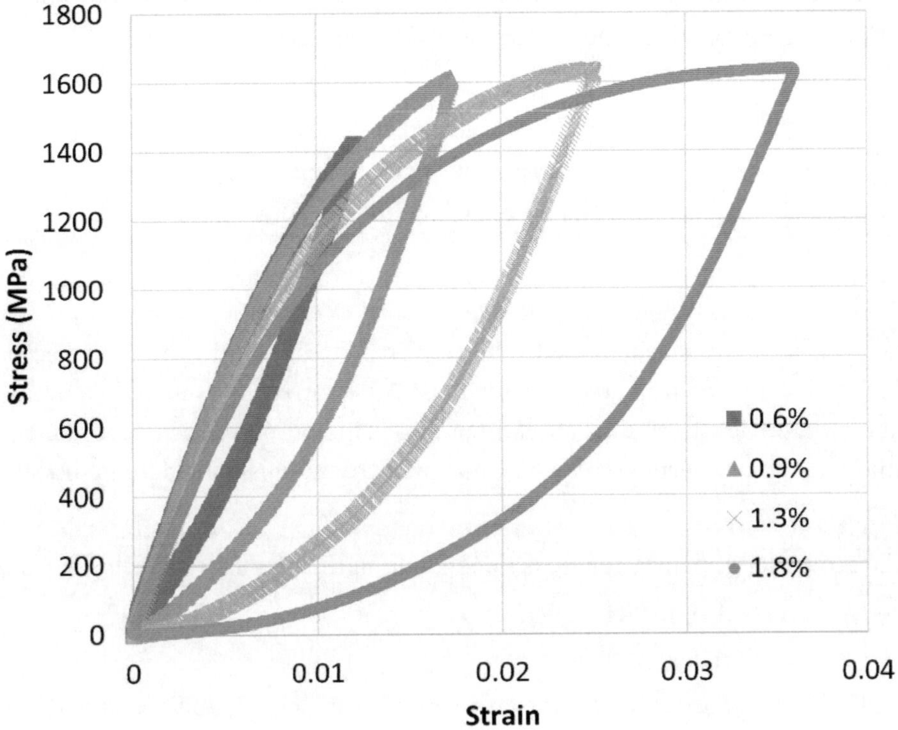

Fig. 4 Non-Masing behaviour of Inconel 718 manufactured by SLM technology.

Table 2. Main results of LCF tests performed on Inconel 718.

Strain amplitude [%]	Stress amplitude [MPa]	Number of cycles to failure [-]
0.2	335.6	50082
0.3	467.1	21839
0.4	594.9	8407
0.6	689.9	1935
0.9	758.9	1063
1.3	762.7	302
1.8	767.2	103

3 Constitutive modelling

3.1 Cyclic plasticity model description

For simplicity, the constitutive model is described for the uniaxial loading case. The main equations of the incremental theory of plasticity follow. The yield condition is defined as

$$f = |\sigma - \alpha| - \sigma_Y - R = 0 \tag{1}$$

where σ is the axial stress, α is the backstress (kinematic variable), R is the isotropic variable and σ_Y is the yield stress. The superposition of six backstress parts ($M = 6$) is considered as in the original Chaboche model, i.e.

$$\alpha = \sum_{i=1}^{M} \alpha^{(i)} \tag{2}$$

whereas the evolution equation of kinematic hardening is considered as

$$d\alpha^{(i)} = C_i d\varepsilon_p - \gamma_i \varphi(p) \alpha^{(i)} dp, \tag{3}$$

where C_i, γ_i are material parameters, $d\varepsilon_p$ is the increment of plastic strain and dp is the accumulated plastic strain increment. The Marquis parameter is dependent on the current value of the accumulated plastic strain

$$\varphi(p) = \varphi_{\infty 1} - (\varphi_{\infty 1} - \varphi_{01})e^{-\omega_1 \cdot p} + \varphi_{\infty 2} - (\varphi_{\infty 2} - \varphi_{02})\, e^{-\omega_2 \cdot p}, \tag{4}$$

Two Marquis terms are used to describe the cyclic hardening/softening behaviour. The parameters φ_{01} and $\varphi_{\infty 2}$ are dependent on the size of the memory surface R_M:

$$\varphi_{01} = A_{01} e^{B_{01}\mathrm{R}_M}, \tag{5}$$

$$\varphi_{\infty 2} = A_{\infty 2} e^{B_{\infty 2}\mathrm{R}_M}, \tag{6}$$

In this work, the memory surface is introduced in the stress space with isotropic behaviour and is calculated from the virtual backstress [5]. Its size corresponds to the maximal absolute value of the virtual backstress for uniaxial loading. This property is in accordance with the Jiang-Sehitoglu memory surface [6]. The isotropic hardening is considered a linear one

$$dR = R_0 dp, \tag{7}$$

but the value of material parameter R_0 is dependent on the memory surface size as follows

$$R_0 = -\langle A_1 \cdot \mathrm{R}_M + A_2 \rangle, \text{ for } \mathrm{R}_M < \mathrm{R}_{ML}, \tag{8}$$

$$R_0 = -(A_3 \cdot \mathrm{R}_M + A_4), \text{ otherwise}, \tag{9}$$

where A_1, A_2, A_3, A_4 are material parameters, $\langle x \rangle = (x + |x|)/2$, and the transition memory size is calculated by

$$\mathrm{R}_{ML} = (A_4 - A_2)/(A_1 - A_3). \tag{10}$$

All material parameters are stated in Table 3. The implementation of the proposed cyclic plasticity model was done in Python for the strain-controlled uniaxial case based on the algorithm presented elsewhere [4].

Table 3. Parameters of the proposed cyclic plasticity model.

Parameter	Value
C_{1-6} [MPa]	426190, 99567, 41526, 25502, 18386, 5494
γ_{1-6}	3333, 1212, 519, 253, 137, 54
σ_Y [MPa]	220
ω_1, ω_2	4.3, 2.2
$\varphi_{\infty 1}, \varphi_{02}$	-0.15, 0.8
A_{01}, B_{01}	0.003056, 0.008469
$A_{\infty 2}, B_{\infty 2}$	0.3216763, 0.0021203
A_1, A_3	0.007763, 0.1351
A_2, A_4 [MPa]	-1.455174, -61.295

3.2 Results of simulations

The first example of stress-strain prediction corresponds to the fatigue test with a constant strain amplitude of 1.8%. Uniaxial hysteresis loops from the whole history of loading are presented in Fig. 5. The prediction of the hysteresis loop shape is good as can be seen from Fig. 6.

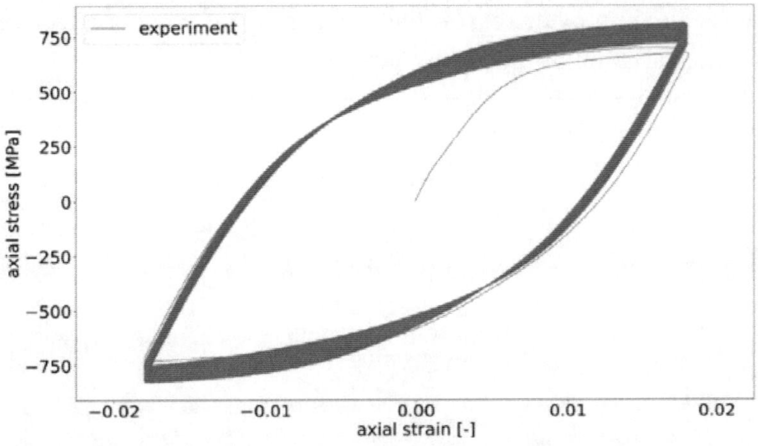

Fig. 5 Uniaxial hysteresis loop for 1.8% of strain amplitude - experiment.

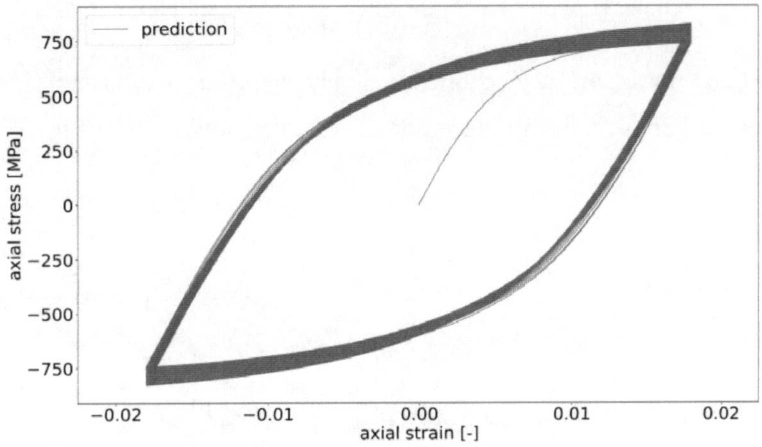

Fig. 6 Uniaxial hysteresis loop for 1.8% of strain amplitude - prediction.

The cyclic hardening/softening behaviour is accurately described for the case of 1.8% strain amplitude (Fig. 7).

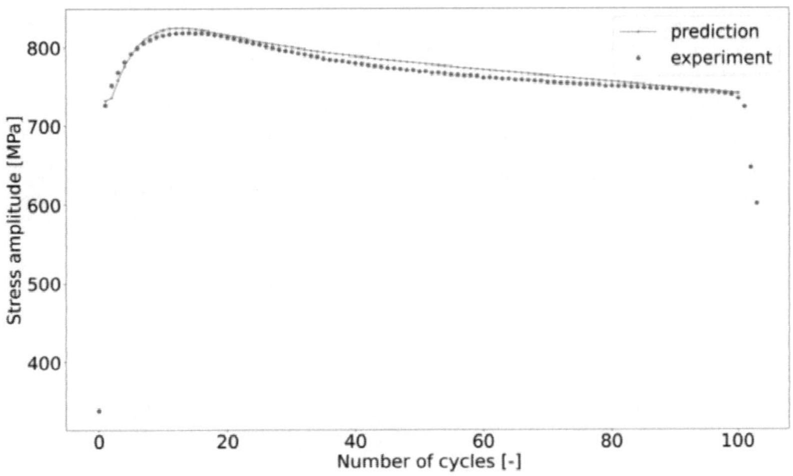

Fig. 7 Results of evolution of stress amplitude with cycling - 1.8% of strain amplitude.

The incremental test performed under strain amplitudes 0.22%, 0.31%, 0.44%, 0.66%, 0.88% and 1.1% serves us as a verification case. From Fig. 8, it is obvious that the highest error occurs for small amplitudes, where the dynamic modulus was higher than E=130000MPa.

Therefore, a possible improvement of the robustness of the material model is in the consideration of the variation of the elastic modulus.

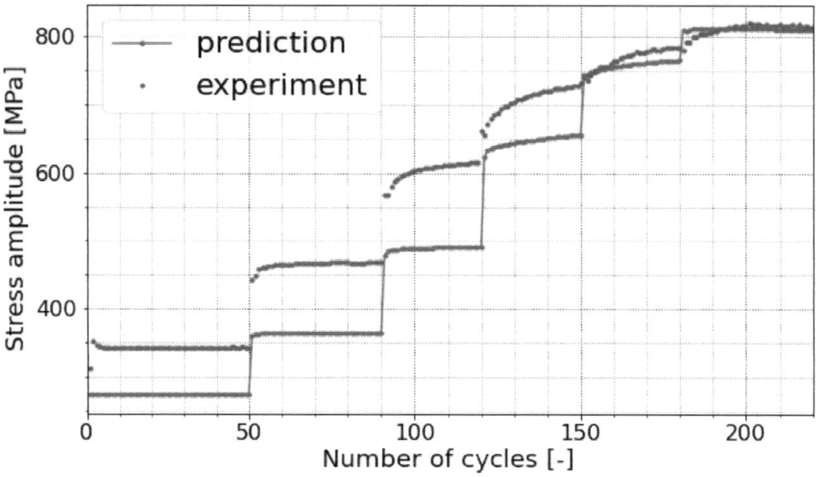

Fig. 8 Results of evolution of stress amplitude with cycling – the incremental test.

Uniaxial hysteresis loops from the incremental test are shown in the Fig. 9. The prediction by proposed material model is very good, see Fig. 10.

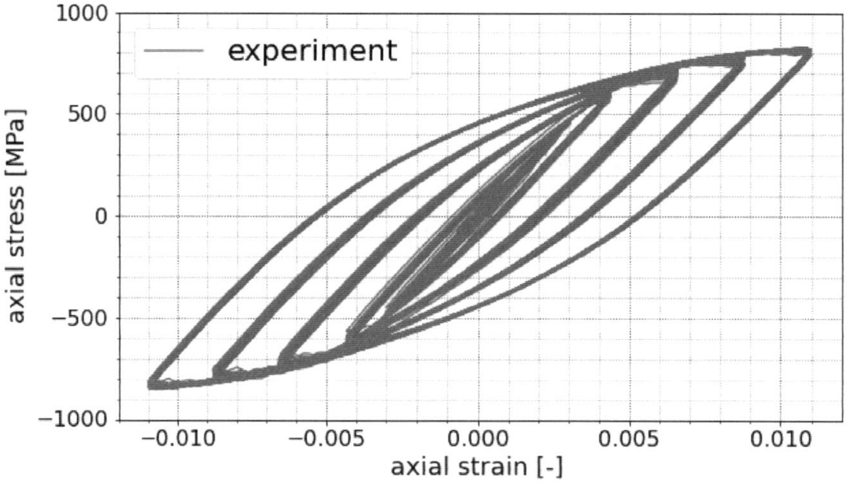

Fig. 9 Uniaxial hysteresis loops from the incremental fatigue test.

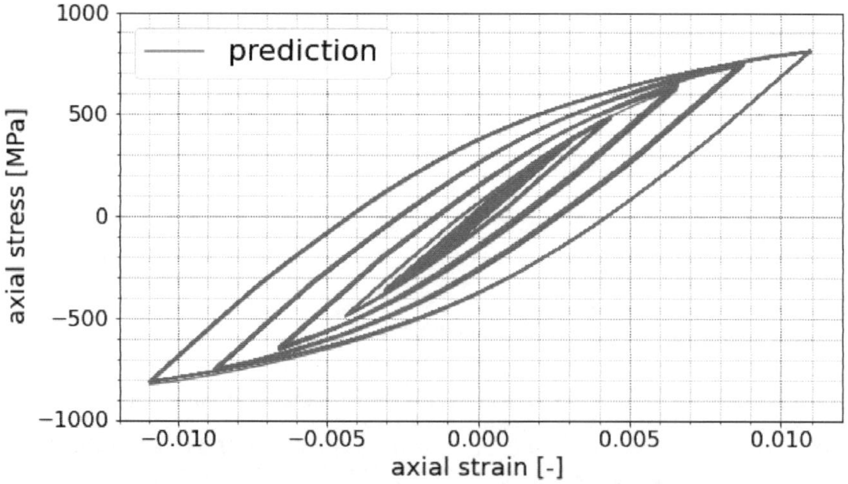

Fig. 10 Prediction of uniaxial hysteresis loops from the incremental fatigue test.

Conclusion/Summary

The article describes selected results of low-cycle fatigue tests performed on as built specimens made of Inconel 718. The material printed with SLM technology shows a complex cyclic hardening/softening and non-Masing behaviour. Dynamic moduli are strain dependent, but for the numerical predictions an average value was considered. A new cyclic plasticity model with a memory surface is applied to describe the complex stress-strain behaviour of the material under constant amplitude loading, as well as in an incremental fatigue test. Very good correlation of predictions with experiments was achieved in the case of hysteresis loop shape and in the evolution of the stress amplitude with cycling.

Acknowledgements

This work is an output of the project Computational modelling of ductile fracture of identical wrought and printed metallic materials under ultra-low-cycle fatigue created with financial support from the Czech Science Foundation under the registration no. 23-04724S and the SGS project supported by the Ministry of Education, Youth and Sports of the Czech Republic under the registration no. SP2023/027.

References

[1] Marchese, G., Atzeni, E., Salmi, A. and Biamino, S. "Microstructure and residual stress evolution of laser powder bed fused Inconel 718 under heat treatments," *J Mater Eng Perform,* **30**, pp. 565–574 (2021).

[2] Trosch, T, Strößner, J, Völkl, R and Glatzel, U. "Microstructure and mechanical properties of selective laser melted Inconel 718 compared to forging and casting," *Mater Lett,* **164**, pp. 428-431 (2016).

[3] Zhang, Q, Ren, P, Tu, X, Dai, Y., Wang, X. and Li, W. "Effect of heat treatment on microstructure evolution and mechanical properties of selective laser melted inconel 718 alloy," *J Mater Eng Perform*, **28**, pp. 5376–5386 (2019).

[4] Halama, R, Markopoulos, A, Jančo, R and Bartecký, M. "Implementation of MAKOC Cyclic Plasticity Model with Memory," *Adv Eng Softw,* **113**, pp. 34-46 (2017).

[5] Halama, R., Fumfera, J., Gál, P., Kumar, T. and Markopoulos, A. "Modeling the Strain-Range Dependent Cyclic Hardening of SS304 and 08Ch18N10T Stainless Steel with a Memory Surface," *Metals,* **9**, 832 (2019).

[6] Jiang, Y. and Sehitoglu, H. "Modeling of cyclic ratchetting plasticity, part I: Development of constitutive relations," *J Appl Mech,* **63**(3), pp. 720-725 (1996).

Uniaxial Compression of Porous Magnetorheological Elastomer

Mohammad Yousef Amer Hdaib[1,a] and Iva Petríková[1,b]

[1]*Technical University of Liberec; Studentská 1402/2, Liberec, Czech Republic;*

[a]*mohammad.yousef.hdaib@tul.cz*, [b]*iva.petrikova@tul.cz*

Abstract: The development and progression of robotics and sensors expand the requirements for intelligent materials. Thus, this study is concerned with the fabrication, material modeling, and experimentation of Magnetorheological elastomer in foam structure (MREF). The MREF comprised a synthetic silicon matrix filled with iron particles. The magnetoelastic and hyperelastic parameters were identified using a modified compressible Mooney-Rivlin model. The engineering stress was derived based on decomposing the constitutive relation into isochoric and volumetric parts. Then the material parameters were identified for uniaxial compression of MREF in the presence and absence of a magnetic field. The model fitted the experimental data accurately with similar behavior reported in the literature.

Keywords: MRE; Porous magnetorheological elastomer; Magneto-hyperelastic model; Highly-compressive MRE.

1 Introduction

The fabricated Magnetorheological elastomeric foam (MREF) consists of a porous rubber matrix filled with carbonyl iron particles. Such Magnetoelastic material has adjustable stiffness with a magnetic field and vice versa. Particularly, the MREF exhibits a noticeable change in magnetization with displacement because it is highly compressive, which makes it suitable for soft robotics like tactile sensors [1].

To estimate the magnetoelastic coupling for MREF and to employ it for finite element analysis, it is crucial to derive a suitable compressive form of engineering stress. For conventional magnetorheological elastomer (MRE), Kankanala et al. [2,3] derived the constitutive relations for the Cauchy stress, magnetic field, and entropy. The authors derived identical constitutive relations based on Eulerian and Lagrangian forms. In addition, they illustrated how magnetization, magnetostriction, and stress behave with displacement and magnetic field. Afterward, Dorfmann and Ogden [4] proposed a more simplified form of the constitutive relations by merging the magnetic body stresses with the Cauchy stress into a total stress tensor ($\boldsymbol{\tau}$). Subsequently, based on the Dorfmann and Ogden derivation of the total stress tensor, Ottenio et al. [5] illustrated the incremental formulation of magnetoelastic coupling and proposed a modified Mooney-Rivlin Hyperelastic model. The proposed model proved numerical stability through literature based on the results acquired by Marvalova et al. [6] and Janbaz et al. [7].

In this study, an isotropic MREF sample was fabricated using a soluble mold, synthetic silicon rubber, and iron particles. The engineering stress for a highly compressive MREF was derived based on the decomposition of the stress tensor into isochoric and volumetric terms [8]. In addition, the derived form of engineering stress was compared with experimental data.

2 Theory

Eq. (1) and Eq. (2) illustrates the constitutive relation for the total stress ($\boldsymbol{\tau}$) and magnetic field ($\mathbf{H}$), respectively. This formation is derived based on the total stress proposed by

Dorfmann and Ogden [4] and its decomposition into isochoric and volumetric parts, as illustrated by Holzapfel [8].

$$\boldsymbol{\tau} = \boldsymbol{\tau}_{\text{vol}} + \boldsymbol{\tau}_{\text{iso}} \tag{1}$$

$$\boldsymbol{\tau} = \frac{\partial \psi_{\text{vol}}(J)}{\partial J}\mathbf{I} + 2J^{-1}\mathbf{F}.\frac{\partial \psi_{\text{iso}}(\bar{\mathbf{C}},\mathbf{B}_{\text{L}})}{\partial \mathbf{C}}.\mathbf{F}^{\text{T}} \tag{2}$$

$$\mathbf{H} = J^{-1}\frac{\partial \psi(\mathbf{C},\mathbf{B}_{\text{L}})}{\partial \mathbf{B}} \tag{3}$$

The partial differentiation in Eq. (2) and (3) can be expanded additively as:

$$\frac{\partial \psi}{\partial \mathbf{C}} = \sum_{i=1}^{6} \frac{\partial \psi}{\partial \mathbf{C}}\frac{\partial \mathbf{C}}{\partial \mathbf{I}_i} \tag{4}$$

$$\frac{\partial \psi}{\partial \mathbf{B}} = \sum_{i=1}^{6} \frac{\partial \psi}{\partial \mathbf{B}}\frac{\partial \mathbf{B}}{\partial \mathbf{I}_i} \tag{5}$$

Considering a quasi-static case and isotropic material, the total stress and magnetic field take the following tensorial forms:

$$\boldsymbol{\tau} = \frac{\partial \psi_{\text{vol}}(J)}{\partial J}\mathbf{I} + 4J^{-1}\left[J^{-2/3}\frac{\partial \psi_{\text{iso}}}{\partial \mathbf{I}_1}\mathbf{C} + J^{-4/3}\frac{\partial \psi_{\text{iso}}}{\partial \mathbf{I}_2}(\mathbf{I}_1\mathbf{C} - \mathbf{C}.\mathbf{C}) + \frac{\partial \psi_{\text{iso}}}{\partial \mathbf{I}_5}\mathbf{B}\mathbf{B}\right] \tag{6}$$

$$\mathbf{H} = 2\left(\frac{\partial \psi_{\text{iso}}}{\partial \mathbf{I}_4}\mathbf{C}^{-1}.\mathbf{B} + \frac{\partial \psi_{\text{iso}}}{\partial \mathbf{I}_5}\mathbf{B}\right) \tag{7}$$

The used energy function was proposed by Ottenio et al. [5] as follows:

$$\psi_{\text{iso}}(\mathbf{F},\mathbf{B}_{\text{L}}) = C_{10}(\bar{I}_1 - 3) + C_{01}(\bar{I}_2 - 3) + \frac{1}{\mu_0}(\alpha I_4 + \beta I_5) \tag{8}$$

$$\psi_{\text{vol}}(J) = \frac{1}{2}\kappa(J-1)^2 \tag{9}$$

The engineering stress and magnetic field are derived by expanding Eq. (6) and (7) for the principal directions with the following considerations: uniaxial boundary conditions $\left(\lambda_1 = \lambda,\ \lambda_2 = \lambda_3 = \sqrt{J/\lambda}\right)$, magnetic field parallel with the loading direction ($B_1 = B,\ B_2 = B_3 = 0$), and engineering (nominal) stress ($\mathbf{T} = J\mathbf{F}^{-1}\boldsymbol{\tau}$). Eq. (10) and (11) illustrate the derived engineering stress and magnetic field, respectively.

$$T_{\text{eng}} = C_{10}\left(\lambda - \frac{J}{\lambda^2}\right)J^{-2/3} + C_{01}\left(1 - \frac{J}{\lambda^3}\right)J^{-1/3} + \beta\frac{B^2}{\lambda\mu_0} \tag{10}$$

$$H = \frac{2B}{\mu_0}\left(\frac{\alpha}{\lambda} - \beta\lambda\right) \tag{11}$$

3 Experimental measurements

The foam MRE sample was fabricated using liquid silicon rubber, iron particles, 3d-printed mold, and soluble mold (sugar) according to the following steps:

- The following raw materials were mixed: silicone, hardener, and iron particles.
- The mixture was vacuumed and poured on the soluble mold (rigid sugar) inside the 3d-printed assembly, as shown in Fig. 1b and c.
- The whole mixture with sugar was left for setting in the fridge for 72 hours. In this context, setting means providing enough time for the mixture to flow inside the small canals between the sugar crystals.
- The mixture was left to polymerize in a vacuum at room temperature.
- The sugar-MRE texture was cut off from the polymerized mixture, as shown in Fig. 1d.
- The sugar-MRE texture was sunk in water for 24 hours, at which the sugar melted, and the MREF sample was acquired. The sample porosity (ϕ) was calculated using the occupied volume by MRE (V_{MRE}), as follows:

$$\phi = 1 - \frac{V_{MRE}}{V_{total}} \tag{12}$$

Fig. 1. Sample fabrication.

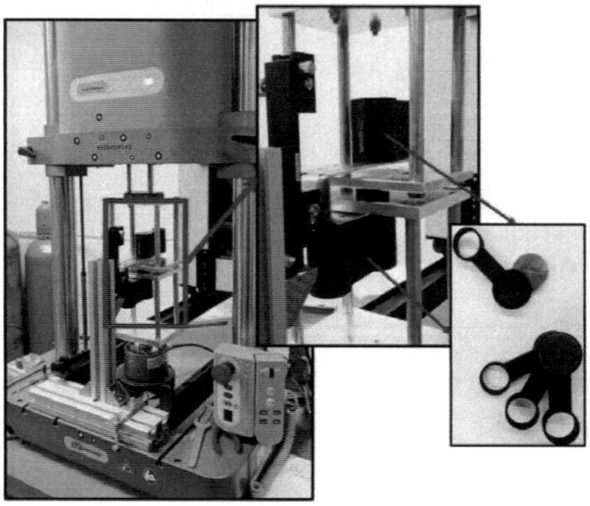

Fig. 2. Experimental setup.

The uniaxial compression was performed using Instron E3000 with grips designed to attach neodymium magnets and prevent them from interfering with the signal of the displacement sensor. The grips are made from Aluminum and brass to avoid interference with the magnetic field. As shown in Fig. 2, the magnetic field can be adjusted by changing the number of magnets.

4 Results

Fig. 3 illustrates a quasi-static uniaxial compression with a 0.003 s^{-1} strain rate. The sample was compressed to a 0.7 stretch (λ) at variable magnetic fields. The dotted curves represent the experimental data, while the solid ones represent the theoretical model (Eq. 10).

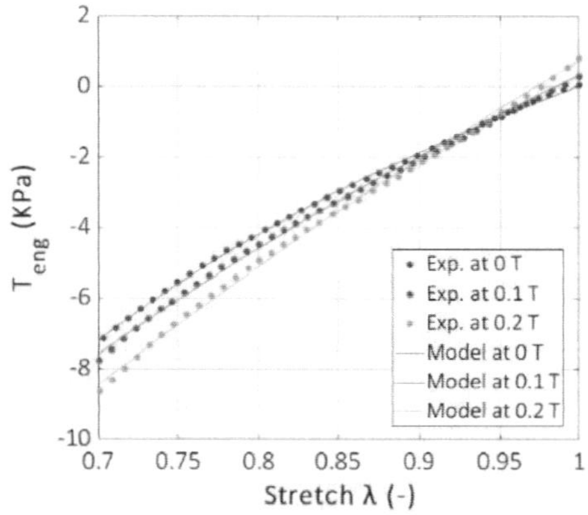

Fig. 3. Theoretical model fitting results with a quasi-static uniaxial compression for MRE foam.

Table 1. Material parameters.

	At 0 [T]	At 0.1 [T]	At 0.2 [T]
C_{01}[Pa]	-847.11	-4853.2	-7314.9
C_{10}[Pa]	7822.7	13914	18511
α [-]	0	0.387	0.399
β [-]	0	0.0406	0.0229

At zero displacements, a positive stress appears, which agrees with the isotropic MRE behavior reported by Kankanala et al. [2]. Moreover, the engineering stress increases with the magnetic field, and the maximum magnetorheological effect equals 8.7% and 20% at 0.1 T and 0.2 T, respectively. For the fitting model, a Poisson ratio of 0.22 was used based on a study performed by Sanborn et al. [9]. The theoretical model predicted the material behavior accurately using parameters shown in Table 1.

Conclusion

In this study, an isotropic MRE foam sample was fabricated using a soluble mold, liquid silicon, and carbonyl iron particles. The engineering stress was derived for compressible MREF. The magneto-hyperelastic parameters were identified using data from uniaxial compression in the presence and absence of a magnetic field.

For the fabrication process, the authors recommend using a less viscous liquid silicon rubber to reduce the setting time and acquire stiffer samples. In contrast, the theoretical model can be improved by including the porosity in the volumetric part of the energy function, which simultaneously treats the fabricated sample as porous and compressible.

Acknowledgment

This publication was written at the Technical University of Liberec as part of the project "Research of advanced materials, and application of machine learning in the area of control and modeling of mechanical systems" nr. SGS-2022-5072 with the support of the Specific University Research Grant, as provided by the Ministry of Education, Youth and Sports of the Czech Republic in the year 2022.

References

[1] Diguet, G., Froemel, J., Muroyama, M. and Ohtaka, K. "Tactile Sensing Using Magnetic Foam," *Polymers*, **14**(4), 834 (2022).

[2] Kankanala, S.V. and Triantafyllidis, N. "On finitely strained magnetorheological elastomers," *J Mech Phys Solids*, **52**(12), pp. 2869-2908 (2004).

[3] Danas, K., Kankanala, S.V. and Triantafyllidis, N. "Experiments and modeling of iron-particle-filled magnetorheological elastomers," *J Mech Phys Solids*, **60**(1), pp. 120-138 (2012).

[4] Dorfmann, A. and Ogden, R.W. "Nonlinear magnetoelastic deformations of elastomers," *Acta Mech*, **167**(1–2), pp. 13-28 (2004).

[5] Otténio, M., Destrade, M. and Ogden, R.W. "Incremental magnetoelastic deformations, with application to surface instability," *J Elast*, **90**(1), pp. 19-42 (2008).

[6] Marvalová, B. and Petríková, I. "Modelling of magneto-mechanical coupling in COMSOL multiphysics," in *Const Models for Rubber IX,* Marvalová B. and Petríková I. ed. (CRC Press, London, 2015), pp. 657–662.

[7] Janbaz, M. and Saeidi Googarchin, H. "Experimental and numerical analysis on magneto-hyper-viscoelastic constitutive responses of magnetorheological elastomers: A characterization procedure," *Mech Mater*, **154**, 103712 (2021).

[8] Holzapfel, G. "Nonlinear Solid Mechanics: A Continuum Approach for Engineering Science," *Meccanica*, **37**(4–5), pp. 489-490 (2002).

[9] Sanborn, B. and Song, B. "Poisson's ratio of a hyperelastic foam under quasi-static and dynamic loading," *Int J Impact Eng*, **123**(1), pp. 48-55 (2019).

UHPC as a Lost Formwork for Bridge Structures

Pavel Horák[1,a], Radoslav Sovják[1,b], Šárka Pešková[1,c] and Petr Vítek[2,d]

[1]*CTU in Prague, Faculty of Civil Engineering, Experimental Centre, Thákurova 7, Prague, Czech Republic;*

[2]*Hochtief CZ a. s., Plzeňská 16/3217, Prague, Czech Republic;*

[a]*pavel.horak@fsv.cvut.cz,* [b]*radoslav.sovjak@fsv.cvut.cz,* [c]*sarka.peskova@fsv.cvut.cz,* [d]*petr.vitek@hochtief.cz*

Abstract: The use of UHPC in bridge structures provides many advantages but also some disadvantages. The main disadvantages are the high technological requirements and the high costs of this material. The main advantages are durability, tensile and compressive strength or ductility. Thus, the ideal application could be creating a precast shell of UHPC that surrounds normal strength concrete (NSC) to ensure durability of bridge structures such as beams or columns. This shell also makes formwork for filling concrete. The predicted behavior of the composite beam and column was verified experimentally.

Keywords: UHPC; Shell; Beam; Column; Shear stress

1 Introduction

The use of UHPC in bridge structures provides many advantages but also some disadvantages. The main disadvantages are the high technological requirements and the high costs of this material. UHPC is thus used mostly for prefabricated parts. The main advantages are durability, tensile and compressive strength or ductility. Thus, the ideal application could be creating a precast shell of UHPC that surrounds normal strength concrete (NSC) in structures to ensure its durability in bridge structures.

The research is focused on two basic types of structures - beams and columns. Samples consist of a lightweight UHPC shell which makes formwork and NSC infill. UHPC shell can be transferred by less load-bearing cranes compared to full-size precast segments. The shell may include conventional reinforcement, fibre reinforcement, and prestressing tendons and the shell itself may be prestressed. Shell beams may be temporarily supported, but it is not necessary. Other benefits besides the absence of conventional formwork are the high environmental resistance of the UHPC shell and the visual quality of UHPC.

Once the infill concrete has hardened, a composite element is created in which both the shell and the filling concrete bear the load. Many structural design options are available to make effective use of the UHPC qualities. An important part of the static calculation is a rheological model that deals with the redistribution of stress between these two materials of different qualities and ages.

The research also investigated the possibility of using secondary materials. Concrete with fluidized bed combustion-based ternary binder (FBC-TB), also called Sorfix binder, which is produced from power plant fly ash, was used as the filling material. The great advantage of this concept is precisely the fact that the filling material is not exposed directly to the environment, because it is protected by a durable shell. If the structure is properly insulated, then water and chemicals should not get to the infill.

The shear strength of the material interface, a composite beam and a composite column both made of UHPC shell and filling concrete were tested experimentally.

2 Methods

To ensure the transfer of shear forces between the shell and the filling concrete at the area of the material interface, two methods are proposed. The first option is creating texture on the shell contact surfaces by inserting a matrix into the formwork. The second option is using steel net as formwork. The choice of the option depends on the structure. In the case of the steel net, reinforcement can pass through this type of formwork. In the case of small elements tested, a texture of protrusions was made by button foil embedded in the formwork for the UHPC shell. Previously presented results showed sufficient shear strength at the interface [1]. The obtained data were used for the numerical model, which was made for beam and column elements.

Two mock-up experiments were arranged as part of this research. The first model represented a bridge beam, which consisted of a UHPC shell and an infill of NSC. The shell had a thickness of 40 mm. The beam was 6.8 m long and was additionally prestressed. The beam span was 6.0 m. The second element was a 2.0 m high column, which consisted of a 40 mm thick rectangular UHPC shell and an infill of NSC.

3 Results

3.1 Beam

To verify the load-bearing capacity and the failure mode, a U-shaped beam was designed. The cross-section and longitudinal section are shown in Fig. 1. Beam was supplemented with blocks at both ends for prestressing anchors and prestressing force distribution. On the inner part of the UHPC shell, a texture of protrusions was made. Two beams of identical dimensions were made. In one of them, half of the cement amount in the filling concrete was replaced with an environmentally friendly binder Sorfix. Loading and failure can be seen in Fig. 2.

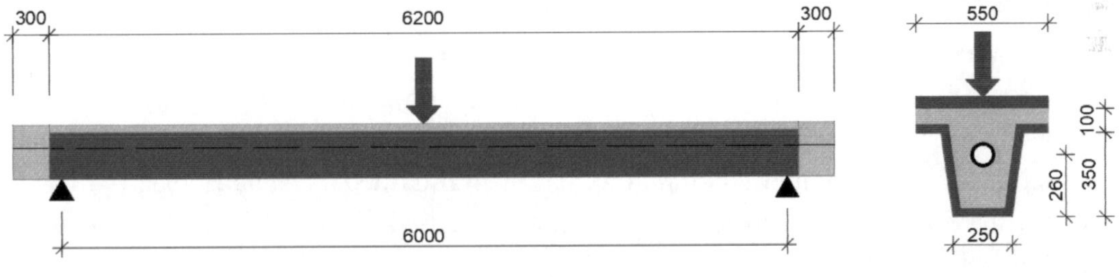

Fig. 1 Composite beam, dark-UHPC, light - normal strength concrete, in the middle is a duct for additional prestressing.

Fig. 2 Beam loading and the failure.

In the experiment, the shell itself was not prestressed. It was reinforced only with fibres and conventional steel reinforcement. The entire composite cross-section was then prestressed with an internal centric cable placed in the duct in filling concrete. The beam was designed to crush the concrete by compression and not to break the prestressing reinforcement during loading. The loading and failure of the beam are shown in Fig. 2. During loading, the first cracks were initiated in the bottom part. Subsequently, a compression failure occurred in the deck slab made of NSC. Then this crack extended and the UHPC shell later separated from the filling concrete.

The dependence of the applied force on the deformation is plotted in Fig. 3. Maximum possible shift of the hydraulic loading machine was approximately 300 mm. Once this deformation was reached, the beam had to be unloaded and an extender was placed. Then the loading continued until a deflection of approximately 500 mm.

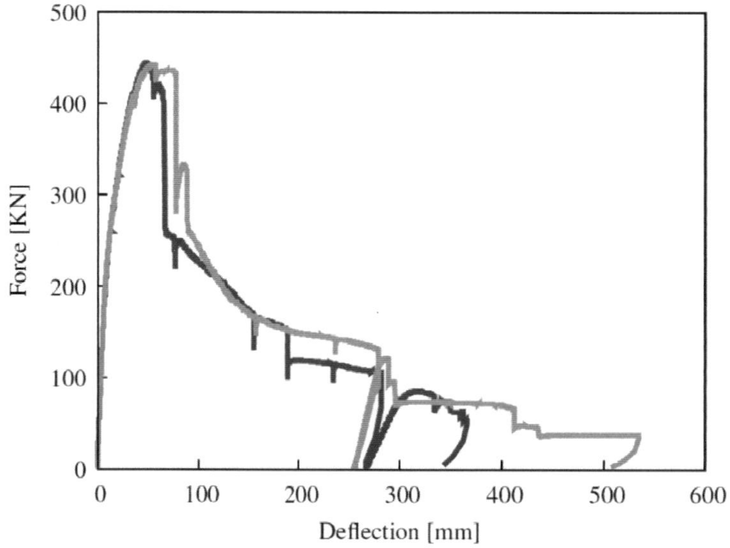

Fig. 3 Deflection of beams versus loading force.

The measured data were very close to the predictions of load-bearing capacity from the numerical analysis. This experiment provided very valuable data about the mode of failure. After reaching a deformation of 300 mm, the beam still provided a residual capacity of about 20% of the maximum load-bearing capacity. In a simplified view, this is about 40% of the design load-bearing capacity.

For large bridge beams, it would be recommended to add more connecting reinforcement to the top plate of the UHPC shell. This would prevent the delamination that can be seen in Fig. 2.

3.2 Column

The column test aimed to verify the whole manufacturing process and then the load-bearing capacity of the element. In this case, a steel net was used for the internal formwork of the shell. The shell was reinforced with longitudinal, transverse and fibre reinforcement. The column shell was concreted in the horizontal position, while the infill was concreted in the vertical position corresponding to the position in the structure. The cross-section and loading of the column are shown in Fig. 4.

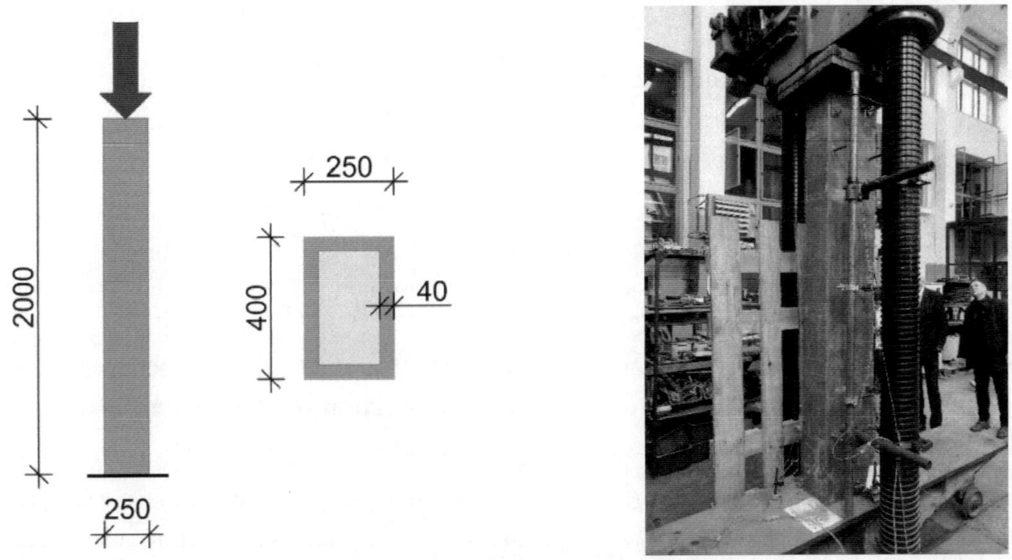

Fig. 4 Arrangement of the experiment.

The column achieved a load-bearing capacity of 4629 kN when shell failure occurred at its top, as shown in Fig. 5. The measured load-bearing capacity was lower than expected. The reason is that the loading method does not guarantee uniform load transfer to the shell and the filling concrete. More load is transferred to the shell due to the higher modulus of elasticity.

Fig. 5 Failure of the composite column.

The experiment has shown that it is necessary to pay attention mainly to the column head in real structures. This part can be connected to the superstructure by bridge bearing or directly or it can be connected to a transverse beam with bridge bearings. It seems more advantageous to create a full-size profile of UHPC in the head of the column or increase the amount of transverse reinforcement in this part of the UHPC shell.

Conclusions

UHPC has many favourable qualities that can be used for larger structures. One possibility is making composite structures combining UHPC shells with conventional concrete. The research proved the applicability of this solution to real bridge structures.

Destructive loading tests demonstrated the efficiency of this approach, the residual load-bearing capacity of elements and the mode of failure. The beam element achieved very good compliance with the predictions. It would be recommended to add more connecting reinforcement to the top plate of the UHPC shell to prevent delamination. The column element also showed a high load-bearing capacity, which could be further increased by more efficient transfer of normal force to the entire cross-section of the column and by more efficient design of transverse reinforcement at the upper part of the column.

When concreting elements made of UHPC, attention must be paid to the quality of the mixture and compliance with the technological rules. We consider it important to monitor the consistency of the mixture and its viscosity. In some cases, the mixture is difficult to work with or segregation of the dispersed reinforcement may occur.

Specimens with filling concrete, where half of the cement amount was replaced by a secondary material binder (fly ash), showed almost the same properties as the specimens with conventional concrete. Due to the properties of UHPC, this approach allows the design of a cement-free filling concrete. This can be the subject of further research.

Acknowledgements

The authors gratefully acknowledge financial support from the project CK02000047-Optimization of bridge construction and durability, using a new composite solution for UHPC and conventional concrete, mineralized admixtures and secondary materials, which was created with the support of the Technology Agency of the Czech Republic. This work was also supported by the Grant Agency of the Czech Technical University in Prague, grant No. SGS SGS23/053/OHK1/1T/11.

References

[1] Horak, P., Pešková, Š., Jogl, M., Sovják, R. and Vítek, P. "Experimental Investigation of Cohesion between UHPC and NSC Utilising Interface Protrusions," *Materials*, **15**(19), 6537 (2022).

Mechanical Properties of 3D Printed Resorbable Material for Manufacturing of Vascular Replacements

Lukáš Horný[1,a], Hynek Chlup[1,b], Katarína Mendová[1,c], Zbyněk Sobotka[1,d], Zdeněk Petřivý[1,e], Jakub Kronek[1,f], Miroslav Kohan[2,g], Tomáš Balint[2,h], and Marek Schnitzer[2,i]

[1]*Czech Technical University in Prague, Faculty of Mechanical Engineering; Technická 4 Prague 160 00, Czech Republic;*
[2]*Technical University of Košice, Faculty of Mechanical Engineering; Letná 9/B 042 00 Košice, Slovakia;*

[a]*lukas.horny@fs.cvut.cz*, [b]*hynek.chlup@fs.cvut.cz*, [c]*katarina.mendova@fs.cvut.cz*, [d]*zbynek.sobotka@fs.cvut.cz*, [e]*zdenek.petrivy@cvut.cz*, [f]*jakub.kronek@fs.cvut.cz*, [g]*miroslav.kohan@tuke.sk*, [h]*tomas.balint@tuke.sk*, [i]*marek.schnitzer@tuke.sk*

Abstract: Modern additive manufacturing methods offer the ability to achieve high geometric complexity, allowing for the production of tubular implants with intricate anatomical details that were previously unattainable. Specifically, the manufacturing process enables the accurate fabrication of structures such as the aortic arch and its branches, which display significant inter-individual geometric variability. In this study, the mechanical properties of a 3D-printed material composed of polylactic acid (PLA) and polyhydroxybutyrate (PHB) were investigated. This material holds the potential for additive manufacturing of such implants. The results of uniaxial tensile tests conducted on PLA-PHB are compared with the mechanical behavior of GORE tubes, currently utilized as vascular grafts in clinical practice. Our findings reveal that, within the range of linear elasticity, the mechanical response of PLA-PHB strips closely resembles that of circumferentially cut GORE-TEX vascular graft strips. Conversely, longitudinally oriented GORE-TEX strips exhibit greater compliance compared to the 3D-printed PLA-PHB material.

Keywords: Additive manufacturing; Bioresorbable material; Elasticity; PLA-PHB; Uniaxial tensile test; Vascular graft

1 Introduction

Bioengineering focused on tissue engineering and tissue replacements development is an emerging field providing innovations in medical technology. Bioengineering uses multidisciplinary concepts of materials science and life sciences that together lead to the development of regenerative treatments and replacements, thereby improving the function of tissues damaged by injury, disease or congenital disorders [1]. In the field of 3D printing and bioprinting of tissue substitutes, materials science focuses on obtaining compatible materials suitable for the fabrication of so-called scaffolds, which are structures whose properties at the macroscopic and microscopic level promote the formation of new tissue and colonization by recipient cells [2]. Natural and synthetic polymers most commonly used for vascular grafts and vascular replacement include collagen, silk, gelatin, expanded polytetrafluoroethylene (ePTFE), polylactic acid (PLA), poly(-caprolactone) (PCL), polyurethane (PU), polyethylene terephthalate (PET), poly(glycolic acid) (PGA), and GORE materials [3-5].

Modern additive manufacturing methods have revolutionized the production of tubular implants by enabling the fabrication of complex geometries that mimic anatomical details [6]. This breakthrough opens up possibilities for the creation of custom-made urethras, esophagi, and blood vessels [6-8]. A notable application is the development of vascular replacements that can accurately replicate intricate structures such as the aortic arch, which includes the brachiocephalic trunk, the left common carotid artery, and the left subclavian artery [8].

The focus of our research is the advancement of 3D printing techniques for vascular replacements. In this the feasibility study, which follows [9,10], the use of a biodegradable printed material composed of polylactic acid (PLA) and polyhydroxybutyrate (PHB) was investigated.

2 Material and methods

Rectangular strips of PLA-PHB were fabricated using 3D printing technology. Preprocessing of the printing process was carried out in Simplify 3D software (Simplify 3D, USA). The filament used for printing was developed in-house and prepared using the Maker Composer 450 series (3devo, Netherlands). Before the actual filament production, the apparatus was cleaned with HDPE cleaning material at 180 °C. The cleaning was followed by extrusion of PLA/PHB + TAC 25% material. The filament composition consisted of 85% PLA and 15% PHB by weight. To enhance the material's properties, 25% TAC (Tricytyl 2-Acetyl Citrate) was incorporated into the final PLA-PHB blend as a plasticizer. Fig. 1 shows the design and dimensional characteristics of the rectangular samples fabricated using Fused Deposition Modeling (FDM) technology, utilizing the DeltiQ2 printer (Trilab, Czech Republic). The temperature of the extruder was 175°C, the speed of the extrusion was 30 mm/s, and the cooling of the extruded melt was 60% (temperature of the substrate was 90°C). The entire sample preparation process was conducted at the Department of Biomedical Engineering and Measurement, Faculty of Mechanical Engineering, Technical University of Košice.

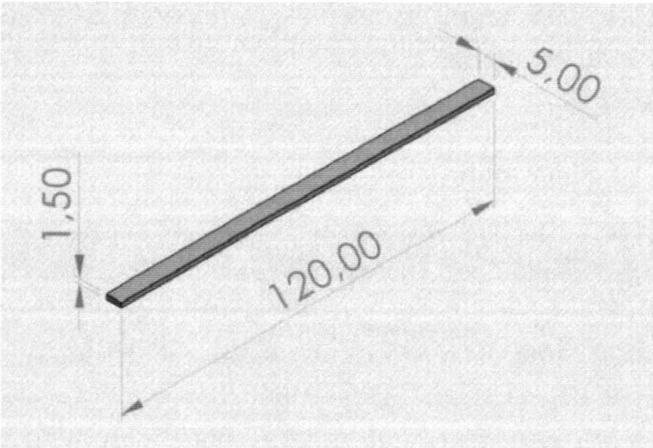

Fig. 1 Design and dimensional characteristics of rectangular samples for mechanical testing.

There is an EN ISO 7198 standard for the marketing of vascular substitutes. This specifies the tests to be carried out on a sample of tubular vascular substitutes. However, our study is at a point where we are still in the process of testing the suitability of PLA-PHB material. For such a stage of development, it is more appropriate to use conventional uniaxial tensile tests to show how the PLA-PHB material actually behaves when compared to a material that has already met EN .ISO 7198.

Cyclic uniaxial tensile tests were conducted at the Department of Mechanics, Biomechanics, and Mechatronics, Faculty of Mechanical Engineering, Czech Technical University in Prague. A Zwick/Roell universal testing machine (Zwick/Roell, Germany), see Fig. 2, specifically designed for testing soft tissues (blood vessels) and polymeric materials, was employed. The testing machine featured electromechanical actuators with a working load range of -200 to 200 N in compression-tension mode. HBM U9C +/- 25N load cells were used in our experiments. The machine offered a displacement range of 0 to 140 mm, with a position resolution of 1 μm. Deformation of the specimens was measured using a built-in video extensometer equipped with a 5 Mpx uEye 3.0 CMOS camera. The applied loading rate was set at 0.2 mm/s. The specimens underwent preconditioning through 6 loading cycles, reaching up to 7 N and then returning to 0 N (not necessarily back to the initial length). All data were recorded on a control PC at a sampling rate of 20 Hz for subsequent post-processing.

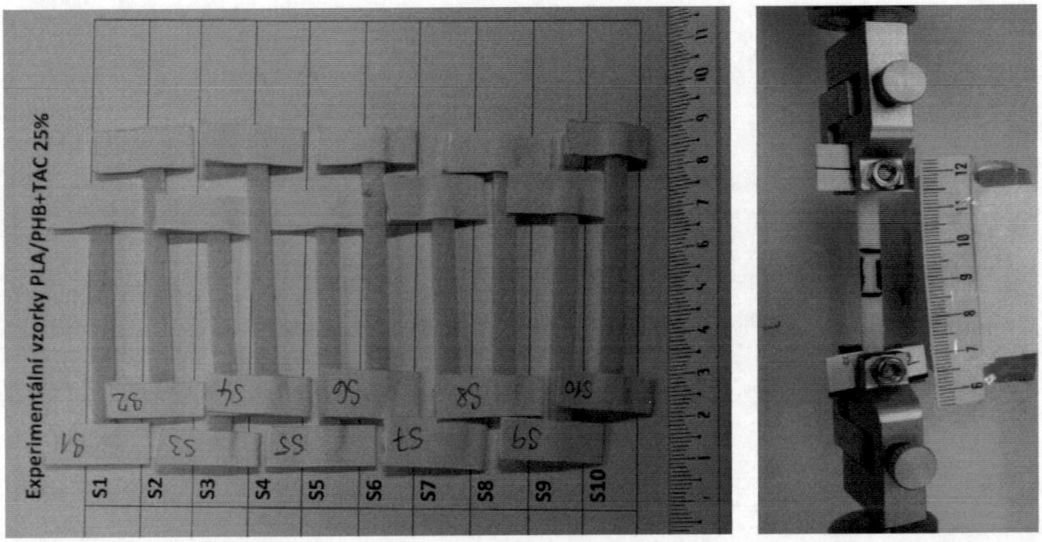

Fig. 2 Rectangular samples of PLA-PHB prior to the tensile testing (left panel) and the sample mounted in the testing machine (right panel).

The mechanical response of the tested specimens was described using the nominal stress $\sigma = F/S_0$, where F represented the recorded loading force, and S_0 denoted the reference cross-sectional area of the specimens. Deformation was quantified as the engineering strain $\varepsilon = l/L - 1$, where L referred to the distance between marks made on the specimen's surface prior to the test, and l represented this distance in the deformed state. The initial Young's modulus E_{ini} was determined by performing linear regression on the recorded responses within the small deformation (linear) regime ($\varepsilon < 0.01$).

In addition to the PLA-PHB strips, mechanical tests were also conducted on commercially available samples of GORE-TEX material. Longitudinal and circumferential strips were prepared from a vascular substitute marketed under the brand name SA1802 (W. L. Gore & Associates, Inc., USA) and subjected to the same uniaxial tensile tests as the PLA-PHB strips.

3 Results and discussion

The stress-strain relationships obtained from the experiments are depicted in Fig. 3 and Fig. 4. Fig. 3 illustrates the responses of PLA-PHB and GORE-TEX materials during the first cycle of loading, while the preconditioned response of the GORE material (6th cycle of loading) is presented in Fig. 4. It was observed that the preconditioned response of PLA-PHB exhibited negligible differences compared to the primary response, hence only the primary response is shown in both figures. However, this was not the case for the GORE material. Its response indicated saturation of hereditary effects, and the response during the 6th cycle was deemed repeatable.

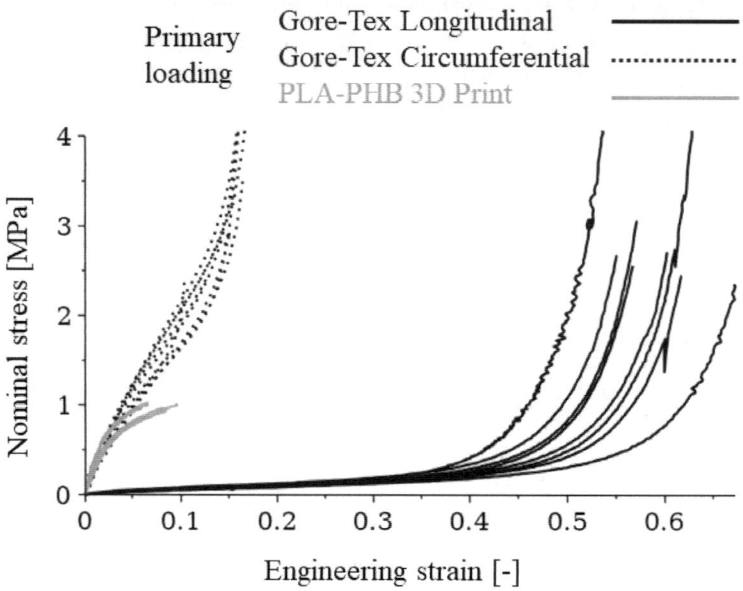

Fig. 3 Stress-strain relationships determined in uniaxial tensile tests for the pristine PLA-PHB and GORE-TEX materials.

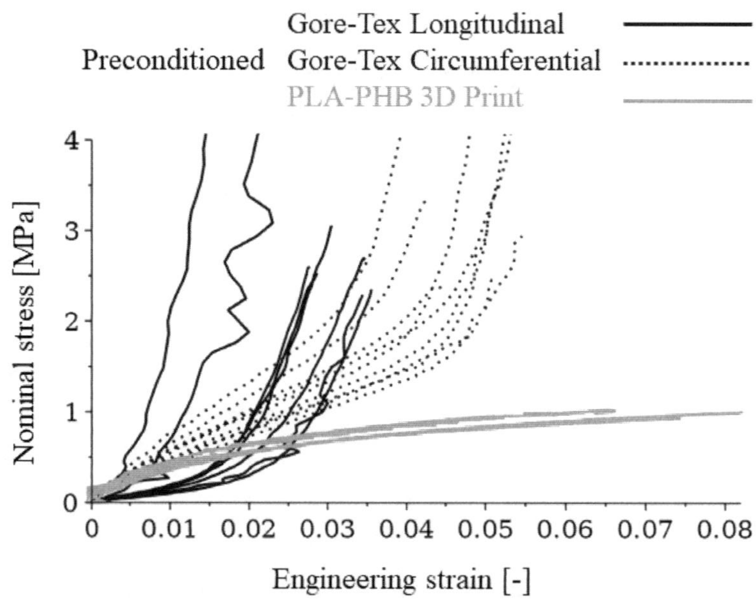

Fig. 4 Stress-strain relationships determined in uniaxial tensile tests for the preconditioned (6th loading cycle) PLA-PHB and GORE-TEX materials.

In contrast to PLA-PHB, where all rectangular samples were obtained from the same printing plane, GORE samples were cut from the tube in two different orientations, leading to anisotropic behaviour. Fig. 3 and Fig. 4 illustrate the stress-strain responses of GORE material in longitudinal and circumferential orientations. It is evident that longitudinal strips of GORE material exhibit significantly higher initial compliance compared to circumferential strips, regardless of the loading history.

The most significant observation from our study is that within the linear regime (small deformations with $\varepsilon < 0.01$), the mechanical response of circumferentially oriented GORE strips closely resembles that of PLA-PHB. The mean initial Young's modulus obtained for PLA-PHB samples was E_{ini} = 39.04 MPa (± 8.27 MPa SD). In comparison, GORE strips exhibited E_{ini} values of 1.49 MPa (± 0.339 MPa), 30.1 MPa (± 5.84 MPa), 41.7 MPa (± 61.0 MPa), and 54.9 MPa (± 10.4 MPa) for the 1st cycle in the longitudinal and circumferential directions, and for the 6th cycle in the longitudinal and circumferential directions, respectively. These results suggest that although there are noticeable differences in the perceived compliance when handling the materials (GORE appearing more compliant), objective measurements indicate that this difference is not significant in the small deformation range. Fig. 5 provides a detailed comparison of the mechanical properties between the two materials through a box plot representation.

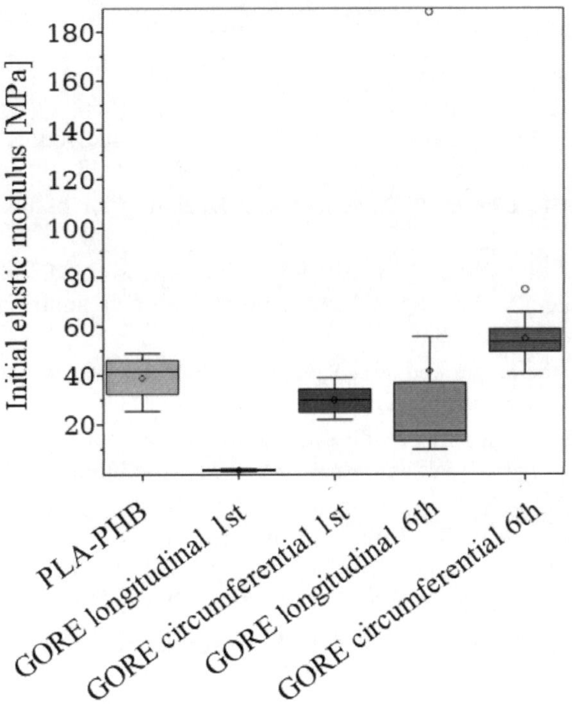

Fig. 5 Box plot comparing the obtained values of the initial Young's modulus of elasticity for PLA-PHB and GORE-TEX materials.

Conclusion

Our study focused on comparing the mechanical properties of PLA-PHB material, which is easily applicable in current 3D-printing methods, and GORE-TEX material used for vascular replacements. Using uniaxial tensile testing, the response of GORE-TEX was found to be significantly non-linear, anisotropic and dependent on the deformation history. This behaviour of GORE-TEX can be attributed to its elastomeric nature, where gradually straightening molecular chains are responsible for its large strain stiffening.

On the other hand, the observed behaviour of PLA-PHB was consistent with the response known for thermoplastics and was not highly dependent on loading history in the small strain regime. Although GORE-TEX is more compliant than PLA-PHB in a certain strain range, in the range of validity of the linear approximation (small strains), the differences between PLA-PHB and GORE were not significant. As GORE-TEX is an established mateiral in the field of vascular replacements, this comparison suggests that a vascular replacement could be developed using PLA-PHB with mechanical properties that match the physiological conditions of blood vessels. However, it is of course clear that further research is needed to verify the long-term mechanical behaviour, biocompatibility, and other properties of the PLA-PHB material to comprehensively evaluate its suitability.

Acknowledgements

We would like to acknowledge the support provided by the Czech Ministry of Education, Youth and Sport in the project LUASK22174 "Bioresorbable materials for additive manufacturing of vascular replacements and their biomechanical characterization." Additionally, we are grateful for the support received from the Slovak Research and Development Agency in the project SK-CZ-RD-21-0056 "Bioresorbable materials for additive manufacturing of vessel substituents and their biomechanical characterization."

References

[1] OLSON, Jennifer L.; ATALA, Anthony; YOO, James J. Tissue engineering: current strategies and future directions. *Chonnam medical journal*, 2011, 47.1: 1-13.

[2] ZASZCZYŃSKA, Angelika, et al. Advances in 3D printing for tissue engineering. *Materials*, 2021, 14.12: 3149.

[3] WANG, Dongfang, et al. Artificial small-diameter blood vessels: Materials, fabrication, surface modification, mechanical properties, and bioactive functionalities. *Journal of Materials Chemistry B*, 2020, 8.9: 1801-1822.

[4] LI, Mei-Xian, et al. Challenges and advances in materials and fabrication technologies of small-diameter vascular grafts. *Biomaterials Research*, 2023, 27.1: 58.

[5] PÁEZ, JM García, et al. Comparison of the mechanical behaviours of biological tissues subjected to uniaxial tensile testing: pig, calf and ostrich pericardium sutured with Gore-Tex. *Biomaterials*, 2003, 24.9: 1671-1679.

[6] Findrik Balogová, A., Hudák, R., Tóth, T., Schnitzer, M., Feranc, J., Bakoš, D. and Živčák, J. "Determination of geometrical and viscoelastic properties of PLA/PHB samples made by additive manufacturing for urethral substitution," *J Biotechnol*, **284**, pp. 123-130 (2018).

[7] Baudis, S., Nehl, F., Ligon, S.C., Nigisch, A., Bergmeister, H., Bernhard, D. and Liska, R. "Elastomeric degradable biomaterials by photopolymerization-based CAD-CAM for vascular tissue engineering," *Biomed Mater*, **6**(5), 055003 (2011).

[8] Marti, P., Lampus, F., Benevento, D. and Setacci, C. "Trends in use of 3D printing in vascular surgery: A survey," *Int Angiol*, **38**(5), pp. 418-424 (2019).

[9] Balint, T., Findrik Balogová, A., Hudák, R., Živčák, J., Schnitzer, M. and Feranc, J. "Production, additive printing and mechanical testing of PLA/PHB material with different concentrations of TAC emollient," in *Proceedings of the 19th IEEE World Symposium on Applied Machine Intelligence and Informatics, SAMI 2021* (Herľany, Slovakia, 2021).

[10] Kohan, M., Lancoš, S., Schnitzer, M., Živčák, J. and Hudák, R. "Analysis of PLA/PHB Biopolymer Material with Admixture of Hydroxyapatite and Tricalcium Phosphate for Clinical Use," *Polymers*, **14**(24), 5357 (2022).

Determination of the Elastic Constants of a Composite Using FE Model Updating Method

Róbert Huňady[1,a], Adam Kaľavský[1,b], Peter Palička[1,c] and Štefan Kender[1,d]

[1]*Technical University of Košice; Letná 1/9, Košice, Slovakia;*

[a]*robert.hunady@tuke.sk*, [b]*adam.kalavsky@continental-corporation.com*, [c]*peter.palicka@tuke.sk*, [d]*stefan.kender@tuke.sk*

Abstract: The paper deals with an alternative approach for determining elastic constants using the FE model update method. The object of interest is a unidirectional carbon fibre reinforced composite plate, which is subjected to experimental modal analysis. The natural frequencies of the freely supported plate are then used in an optimization process that aims to tune the homogenized numerical model of the plate to achieve modal parameters agreement. The material constants obtained in this way are compared with the results of mechanical tests.

Keywords: Elastic constants; FE model updating; carbon fibre composite; modal analysis

1 Introduction

Knowledge of the material properties is crucial for describing its behaviour in terms of elastic response, plastic response, failure or fatigue. There are several methods for determining material constants and mechanical properties, which are mainly based on material tests. Many of them are standardised. The implementation of these tests can usually be time and cost consuming, as measurements need to be performed on a larger number of samples in order to increase statistical precision. The determination of elastic constants is a relatively simple matter, especially for homogeneous isotropic materials. However, in the case of orthotropic materials, such as fibre-reinforced composites, the situation becomes more complex as the number of constants increases with the degree of orthotropy. One alternative to the material tests required to determine elastic moduli, shear moduli and Poisson's ratios is the FE model updating method [1]. Its goal is to optimize the unknown parameters of the FEM model so that its response matches the measured data. The objective function can be the displacement or stress at a certain point, or the natural frequencies of the model, etc. This is a method that has wide application in practice. It can be used in dynamic analysis to identify damping in the structure [2]. The paper [3] presents the use of the FE model updating method to determine the damage of a structure based on the PSD of the vibration response obtained by measurement. Teughels and De Roeck used measured modal parameters to identify damage to the highway bridge structure [4]. A similar approach was used by the authors in [5] to determine the bond stiffness parameters. They also demonstrated the possibility of replacing the part of the numerical model at the location of the structure partition with a flexible bond (bushing) of equivalent stiffness. The method is also applicable to the refinement of various material parameters of composites or their components [6-8]. In the study [9], the authors tuned a material model of a 3D printed structure using a genetic algorithm (GA), trying to capture the relaxation and creep of the material. The experimental-numerical approach can also be used to determine residual stresses using the inverse algorithm, as shown in [10].

In this work, the material constants of a unidirectional carbon fibre reinforced composite plate are determined using FE model updating method. Basis for this is the measured natural frequencies and plate oscillation shapes. The basis is an experimental modal analysis of the plate. To verify the approach used, the material constants are also determined by mechanical tests. The results are compared with each other and discussed.

2 Composite sheet and tested specimens

A vacuum bag moulding method combined with hand lay-up was used to produce the composite sheet. Using this method, the pressure inside the bag is reduced and the external pressure is increased, thereby removing excess air and resin. This method helps to produce laminates with uniform consolidation, complete wetting of the fibres and better core bonding. The composite material consisted of epoxy resin (MGS L258) and unidirectional carbon fibers (200 g/m^2, 50 K) that were stacked in eight layers. Each layer was separately saturated with resin to achieve homogeneous properties through the thickness. The resulting nominal sheet thickness was approximately 2 mm. The vacuuming process can be seen in Fig. 1a. Subsequently, the composite sheet was cured in an oven for 12 hours at 60 °C. After curing, specimens for mechanical testing and modal analysis were made from the sheet by milling (Fig. 1b).

(a)  (b) 

Fig. 1 (a) vacuum bag moulding process, (b) milling of specimens.

3 Constitutive model of the composite

The unidirectionally reinforced composite represents a transversely isotropic material. This material is symmetric about an axis that is normal to a plane of isotropy (see Fig. 2). This transverse plane has infinite planes of symmetry and thus, within this plane, the material properties are same in all directions. This is a special case of orthotropy, where 5 independent constants in the elasticity tensor are needed to describe the linear elastic behaviour of the material.

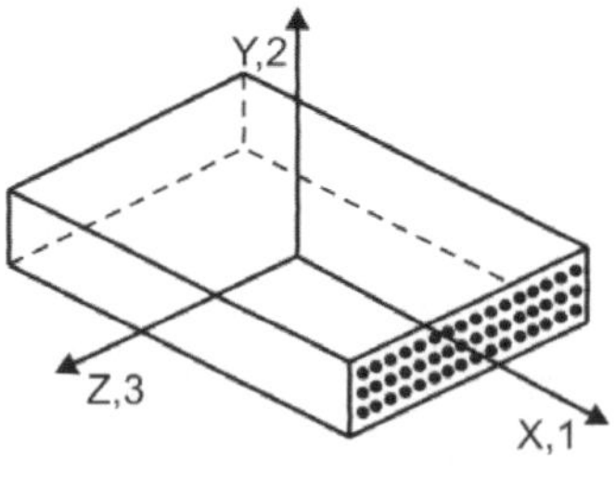

Fig. 2 Transversely isotropic material.

Assuming the 2–3 plane to be the plane of isotropy, transverse isotropy requires that $E_2 = E_3 = E_p$, $\nu_{12} = \nu_{13} = \nu_{tp}$, $\nu_{21} = \nu_{31} = \nu_{pt}$, $G_{12} = G_{13} = G_t$, where p and t stand for "in-plane" and "transverse," respectively. Thus, while ν_{tp} has the physical interpretation of the Poisson's ratio that characterizes the strain in the plane of isotropy resulting from stress normal to it, ν_{pt} characterizes the transverse strain in the direction normal to the plane of isotropy resulting from stress in the plane of isotropy. In general, the quantities ν_{tp} and ν_{pt} are not equal and are related by $\frac{\nu_{tp}}{E_t} = \frac{\nu_{pt}}{E_p}$. The stress-strain laws reduce to [11]

$$\begin{Bmatrix} \varepsilon_{11} \\ \varepsilon_{22} \\ \varepsilon_{33} \\ \gamma_{12} \\ \gamma_{13} \\ \gamma_{23} \end{Bmatrix} = \begin{bmatrix} 1/E_t & -\nu_{pt}/E_p & -\nu_{pt}/E_p & 0 & 0 & 0 \\ -\nu_{tp}/E_t & 1/E_p & -\nu_p/E_p & 0 & 0 & 0 \\ -\nu_{tp}/E_t & -\nu_p/E_p & 1/E_p & 0 & 0 & 0 \\ 0 & 0 & 0 & 1/G_t & 0 & 0 \\ 0 & 0 & 0 & 0 & 1/G_t & 0 \\ 0 & 0 & 0 & 0 & 0 & 1/G_p \end{bmatrix} \begin{Bmatrix} \sigma_{11} \\ \sigma_{22} \\ \sigma_{33} \\ \sigma_{12} \\ \sigma_{13} \\ \sigma_{23} \end{Bmatrix}, \tag{1}$$

where $G_p = E_p / 2(1+\nu_p)$. In the transversely isotropic case the stability relations for orthotropic elasticity are

$$\begin{aligned} &E_p, E_t, G_p, G_t > 0, \\ &|\nu_p| < 1, \ |\nu_{pt}| < \sqrt{E_p / E_t}, \ |\nu_{tp}| < \sqrt{E_t / E_p}, \\ &1 - \nu_p^2 - 2\nu_{pt}\nu_{tp} - 2\nu_p\nu_{pt}\nu_{tp} > 0. \end{aligned} \tag{2}$$

4 Mechanical tests

The elastic constants of the composite material were determined based on mechanical tensile and shear tests according to the relevant standards [12-14]. Loads were measured by force sensors integrated in the test machines. A Dantec Dynamics Q-400 3D correlation system was used to measure strain components (Fig. 3).

(a)

(b)
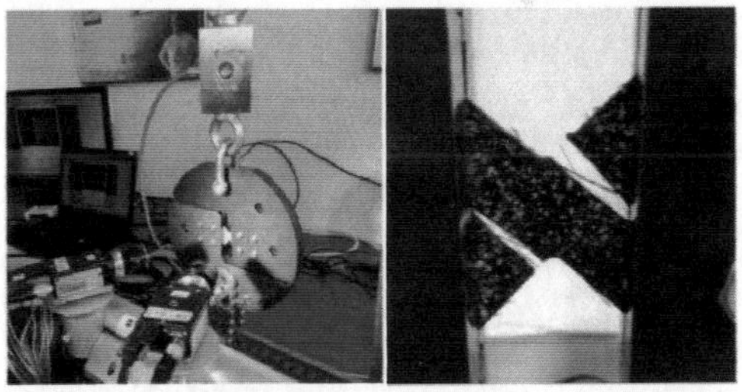

Fig. 3 Testing of material properties in tension (a) and shear (b).

From the measured data, elastic moduli, shear moduli and Poisson's ratio values were calculated using well-known procedures. To increase the statistical accuracy, 5 specimens were tested and the calculated values were averaged. The resulting material constants are shown in Table 1.

Table 1. Material constants of the composite obtained by mechanical tests.

E1 [MPa]	E2 [MPa]	G12=G13 [MPa]	G23 [MPa]	Nu12 [-]
120442	5779	3492	-	0.289

5 Experimental modal analysis

The test object was a rectangular plate with dimensions of 200.6 x 130.5 mm, whose average thickness was 1.969 mm. The plate was cut from the composite sheet at an angle of 40°. The mass of the plate determined by weighing was 66.44 g. The density of the homogenized plate was determined by calculation to be $1.289 \cdot 10^{-9}$ t/mm^3. The plate was placed on thin elastic rubber bands so that the conditions for a free fit were met. 3D high-speed correlation system Q-450 Dantec Dynamics was used to measure vibration response (Fig. 4). A random stochastic pattern was created on the surface of the slab for the purpose of correlation measurements.

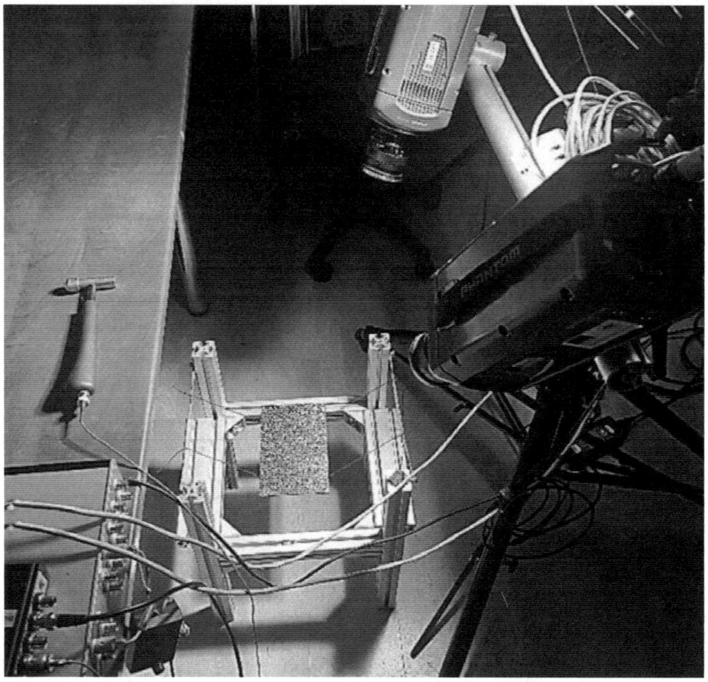

Fig. 4 Experimental configuration for measuring modal parameters of a composite plate.

The structure was excited with a Bruel&Kjaer Type 8206 modal hammer with a single impact on the bottom of the plate. The modal parameters (Fig. 5) were extracted using DICMAN 3D software [15,16]. Ten modes were identified whose frequencies were later used in the optimization process of FEM model.

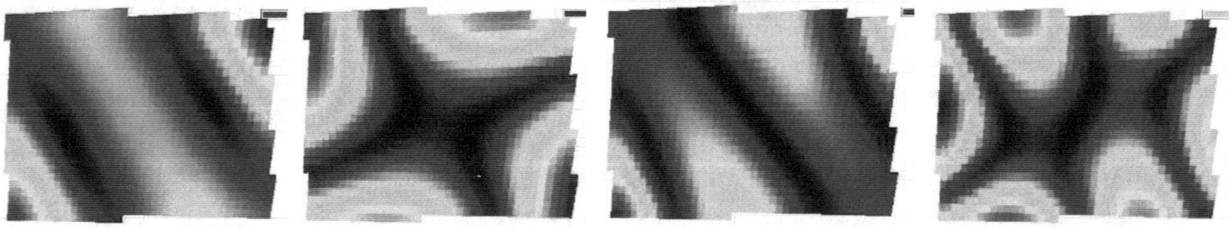

1. mode: 110.4 Hz 2. mode: 197.7 Hz 3. mode: 281.3 Hz 4. mode: 452.2 Hz

Fig. 5 The first four mode shapes and natural frequencies of the composite plate.

6 FE model updating

The FEM model of the plate was created in Abaqus/CAE. The plate was modelled as a homogenised shell, so a *"Lamina"* type material model was used to define its elastic properties. The plate was considered to be free, i.e. without any constraints. A linear perturbation procedure of type "*Frequency*" was used to calculate the natural frequencies and mode shapes. Rigid body modes have been excluded. The mesh was formed by 260 linear quadrilateral elements of type S4R with 294 nodes (Fig. 6).

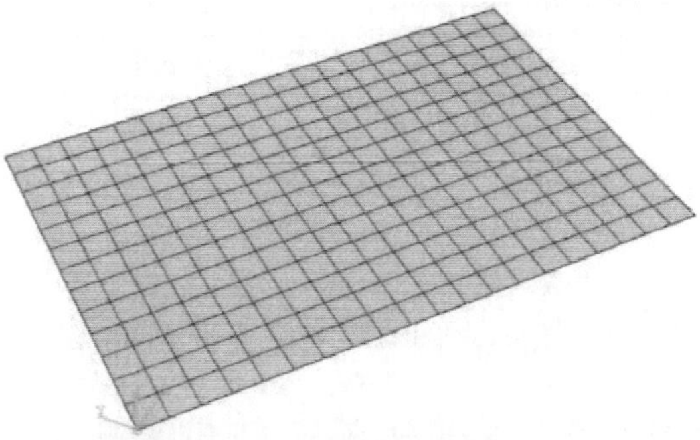

Fig. 6 FE model of the plate.

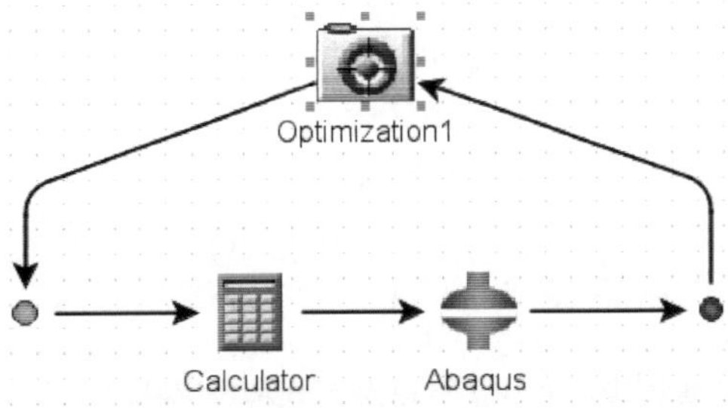

Fig. 7 Diagram of the optimization process.

Within the definition of geometric and material properties, the average density and thickness of the plate were considered according to Section 5. The material coordinate system was rotated at an angle of 40°. Other material constants were assumed to be unknown. Their values were determined by optimization in Isight. The aim of the optimization was to tune the model so that its natural frequencies matched the measured ones. A block diagram of the optimization process is shown in Fig. 7. As an optimization algorithm was used Neighborhood Cultivation Genetic Algorithm (NCGA), which is based on Pareto optimal solution [17]. Block 'Calculator' verified the fulfilment of stability conditions (2). The FEM model was included in the 'Abaqus' block. Five parameters were optimized: E1, E2, Nu12, G12=G13, G23. These parameters were defined as design variables. Their initial bounds are shown in Table 2. The natural frequencies determined by the measurements were used as objective functions (see Table 3). For the first five, optimization constraints were at the same time defined (Table 4).

Table 2. Initial and final lower and upper bounds of design variables.

Name	Initial Lower Bound	Initial Upper Bound	Final Lower Bound	Final Upper Bound
Elastic_LAMINA_1_E1	2000	200000	85000	120000
Elastic_LAMINA_1_E2	200	20000	4500	7500
Elastic_LAMINA_1_G23	200	20000	1500	7500
Elastic_LAMINA_1_Nu12	-1	1	0.1	0.35
Elastic_LAMINA_1_G13	200	20000	1500	5000

Table 3. Objective functions.

Name	Direction	Target
Step_1__History__EIGFREQ[0,1]	target	110.4
Step_1__History__EIGFREQ[1,1]	target	197.7
Step_1__History__EIGFREQ[2,1]	target	281.34
Step_1__History__EIGFREQ[3,1]	target	425.2
Step_1__History__EIGFREQ[4,1]	target	508.2
Step_1__History__EIGFREQ[5,1]	target	577.6
Step_1__History__EIGFREQ[6,1]	target	741.5
Step_1__History__EIGFREQ[7,1]	target	831.15

Table 4. Optimization constraints.

Name	Lower Bound	Upper Bound
Step_1__History__EIGFREQ[0,1]	105.0	115.0
Step_1__History__EIGFREQ[1,1]	190.0	210.0
Step_1__History__EIGFREQ[2,1]	270.0	290.0
Step_1__History__EIGFREQ[3,1]	415.0	435.0
Step_1__History__EIGFREQ[4,1]	490.0	520.0

It should be noted that the optimization process was run four times, with the interval of the design variables being narrowed after each run. The boundaries of the last run are shown in Table 2. The elastic constants determined by Isight are listed in Table 5.

Table 5. Material constants of the composite obtained by optimization in Isight.

E1 [MPa]	E2 [MPa]	G12=G13 [MPa]	G23 [MPa]	Nu12 [-]
115750	5367	3289	3603	0.309

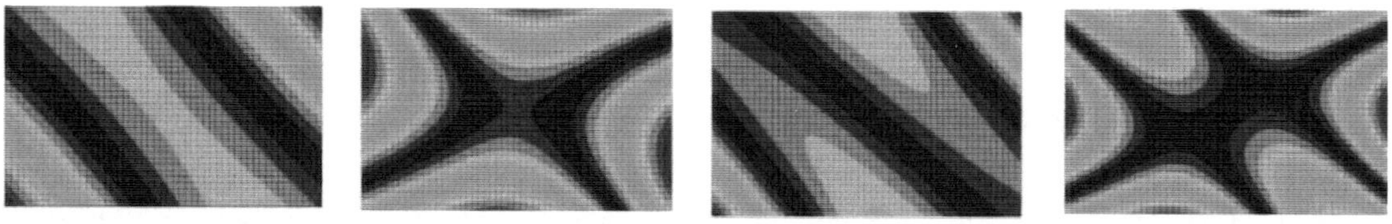

1. mode: 115.7 Hz   2. mode: 203.8 Hz   3. mode: 286.6 Hz   4. mode: 435.5 Hz

Fig. 8 The first four mode shapes and natural frequencies of the optimized FE model of the composite plate.

Fig. 8 shows the mode shapes and natural frequencies of the FE model of the homogenised plate with the optimized material constants. Table 6 shows the percentage differences in the values of elastic constants and natural frequencies determined numerically and by the measurement. The measured data are considered as reference data. Comparison of the simulation and experimental results shows very good agreement, indicating high estimation accuracy.

Table 6. Percentage differences between numerical estimation and experimental measurement.

E1	E2	G12=G13	Nu12	1. mode	2. mode	3. mode	4. mode
3.9 %	7.1 %	5.8 %	-6.9 %	-4.8 %	-3.1 %	-1.9 %	-2.4 %

Conclusion

The paper presented an alternative method for determining the material constants of a composite material, which is based on the FE model updating method. The elastic parameters of the homogenized numerical model are optimized so that its natural frequencies match the frequencies determined by the experimental modal test. The accuracy of the estimation depends on several factors such as the dimensions and shape of the structure, the average density of the material, the way the structure was supported during modal analysis (bond stiffness), which must be known. If the orientation of the fibres is not known, it can be included as a design variable, in the same way the density value can be refined. As shown, the results obtained by FE model updating are in reasonable agreement with the results of the mechanical tests. With respect to the course of the optimization process, it should be noted that the influence of the Poisson's ratio on the natural frequencies is not very pronounced. Its value could be refined by including a MAC criterion comparing the mode shapes of the structure.

Acknowledgements

This paper was supported by the Ministry of Education of Slovakia Foundation under grant project VEGA 1/0500/20 and VEGA 1/0516/22.

References

[1] Huňady, R., Pavelka, P. and Lengvarský, P., "Homogenization of material properties of a combined rubber/steel sheet using modal analysis," in *Experimental Stress Analysis 2019 - Conference Proceedings*, Luhačovice, Czech Republic, 2019 (Czech Society for Mechanics, 2019), pp. 146-150.

[2] Arora, V., Singh, S.P. and Kundra, T.K., "Comparative study of damped FE model updating methods," *Mech Syst Signal Pr*, **23**(7), pp. 2113-2129 (2009).

[3] Pedram, M., Esfandiari, A. and Khedmati, M.R., "Damage detection by a FE model updating method using power spectral density: Numerical and experimental investigation," *J Sound Vib*, **397**, pp. 51-76, (2017).

[4] Teughels, A. and De Roeck, G., "Structural damage identification of the highway bridge Z24 by FE model updating," *J Sound Vib*, **278**(3), pp. 589-610, (2004).

[5] Huňady, R. Lengvarský, P., Pavelka, P., Kaľavský, A. and Mlotek, J., "Stiffness estimation and equivalence of boundary conditions in FEM models," *Applied Sciences*, **11**(4), pp. 1-14, 1482 (2021).

[6] Cunha, J. and Piranda, J., "Application of model updating techniques in dynamics for the identification of elastic constants of composite materials," *Compos Part B-Eng*, 30(1), pp. 79-85, (1999).

[7] Huňady, R., Pavelka, P., Lengvarský, P. and Kaľavský, A., "The estimation of elastic material constants of unidirectional fibre-reinforced composite based on modal analysis," in *Proceedings of the MMaMS 2019*, Sromowce Niżne, Poland, 11–13 September 2019 (TUKE 2019), pp. 1–9.

[8] Bagha, A.K., Gupta, P. and Panwar, V., "Finite element model updating of a composite material beam using direct updating method," *Mater Today-Proc*, **27**, Part 3, pp. 1947-1950, (2020).

[9] Fusek, M., Paška, Z. and Rojiček, J. "Identification of material parameters and material model for 3D printed structure", in *Proceedings of Experimental Stress Analysis - 58th International Scientific Conference*, EAN 2020, pp. 87-92.

[10] Pecenka, L., Halama, R., Fojtík, F., Paška, Z. and Rojícek, J. "Hybrid method suitable for measurement of residual stresses by hole-drilling technique", in *Proceedings of Experimental Stress Analysis - 57th International Scientific Conference*, EAN 2019, pp. 384-389.

[11] Abaqus Documentation, Dassault Systèmes, 2015.

[12] ČSN EN 2561: Aeronautics and astronautics - Carbon fibre reinforced plastics - Unidirectional composites - Tensile test parallel to fibre direction (in Czech).

[13] ČSN EN 2597: Aeronautics and astronautics - Carbon fibre reinforced plastics - Unidirectional composites - Tensile test perpendicular to fibre direction (in Czech).

[14] ASTM D7078: Standard Test Method for Shear Properties of Composite Materials by V-Notched Rail Shear Method.

[15] Huňady, R. and Hagara, M., "A new procedure of modal parameter estimation for high-speed digital image correlation," *Mech Syst Signal Pr*, **93**, pp. 66-79 (2017).

[16] Huňady, R., Pavelka, P. and Lengvarský, P., "Vibration and modal analysis of a rotating disc using high-speed 3D digital image correlation," *Mech Syst Signal Pr*, **121**, pp. 201-214 (2019).

[17] Watanabe, S., Hiroyasu, T. and Miki, M., "NCGA: Neighborhood Cul-tivation Genetic Algorithm for Multi-Objective Optimization Problems," in *Proceedings of the Genetic and Evolutionary Computation Conference,* GECCO Late Breaking Papers (2002), pp. 458-465.

Biodegradable vs. Conventional Implants in Skeletal Traumatology: A Study of Load Capacity in the Inner Ankle Region

Aleš Jíra[1,a] and Jiří Dupal[2, 3b]

[1]*Czech Technical University in Prague, Faculty of Civil engineering, Department of Mechanics; Thákurova 7, Prague 6, 166 29, Czech Republic;*
[2]*Department of Traumatology, Faculty of Medicine and Dentistry, Palacky University Olomouc, Hnevotinska 976/3, Olomouc 775 15, Czech Republic;*
[3]*Orthopaedic Department of Sumperk Hospital, Nerudova 640/41, Sumperk, 787 01, Czech Republic;*

[a]*jira@fsv.cvut.cz,* [b]*jiri.dupal@fnol.cz*

Abstract: This study investigates the use of biodegradable implants in skeletal traumatology. The focus is on evaluating the cyclic loading of intraosseous implants in the inner ankle region and comparing the load capacity of conventional and biodegradable implants in healthy tissue. Mechanical testing and retrospective analysis were conducted on chromium nickel and magnesium implants to assess their strength and stability. Laboratory testing involved simulating ankle fractures and analyzing the performance of different implants. The results show significant differences in fatigue capacity, with the biodegradable implants demonstrating lower ultimate stress but sufficient load-bearing capacity under actual stress conditions during the healing process. These findings contribute to our understanding of implant performance in skeletal traumatology and highlight the potential of biodegradable implants as a viable alternative in reducing invasiveness and improving patient outcomes.

Keywords: Biodegradable implants; Skeletal traumatology; Cyclic loading; Ultimate fatigue capacity

1 Introduction

In the field of skeletal traumatology, when conservative treatment is not feasible, the primary objective is to minimize the invasiveness of surgical interventions. Presently, in addition to conventional implants, absorbable and biodegradable implants can be utilized for osteosynthesis. The integration of these absorbable materials into the therapeutic procedure mitigates certain side effects associated with osteosynthesis, particularly the need for a secondary surgical intervention for implant removal. This considerably reduces the overall surgical risks, such as general anesthesia, infection, embolism, and also alleviates the economic burden on the healthcare system. Consequently, the utilization of biodegradable and resorbable materials can be perceived as a minimally invasive approach [1].

Certain types of biodegradable materials possess the capability to degrade and be absorbed by the body without adversely affecting the bone tissue. Their crucial mechanical property lies in maintaining static stability, ensuring the strength of the implant over a specific duration, particularly during the healing of bone tissue (e.g., fracture healing). During the initial phase, the fracture stability is upheld by the osteosynthetic material, enabling bone healing to occur. Subsequently, the integrity and stability of the bone tissue are restored (bone callus is formed), and it is during this phase that the process of biodegradation accelerates [2]. The specific type and chemical properties of biodegradation are influenced by the composition of the implant itself.

Currently, the predominant material used for biodegradable implants is poly-L-lactide (PLLA) due to its exceptional mechanical properties, stability, natural biocompatibility, and, importantly, ease of processing [3]. However, the degradation process of PLLA in the body takes approximately 14 to 36 months, which does not align with the timeframe required for the healing of living bone (8 to 12 weeks) [4]. Moreover, the acidic by-products resulting from PLLA degradation often incite a local inflammatory response that hinders tissue regeneration [4]. PLLA also lacks osteogenic activity and can even impede osteogenic proliferation and differentiation due to its hydrophobic nature [4]. Consequently, alternative materials with improved characteristics are actively sought in the field of orthopedics and skeletal traumatology.

One alternative is the utilization of novel polymeric materials such as polyglycolide (PGA), polycaprolactone (PCL), or their combination with PLLA [5]. These materials exhibit faster degradation rates and elicit more favorable biological responses compared to PLLA. Additionally, they demonstrate the ability to promote osteogenesis and facilitate bone regeneration [6]. Another option involves the use of osteosynthetic materials based on hydroxyapatite (HA) or tricalcium phosphate (TCP), which significantly enhance ossteointegration and can expedite bone healing. However, TCP's biomechanical properties render it brittle and prone to disintegration. Studies indicate that this material is not advantageous for osteosynthesis and is instead primarily employed for bone cyst filling [7].

Another category of materials applicable in bone surgery comprises biodegradable implants based on magnesium alloys. The initial attempts to employ magnesium in bone surgery date back to the early 19th century, exemplified by Lambotte's pioneering operation [8]. The notable advantage of magnesium alloys lies in their modulus of elasticity and strength, which closely match those of bone tissue, unlike common materials used in osteosynthesis such as stainless steel or titanium. For instance, the modulus of elasticity of bone ranges between 3 and 20 GPa, while that of magnesium falls within 41-45 GPa, and titanium alloy exhibits a range of 110-117 GPa [9].

Due to the desirable properties and biodegradability of magnesium alloys, it is possible to minimize undesirable effects such as stress shielding, soft tissue irritation, loosening, or osteopenia that can occur with standard implants [9]. One notable innovative material based on magnesium alloys is an osteosynthetic material called Magnezix (Syntellix AG, Hannover, Germany). Introduced in 2013, Magnezix was the first magnesium implant approved for human use [10]. The chemical composition of this magnesium alloy, known as MgYREZr or "WE43," comprises magnesium, zirconium, and rare earth elements (REEs) such as yttrium, gadolinium, and others [7]. Through this metallurgical process and alloying, the material achieves high strength and a certain degree of stability during degradation via corrosion processes [11].

Upon implantation into bone tissue, Magnezix undergoes resorption and degradation through corrosion, releasing hydrogen gas according to the equation: $Mg + 2H2O \rightarrow Mg(OH)2 + H2$ [9]. Subsequently, it is gradually replaced by new bone tissue. Studies have also demonstrated a certain degree of osteoinductive potential associated with this material. Complete degradation of the material occurs within one year, although radiographic evidence of degradation can be observed earlier [12,13]. Magnezix implants are available in four types: compression screws (Magnezix CS/CSc), pins (Magnezix Pin), cortical screws (Magnezix CBS), and implants for bone fusion (Magnezix StarFuse).

Therefore, this study investigates the cyclic loading of ankle implants and compares the load-bearing capacity of conventiona) and biodegradable implants in relation to the physiological loading experienced by healthy tissue.

2 Methodology

2.1 Preparation of Specimens and Implantation of Fixation Screws

In our laboratory study, we conducted a comparative analysis of the stability and properties between Magnezix CBS screws and standard cortical screws (ISO 58321E implant steel) for medial malleolus fractures of the ankle. Medial malleolar fractures are very common among the adult population and constitute approximately 9% of all fractures [14]. Surgical treatment, specifically open reduction and internal fixation, is widely recommended in recent literature for dislocated ankle fractures [13]. The common technique employed involves operative fixation using two cannulated screws for medial malleolus fractures [15]. However, other studies have indicated that the stability of medial malleolus osteosynthesis performed with fully pre-drilled cortical screws did not significantly differ [16].

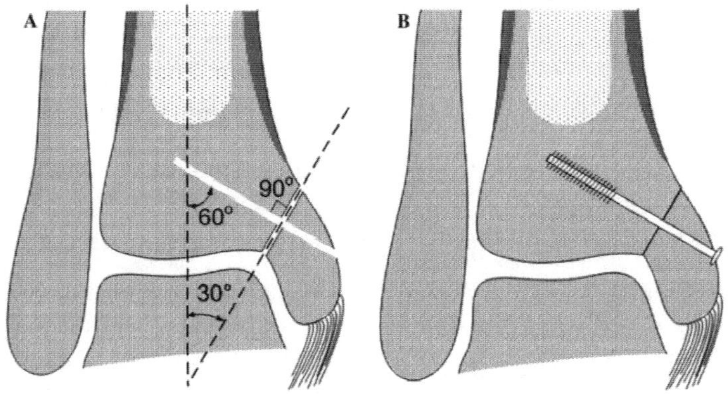

Fig. 1 A schematic representation of the osteotomy and osteosynthesis procedure.

Cadaveric bone, specifically the fresh-frozen distal third of the tibia, was utilized for this study. X-rays were conducted to confirm the absence of bone defects or traumatic focal changes in the tibial skeleton. The oblique osteotomy technique (Fig. 1) was selected as it closely simulates the characteristics of medial malleolus fractures and also serves as an approach to the talus articular surface. The osteotomy line was guided between the tibial plateau cartilage and the articular facet of the medial malleolus. Pre-drilled holes for future insertion of cortical screws were created using a 2.7 mm drill bit at a 60° angle relative to the tibia's long axis. Subsequently, the medial malleolus osteotomy was performed using an oscillating saw with a thin saw blade (0.4 mm width) to minimize bone loss. The saw osteotomy was guided at a 30° angle to the tibia's long axis, near the subchondral bone, and completed with an osteotome and thin chisel [12].

Next, osteosynthesis was carried out using two cortical screws, 40 mm in length and 3.5 mm in diameter. Each preparation was fixed using either two standard self-tapping cortical screws or two Magnezix CBS screws (38 or 40 mm length depending on the preparation size) with a diameter of 3.5 mm. The osteosynthesis was examined, and anteroposterior and lateral projection radiographs were obtained (Figs. 2 and 3).

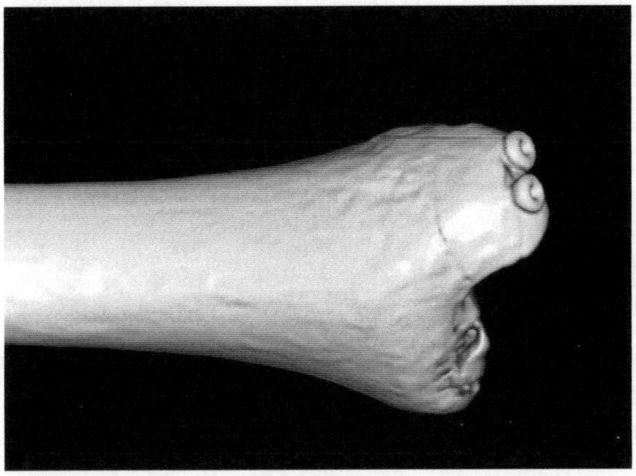

Fig. 2 CT scan of applied screws into the tibial bone.

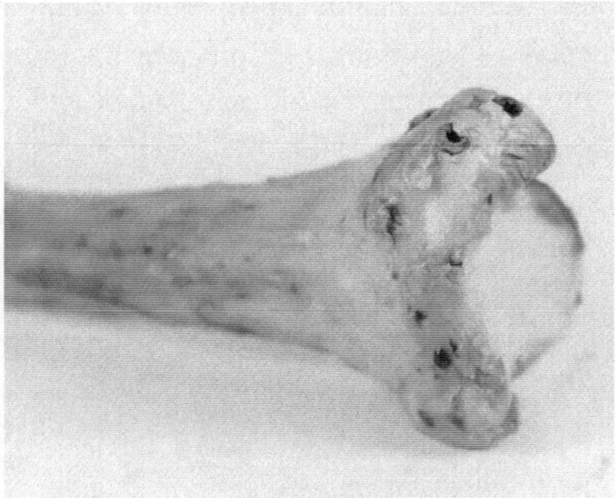

Fig. 3 Bone specimen with implanted screws.

2.2 Mechanical Testing of Biological Samples

The cyclic testing itself was conducted using a LiTeM machine at a frequency of 4 Hz. The amplitude of the applied forces was selected in accordance with previous studies by other authors [17,18], ranging from F_{min} = 10 N to F_{max} = 50 + X*25 N, where X represents the cycle number (0, 1, 2, ...). Each cycle consisted of 10,000 repetitions. The number of cycles continued until mechanical failure of the implants (Fig. 4) or the attainment of the maximum displacement of bone fragments, particularly in the case of chromium-nickel implants.

During the mechanical testing, the sample was placed in a custom-made holder and fixed using three screws at the epiphyseal region and a pair of screws at the diaphyseal region. The securely fastened sample within the inner housing was then stabilized in the optimal position using an outer housing and inserted into the jaws, ensuring overall system stability (Fig. 5).

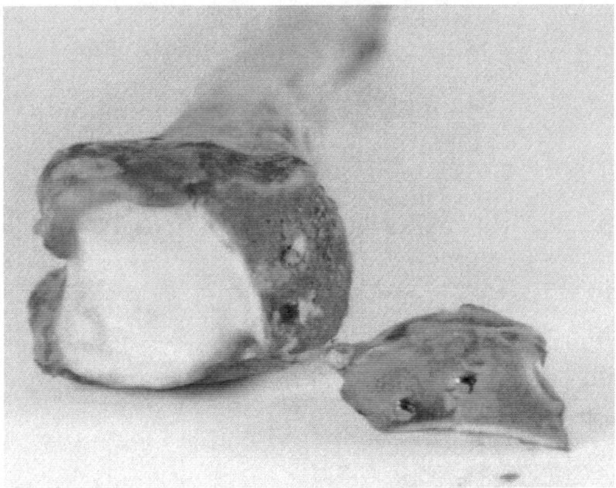

Fig. 4 Final destruction of a magnesium implant.

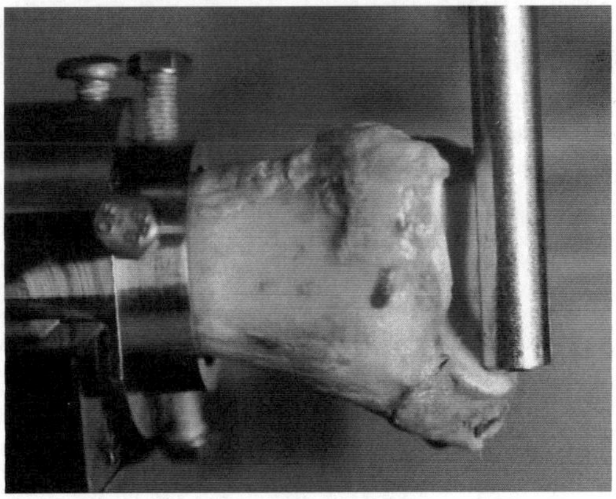

Fig. 5 Picture of real loading in the LiTeM machine.

Loading was generated based on a sinusoidal function, as depicted in Fig. 6. The mean value of the load was automatically increased during each loading cycle, corresponding to the starting position for the respective cycle. The loading direction is derived from studies focusing on fractures in the ankle region and considers the risk loading pattern [19].

3 Results

Mechanical testing revealed significant differences in the ultimate strength of fixation screws. As the testing was conducted on non-living specimens, the resulting values relate solely to the mechanical properties of the implants and the quality of the corresponding bone material. The obtained graphs illustrate an increasing displacement of diverging fragments, attributed to partial compression and bending of the screw, resulting in progressive separation along the osteotomy line.

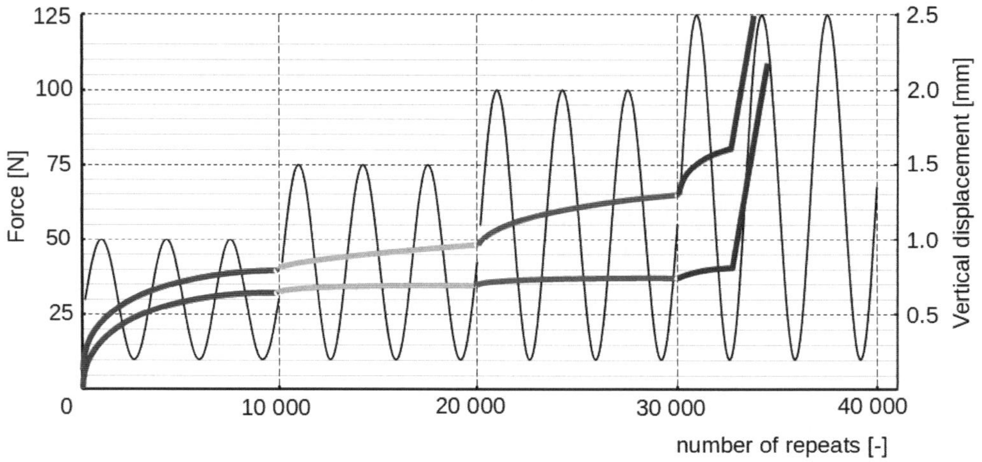

Fig. 6 A diagram depicting the cycles in relation to the applied load.

Fig. 6 presents a diagram depicting the cycles in relation to the applied load. The gradual increase in deformation is evident from Fig. 6, represented by the colored curve. This displacement was calculated based on local minima and maxima of the displacement function obtained from the load machine's displacement sensor. It should be noted that this displacement does not represent the absolute separation perpendicular to the osteotomy plane (Fig. 7). The analysis encompassed a total of six specimens, with three utilizing biodegradable screws (samples 1-3) and three employing chrome-nickel screws (samples 4-6).

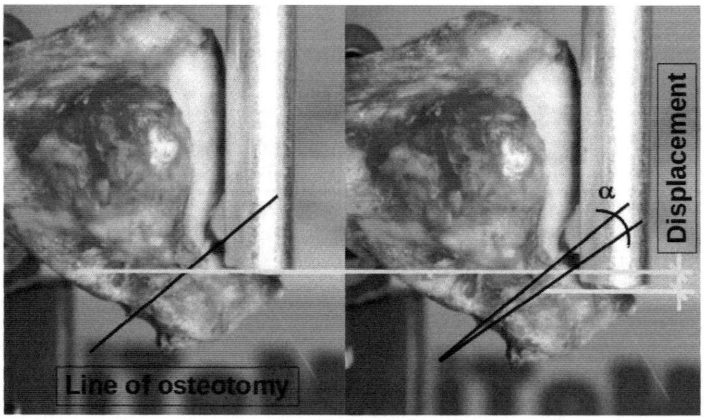

Fig. 7 The photographs of the loaded samples, indicating the osteotomy line, displacement, and the angle α representing the deviation of the fragment from the osteotomy line. The left image depicts a lower applied force compared to the right image, hence the noticeable non-zero displacement.

Table 1 summarizes the results of cyclic loading. For each cycle (10,000 repetitions), the maximum displacement value at the beginning and end of the respective cycle is provided. The table also presents the difference value between the initial and final displacement values in the respective cycle. However, the average value supplemented with the standard deviation is only informative due to the significant variation in the density of biological material.

Table 1. The final value for each sample represents the overall displacement for that particular specimen. For each cycle, the displacement values at the beginning and end of the cycle are provided.

material	sample	cycles 1	2	3	4	5	6
		Displacement [mm]					
Magnezix	1	1.08 1.67	1.91 1.96	2.84 3.49	* after ~ 21800 cycles		
	2	0.48 0.75	1.07 1.17	1.23 1.38	1.43 1.68	* after ~ 32500 cycles	
	3	1.05 1.28	1.61 1.81	2.53 3.06	* after ~ 21400 cycles		
average difference		0.36 ± 0.16	0.12 ± 0.06	0.44 ± 0.21	0.25 ± x		
Stainless steel	4	0.39 0.78	1.05 1.27	1.54 1.83	2.13 2.52	2.83 4.14	4.73 5.32
	5	0.75 2.45	2.97 5.89	# after ~ 6800 cycles			
	6	1.22 2.19	2.87 6.43	6.82 8.27	# after ~ 12400 cycles		
average difference		1.12 ± 0.67	2.23 ± 1.45	0.87 ± 0.58	0.39 ± x	1.31 ± x	0.59 ± x

* – implant failure due to fracture; # – failure due to excessive displacement

Results demonstrate that the evaluation of biological specimens requires a comprehensive approach, taking into account the varying quality of bone. Biodegradable samples 1 and 3 endured a similar number of cycles, with failure (Fig. 4) occurring at a displacement of approximately 3 mm. Sample 2, implanted in denser bone, exhibited significantly lower displacement values and a higher maximum number of repetitions. Greater differences were observed with the conventional screw application. Sample 6 failed after 20,000 repetitions, where no fatigue fracture of the screw occurred, but rather the loosening of the implant within the bone (Fig. 8). Similar failures were observed with samples 4 and 5, resulting in loss of screw fixation. Overall, implant failure in all cases occurred at a displacement of ~3 mm, with biodegradable implants achieving this displacement consistently over the course of repetitions.

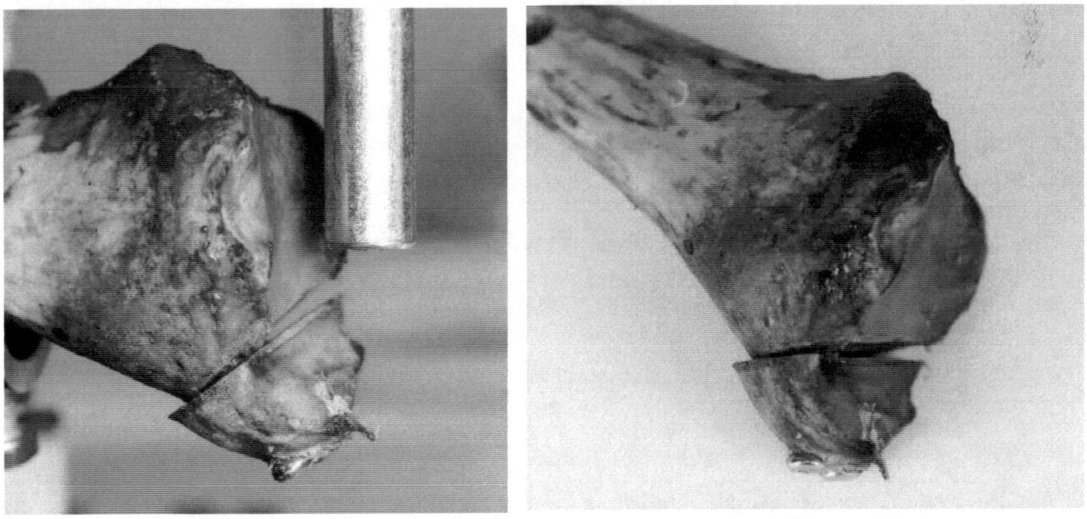

(a) (b)

Fig. 8 Sample 4 after failure of the fixation in the osteotomy line, leading to implant embedding into the bone and significant fragment collapse.

Conclusion

This study revealed a significant difference in the fatigue strength between the conventional and biodegradable implants. The biodegradable implant withstands substantially lower fatigue loads, but considering the actual load at the application site, its strength is sufficient throughout the healing process. Furthermore, the lower material stiffness positively influences the degree of surrounding bone tissue compression, resulting in less pronounced displacements compared to conventional implants. During the healing period, bone stress is lower compared to the physiological state, and the lower implant strength proves to be adequate. It can be assumed that stress gradually transfers from the implant to the bone during the healing process. Biodegradable screws present an intriguing alternative, but a larger sample size would be beneficial for further study. To confirm this study, it will be necessary to expand the number of samples tested. However, considering the availability of biological samples, this is a long-term process.

Acknowledgements

The authors would gratefully like to acknowledge the financial support provided by the Faculty of Civil Engineering, CTU, Prague, project n. SGS23/152/OHK1/3T/11.

References

[1] Heye, P., Matissek, C., Seidl, C., Varga, M., Kassai, T., Jozsa, G. and Krebs, T. "Making Hardware Removal Unnecessary by Using Resorbable Implants for Osteosynthesis in Children," *Children*, **9**(4), 471 (2022).

[2] Technology – Syntellix AG, *Available from* https://www.syntellix.de/en/products/technology.html, *Assessed:* 2023-03-27.

[3] Liu, S., Qin, S., He, M., Zhou, D., Qin, Q. and Wang, H. "Current applications of poly(lactic acid) composites in tissue engineering and drug delivery," *Compos Part B-Eng,* **199**, 108238 (2020).

[4] Shuai, C., Zan, J. Qi, F., Wang, G., Liu, Z., Yang, Y. and Peng, S. "nMgO-incorporated PLLA bone scaffolds: Enhanced crystallinity and neutralized acidic products," *Materials & Desing*, **174**, 107801 (2019).

[5] Walton, M. and Cotton, N.J. "Long-term *in vivo* Degradation of Poly- L -lactide (PLLA) in Bone," *J Biomater Appl*, **21**, pp. 395-411 (2007).

[6] Wu, L., Feyerabend, F., Schilling, A.F., Willumeit-Römer, R. and Luthringer, B.J.C. "Effects of extracellular magnesium extract on the proliferation and differentiation of human osteoblasts and osteoclasts in coculture," *Acta Biomater*, **27**, pp. 294-304 (2015).

[7] Grün, N.G., Holweg, P.L., Donohue, N., Klestil, T. and Weinberg, A.M. "Resorbable implants in pediatric fracture treatment," *Innov Surg Sci*, **3**, pp. 119-125 (2018).

[8] Witte, F. "Reprint of: The history of biodegradable magnesium implants: A review," *Acta Biomater*, **23**, pp. 28-40 (2015).

[9] Walker, J., Shadanbaz, S., Woodfield, T.B.F, Staiger, M.P. and Dias, G.J. "Magnesium biomaterials for orthopedic application: A review from a biological perspective: Magnesium Biomaterials for Orthopedic Application." *J Biomed Mater Res B Appl Biomater*, **102**, pp. 1316-1331 (2014).

[10] Biber, R., Pauser, J., Geßlein, M. and Bail, H.J. "Magnesium-Based Absorbable Metal Screws for Intra-Articular Fracture Fixation," *Case Rep Orthop*, **2016**, 9673174 (2016).

[11] Kopp, A., Fischer, H., Soares, A.P., Schmidt-Bleek, K., Leber, C., Kreiker, H, Duda, G., Kröger, N., van Gaalen, K., Hanken, H., Jung, O., Smeets, R., Heiland, M and Rendenbach, C. "Long-term in vivo observations show biocompatibility and performance of ZX00 magnesium screws surface-modified by plasma-electrolytic oxidation in Göttingen miniature pigs," *Acta Biomater*, **157**, pp. 720-733 (2023).

[12] van Bergen, C.J.A, Tuijthof, G.J.M., Sierevelt, I.N. and van Dijk, C.N. "Direction of the oblique medial malleolar osteotomy for exposure of the talus," *Arch Orthop Trauma Surg*, **131**, pp. 893-901 (2011).

[13] Schatzker, J. and Tile, M. *The Rationale of Operative Fracture Care*. (3 ed., Berlin Heidelberg: Springer; 2005).

[14] Court-Brown, C.M. and Caesar, B. "Epidemiology of adult fractures: A review," Injury, 37, pp. 691-697 (2006).

[15] Femino, J.E., Gruber, B.F. and Karunakar, M.A. „Safe zone for the placement of medial malleolar screws," *J Bone Joint Surg Am*, **89**, pp. 133-138 (2007).

[16] Parker, L., Garlick, N., McCarthy, I., Grechenig, S., Grechenig, W. and Smitham, P. "Screw fixation of medial malleolar fractures: A cadaveric biomechanical study challenging the current AO philosophy," *Bone Jt J*, **95**-B, pp. 1662-1666 (2013).

[17] Burkhart, K.J., Nowak, T.E., Appelmann, P., Sternstein, W., Rommens P.M. and Mueller, L.P. "Screw fixation of radial head fractures: Compression screw versus lag screw—A biomechanical comparison," *Injury, Int J Care Injured,* **41**, pp. 1015-1019 (2010).

[18] Wagner F.C. et al, Post, A., Yilmaz, T., Maier, D., Neubauer, J., Feucht, M.J., Südkamp, N.P. and Reising, K. "Biomechanical comparison of biodegradable magnesium screws and titanium screws for operative stabilization of displaced capitellar fractures," *Biomed Eng online*, **12(**62), pp. 1-8 (2013).

[19] YDE, J. "The Lauge Hansen classification of malleolar fractures," *Acta Orthopaedica Scandinavica*, **51**(1), pp. 181-192 (1980).

Measurement Device for Experimental Investigation of Bolted Joint Diagram in Applications with Non-Metallic Materials

Radka Jírová[1,a], Lubomír Pešík[1,b] and Anna Krobotová[1,c]

[1]*Technical University of Liberec; Studentská 2, 460 01 Liberec, Czech Republic;*

[a]*radka.jirova@tul.cz*, [b]*lubomir.pesik@tul.cz*, [c]*anna.krobotova@tul.cz*

Abstract: The present design of industrial machines is currently constructed with a wide range of non-ferrous and non-metal materials. The connection of machine parts of materials with non-standard strength properties provides specific demands to bolt connections concerning securing the proper preload. To ensure sufficient preload, specially shaped washers can be used. In the design process, the stiffness of a particular shape of the washer must be tested. Therefore, the article describes an innovative measurement device that enables testing washers for connections of various non-ferrous and non-metal materials. The laboratory testing process is shown in the specific example of the washer used in industrial practice.

Keywords: Bolted joint diagram; Stiffness measurement; Washer

1 Introduction

The reliability of bolted connections is currently very important in modern machine design. The connected components are more often produced of non-ferrous and non-metallic materials or composites. The demands on modern machine design regarding economic efficiency, energy consumption, weight, etc., are increasing, and traditional materials are being replaced with a wide range of applications [1,2]. While bolted connections in conventional steel structures are reliable enough, the situation in the case of modern materials is entirely different. The main reason for the reduction in reliability is the loss of the preload, which ensures the strength and tightness of the connection. This phenomenon occurs due to the seating of connected components or thermal or vibration conditions that are more problematic in the case of non-ferrous and non-metallic materials [3-8]. For these reasons, the design of the bolted connection itself regarding the bolted joint diagram of connecting and connected components is of great importance.

This problem can be solved by using a special washer that ensures sufficient preload in the bolted connection even after the seating of connected components or under thermal or vibration conditions [9,10]. In order to verify the designed shapes of washers, a special measurement device was proposed for testing the force and deformation conditions in bolted connections with non-ferrous and non-metallic materials. The article describes the special measurement device and shows its function through an example from an industrial practice.

2 Measurement Device

In order to verify the proposed washer shape, the special measurement device was designed. Using the measurement device and by analysing the measured force and deformations of the washer, the graph of stiffness characteristics of the washer testing specimen can be obtained.

The washer testing specimen can be loaded and unloaded through the pressure base plate using a pressure bolt with trapezoidal thread. The pressure base plate is guided in the measurement device by linear rolling systems fixed to the frame of the device (Fig. 1).

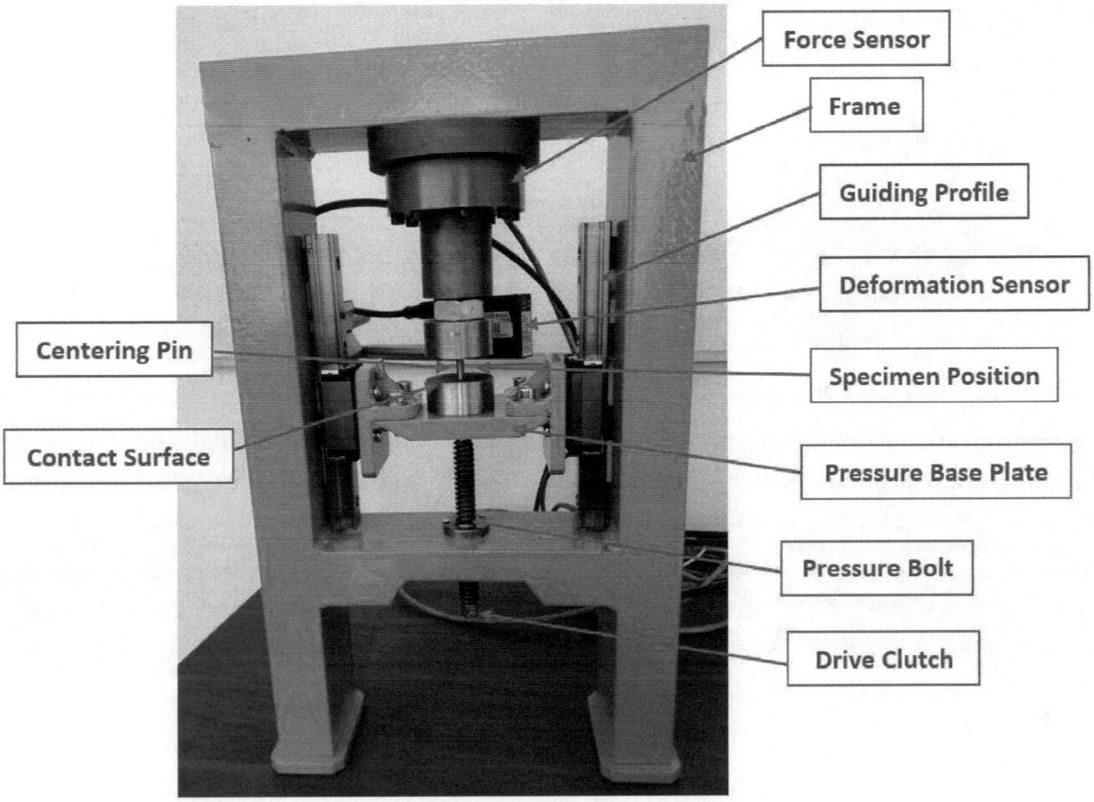

Fig. 1 Measurement device.

The measurement device disposes of changeable contact surfaces for placing the washer testing specimen. The changeable contact surfaces can be made from various non-ferrous and non-metallic materials, including composites. Changeable survaces, as well as, washer testing specimens are centred to the axis of the pressure bolt through centering pin with hardened surface.

A GTM Series K force sensor with a nominal force of 10 kN and a Micro-Epsilon optoNCDT1402 laser deformation sensor with a range of 20 mm was employed at the measurement device. The parameters of the measurement device are summarised in Table 1.

Table 1. The parameters of measurement device.

Maximum Tightening Torque [Nm]	Maximum Axial Force [kN]	Maximum Lift [mm]	Alignment [mm]	Deformation Measurement Sensitivity [mm]
40	10	100	0.5	0.01

3 Example from Industrial Practice

The measurement device was tested on the practical example of the washer in the shape of a disc spring. The washer is used in the bolted connection of batteries cooling system in electric cars. In the cooling system, the coolant distribution is made of aluminium alloy.

Fig. 2 shows the cooling system assembly that is mounted by the bolt with the special washer. The cooling system assembly is realised with washers provided by two producers, Brugola OEB Industriale S.p.A. (Brugola) and Fontana Gruppo (Fontana).

The dimensions of washer testing specimens are shown in Fig. 3. The washer is designed for bolted connection with bolts of M6 size.

Fig. 2 Assembly of cooling system.

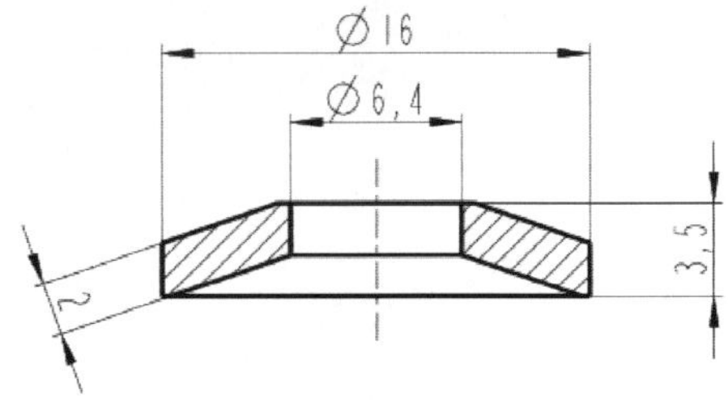

Fig. 3 Dimensions of washer testing specimens.

The testing process of the washers of both producers was processed through the measurement device with a measurement unit Dewesoft Sirius R2DB and software Dewesoft X. Sampling frequency of measurement was set to 100Hz.

Fig. 4 shows the results of the measured axial force and deformation of each washer. Differences in results of stiffness Brugola and Fontana washers are related to their production process. Output stiffness is dependent particularly on the accuracy of dimensions and material properties, especially the heat treatment process of spring steal.

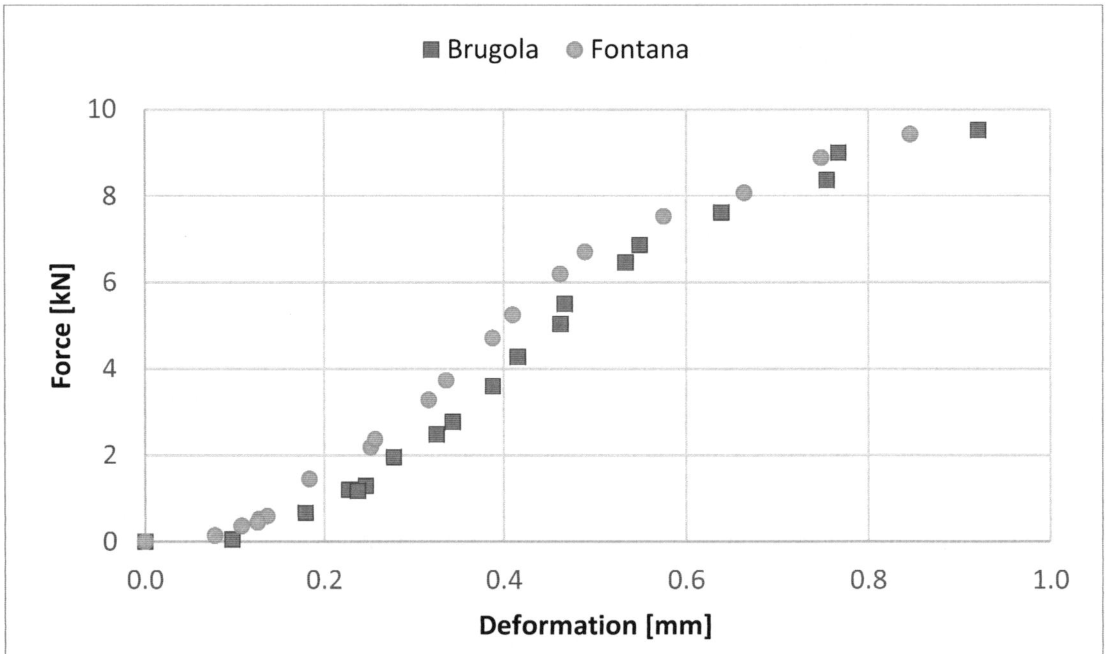

Fig. 4 Washer stiffness.

Conclusions

The article described the innovative measurement device that enables testing and evaluating the stiffness of specially shaped washers for bolted connections used in the construction of non-ferrous and non-metal materials. The measurement device uses the force sensor to detect the axial force and the laser position sensor to measure the axial deformation of the washer. The obtained data can be next composed to the graph of washer stiffness as the dependence of the axial force on the deformation.

Acknowledgements

This work was supported by the Student Grant Competition of the Technical University of Liberec under project No. SGS-2022-5047.

References

[1] Maláková, S., Sivák, S., Guzanová A., and Grega R. "Strength Calculation of Fixed Joints Applied in Passenger Cars," *AD Alta-Journal of Interdisciplinary Research*, **11**(1), pp. 423-427 (2021).

[2] Kanaval, J., Cézová, E. and Starý, F. "Screw Connections with Application of Modern Insurance Sealants and Adhesives Analysis," in *Proceedings of the 59th International Conference of Machine Design Departments (ICMD)*, Demanovska Dolina, Slovakia, 2021.

[3] Jaglinski, T., Nimityongskul, A., Schmitz, R. and Lakes, R.S. "Study of Bolt Load Loss in Bolted Aluminium Joints," *J Eng Mater-T ASME*, **129**(1), pp. 48-54 (2007).

[4] Kober, M. and Kühhorn, A. "Development of Optimized Washer-Shape for Damage-Free Load Application to Composites," *Compos Struct*, **92**(8), pp. 1985-1991 (2010).

[5] den Otter, C. and Maljaars J., "Preload Loss of Stainless Steel Bolts in Aluminium Plated Slip Resistant Connections," *Thin Wall Struct*, **157**, 106984 (2020).

[6] Hu, J., Zhang, K., Cheng, H. and Qi, Z. "Mechanism of Bolt Pretightening and Preload Relaxation in Composite Interference-Fit Joints under Thermal Effects," *J Compos Mater*, **54**(30), pp. 4929-4946 (2020).

[7] Caccese, V., Berube, K.A., Fernandez, M., Melo, J.D. and Kabche, J.P. "Influence of Stress Relaxation on Clamp-up Force in Hybrid Composite-to-Metal Bolted Joints," *Compos Struct*, **89**(2), pp. 285-293 (2009). ISSN 0263-8223.

[8] Peth, J. and Friedrich, C. "Design Influences of Preload Relaxation Behaviour in Bolted Joints Using Aluminium Parts," in *Proceedings of the 6th International Conference on Integrity-Reliability-Failure IRF2018*, Lisabon, Portugal, 2018.

[9] Jírová, R. and Pešík, L. "Analysis of Screw Connection in Air Conditioning Systems," *Scientific Journal of Silesian University of Technology-Series Transport*, **144**, pp.43-54 (2022).

[10] Mantič M., Kuľka J., Grega R., Virostko M. and Kopas M., "Application of Sensors for Incorrect Behavior Identification in a Transport System," *Sensors*, **23**(3), 1635 (2023).

Thermal Actions on Final Lining of the Mezno Tunnel

Martin Jonáš[1,a] and Jan Zatloukal[1,b]

[1]*Experimental Centre, Faculty of Civil Engineering, Czech Technical University in Prague Thákurova 7, 166 29, Praha 6 – Dejvice, Czech Republic;*

[a]*martin.jonas@fsv.cvut.cz*, [b]*jan.zatloukal@fsv.cvut.cz*

Abstract: Temperature measurement in the Mezno tunnel was designed to obtain data about thermal actions reached in tunnel linings. After a period of operation, the initial data has been measured and evaluated. The evaluation presented the temperature differences across the tunnel lining in various spots during the winter and spring periods, identified the maximal thermal actions reached, compared them with available prescripts for static design, and discussed the suitability of further thermal monitoring.

Keywords: Thermal actions; Thermal monitoring; Tunnel; Tunnel lining; Temperature

1 Introduction

Temperature measurement in monolithic concrete structures might be beneficial for investigating their behaviour. Thermal actions on rigid structures may induce high internal forces, potentially resulting in an oversized design if the applied thermal actions exceed those encountered in reality. Hence, a proper understanding of the actual thermal actions on tunnel linings is beneficial. The current situation involves multiple tunnel structures under construction in the Czech Republic, where a previously developed monitoring system has been installed in the secondary linings of these tunnels. The motivation for monitoring tunnel linings arises from the issue of thermal actions, which are relatively insufficiently described in standards. This motivation stems from earlier measurements conducted in the Klimkovice tunnel [1] and Valík tunnel [2], revealing a significant discrepancy with the values prescribed by the TKP ČD [3]. For a brief illustration, technical conditions prescribe only a simple temperature difference on the tunnel lining of 10 °C, whereas the measured differences were only up to 3 °C [1]. Additionally, it might be suitable to consider the division of temperature in the lining as non-linear to model realistic behaviour. However, the methodology used for these measurements cannot encompass the entire spectrum of thermal actions [2].

The monitoring system has been installed or is in the process of installation in three tunnels, with proposals for monitoring other structures in the project. These monitoring systems will continuously measure temperatures and deformations in the lining since the completion of tunnel construction. An integral part of the monitoring system's development is its optimization and adaptation for any other use in construction practice where measurements are appropriate.

The second part of monitoring thermal actions on tunnel linings involves the continuous evaluation of the measured values and the preparation of an overall assessment of the data obtained from these measurements. A method for evaluating long-term measurement data will be proposed to better describe the thermal actions on the linings. The obtained description can be further processed for adjustments to the documents used in the static design of tunnel linings.

2 Monitoring in the Mezno tunnel

The monitoring system in the tunnel has been installed based on the general conditions defined at the onset of the tunnel lining monitoring project [4]. In the Mezno tunnel, there are two measuring profiles. The first profile is located in the cut-and-covered part in block No. 1, approximately 10 meters from the eastern portal of the tunnel. The second measurement profile is situated in the bored part of the tunnel in block No. 15, approximately 180 meters from the eastern portal of the tunnel.

There are two temperature measurement points (as depicted in Fig. 1) within each profile. The first one is positioned at the top of the tunnel vault, while the second one is in the side wall of the tunnel lining. In each location, a set of thermometers is installed at specific positions to cover the entire thickness of the tunnel lining (consisting of 6 or 7 thermometers at specific depths, depicted in the result charts). Measurements are taken using resistive thermometers PT 1000 at a one-minute frequency, which can be adjusted using the remote control of the system.

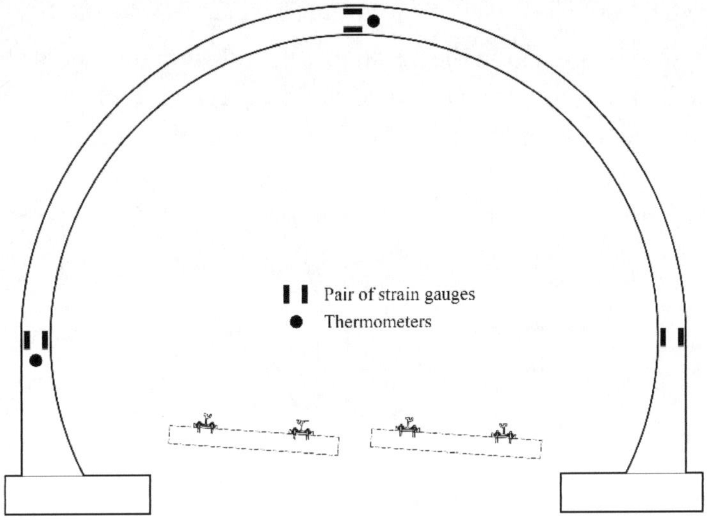

Fig. 1 Scheme of probe set positions in one measurement profile.

Measurements from the Mezno tunnel are conducted by an automatic data acquisition system, which relies on the Raspberry Pi computer and operational modules for the sensors in use. The system has been operational since the beginning of December 2022, allowing for the evaluation of extreme temperature conditions during the winter and spring.

3 Evaluation of thermal actions

The measurements were evaluated with regard to the original purpose of the monitoring. The main emphasis was placed on the extreme linear and non-linear characteristics of the temperature change component (through the lining thickness) and the uniform temperature change component. These values were evaluated with respect to their location – the portal and central part of the tunnel – and the period in which they occurred. From a temporal point of view, it was appropriate to establish these characteristics in seasons and extremes for the entire period of measurement.

According to the period of system operation, data can be basically divided into the winter and spring period. The overall measured data covered the period from December 5th, 2022, to May 30th, 2023. The boundary between the winter and spring period could not be precisely determined due to unstable weather conditions. However, this situation was not problematic, as only periods with maximal or minimal temperatures were chosen for evaluation.

These time periods were selected based on air temperatures measured in the locality of Mezno village, which were extracted from the weather archive [5]. According to the weather archive, three periods during the winter and two periods during the spring were chosen.

4 Winter seasons

During the winter, three specific periods with low temperatures and relatively high temperature changes during the day occurred. The first period was set from December 10th to December 22nd, 2022, the second one from February 3rd to February 11th, 2023, and the last one from February 24th to March 2nd, 2023. However, in contrast to previous winters, air temperatures did not reach such low values. The temperature change component has been evaluated at all four spots for these periods, and extreme situations were chosen for presentation.

4.1 Portal profile

The first evaluated spot is situated in the portal profile, down in the side wall, approximately 1.5 – 2 m from the bottom of the tunnel. Maximal differences reached values up to 5.1 °C. The non-linear component of temperature change is presented in the chart in Fig. 2.

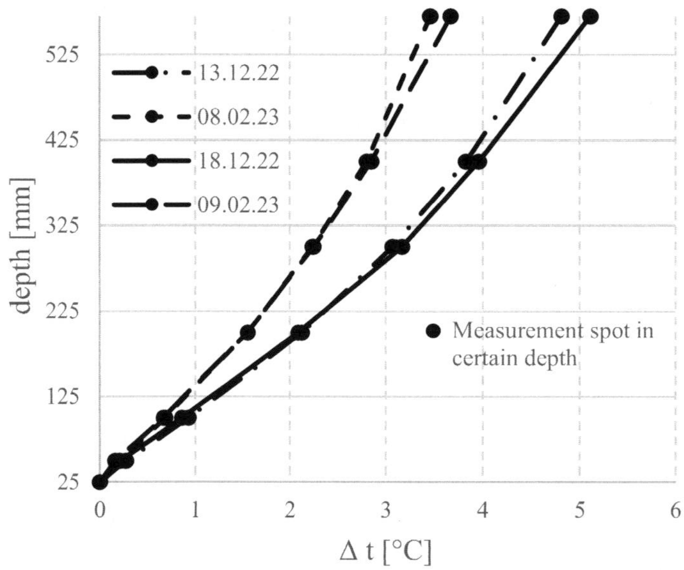

Fig. 2 Extreme temperature differences during the winter across the thickness of tunnel lining in the spot in the side wall of lining, near the portal.

The second evaluated spot is situated in the portal profile, at the top of the tunnel vault. Maximal differences reached values up to 5.2 °C, which are almost identical to those in the bottom part. The non-linear component of temperature change is presented in the chart in Fig. 3.

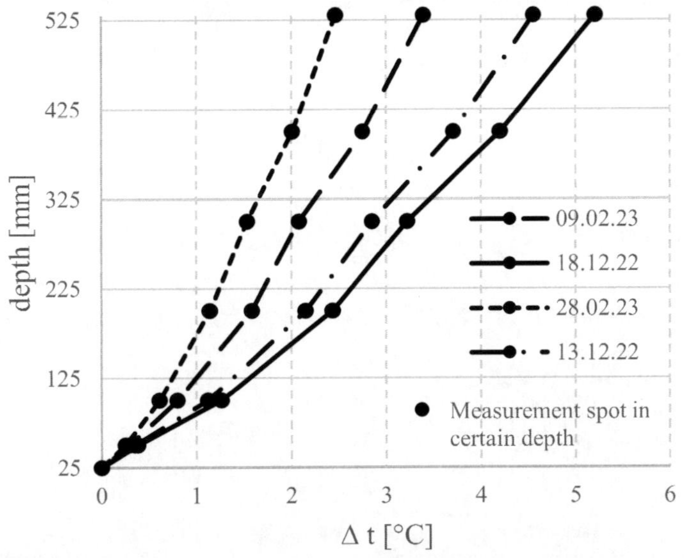

Fig. 3 Extreme temperature differences reached during the winter across the thickness of tunnel lining in the spot in the top of the tunnel lining vault, near the portal.

4.2 Mid-part profile

The spot in the mid-part profile is situated analogously to the portal profile, down in the side wall. Maximal differences reached values up to 4.8 °C. The non-linear component of temperature change is presented in the chart in Fig. 4.

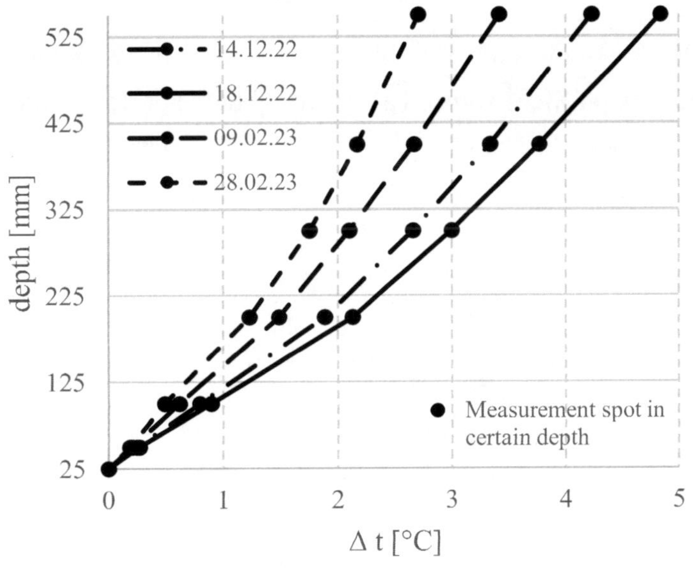

Fig. 4 Extreme temperature differences during the winter across the thickness of lining in the spot in the side wall of lining.

The maximal temperature differences in the spot at the top of the tunnel vault reached values up to 3.3 °C, which is notably less than in the bottom part. The non-linear component of temperature change is presented in the chart in Fig. 5.

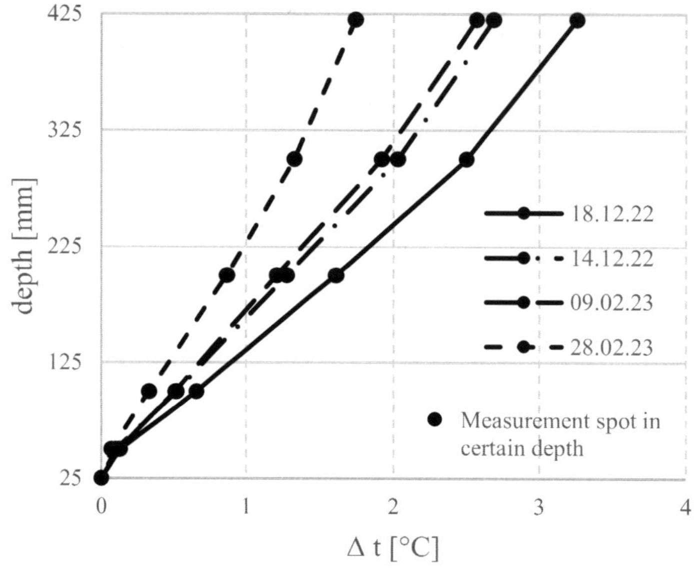

Fig. 5 Extreme temperature differences reached during the winter across the thickness of tunnel lining in the spot in the top of the tunnel lining vault.

5 Spring seasons

Spring was characterized by low air temperatures, which only rose in May, featuring two stable periods with more significant daily air temperature changes. The first one was set from May 6th to May 11th, and the second one from May 20th to May 30th, 2023. Extreme situations and maximal temperature changes are presented.

5.1 Portal profile

The spot in the mid-part profile is situated analogously to the portal profile, down in the side wall. Maximal differences reached values up to 4.9 °C. The non-linear component of temperature change is presented in the chart in Fig. 6.

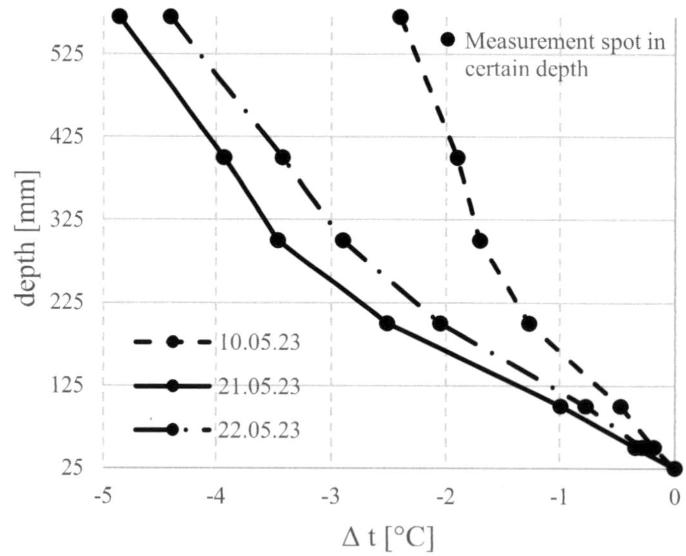

Fig. 6 Maximal temperature differences during the spring across the thickness of the lining in the spot in the side wall of lining, near the portal.

The maximal temperature differences in the spot at the top of the tunnel vault reached values up to 4.9 °C, the same as in the bottom part. The non-linear component of temperature change is presented in the chart in Fig. 7.

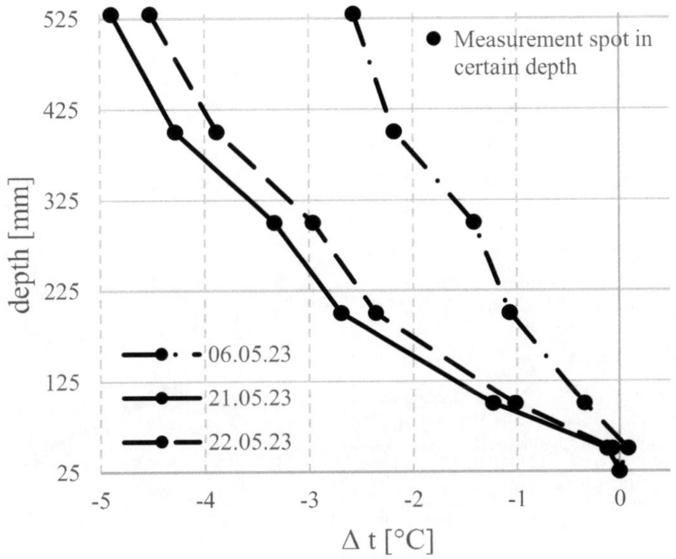

Fig. 7 Extreme temperature differences reached during the spring across the thickness of tunnel lining in the spot in the top of the tunnel lining vault, near the portal.

5.2 Mid-part profile

The spot in the mid-part profile is situated analogously to the portal profile, down in the side wall. Maximal differences reached values up to 4.1 °C. The non-linear component of temperature change is presented in the chart in Fig. 8.

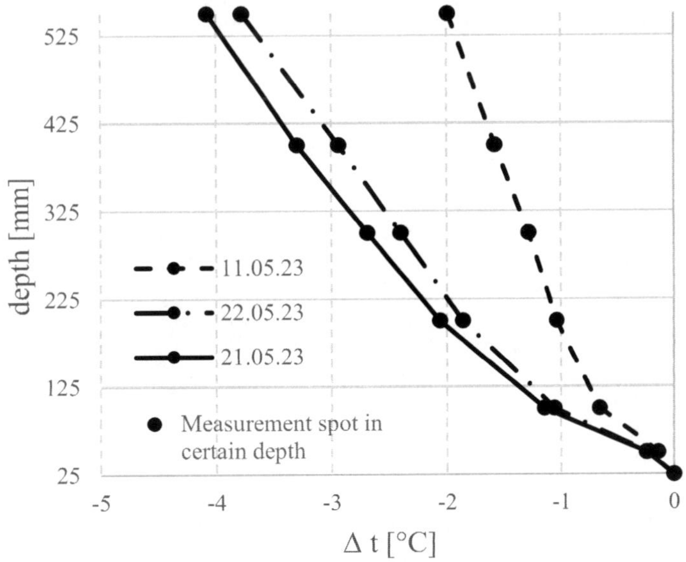

Fig. 8 Maximal temperature differences during the spring across the thickness of the lining in the spot in the side wall of lining.

The maximal temperature differences in the spot at the top of the tunnel vault reached values up to 2.9 °C, which is considerably less than in the bottom part. The non-linear component of temperature change is presented in the chart in Fig. 9.

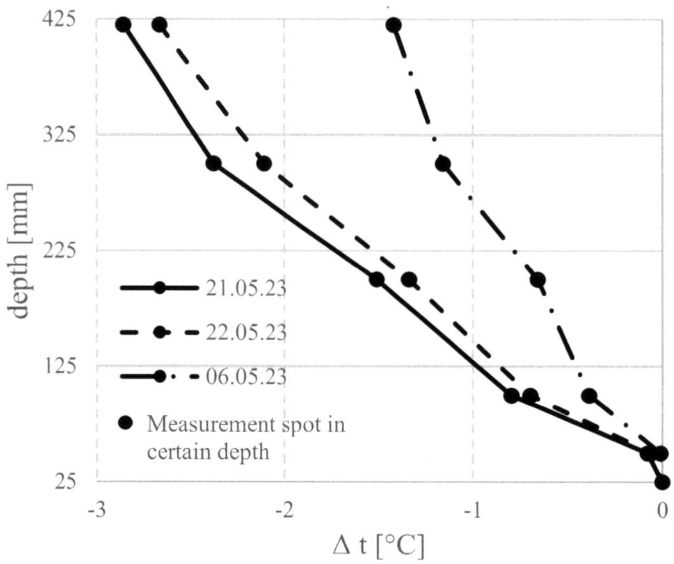

Fig. 9 Extreme temperature differences reached during the spring across the thickness of tunnel lining in the spot in the top of the tunnel lining vault.

Conclusion

The evaluation of thermal actions on the tunnel lining in the Mezno tunnel after half a year of monitoring operations brought specific new information. Firstly, temperature differences across the thickness of the lining are within the expected values. Therefore, the presumption of overestimation of thermal actions by the TKP ČD prescript may be considered confirmed. However, for setting the thermal actions, further measurement and evaluation are necessary.

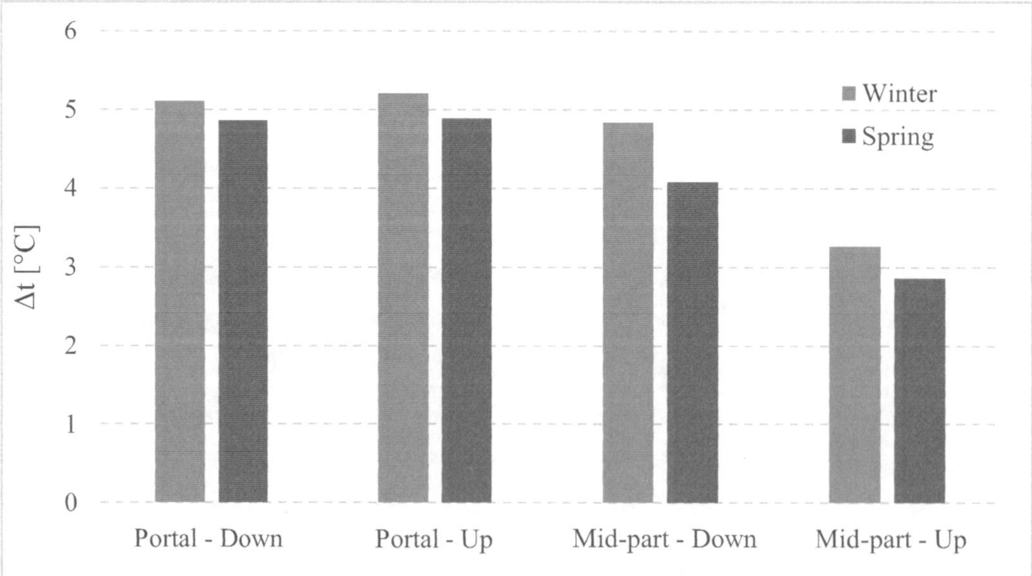

Fig. 10 Comparation of temperature changes reached in the measured spots in winter and spring season.

According to the comparison depicted in the chart in Fig. 10, there are a few differences that deserve discussion. Firstly, the temperature differences reached in the spring season were slightly lower than those reached in the winter. This should be considered as a specificity of the period during which the measurement was taken. These differences could be evaluated only statistically after more years of monitoring operation.

Secondly, a slight difference is observed between the portal part and the middle part of the tunnel. These differences should be confirmed by statistical evaluation; the assumption cannot be made at this point. However, in the case of a significant difference between the upper measurement spot in the mid-part and the other spots, the reason could be sought in the different thickness of the lining in this spot. The lowest point in the measurement is 425 mm deep, which is about 100-140 mm shallower than the deepest points in other spots. Therefore, the thermal behaviour at this point is more variable than in more deep-seated points in other spots of monitoring.

The maximal temperature difference reached 5.2 °C. As mentioned earlier, this value falls within the expected range. Continuous measurement of thermal actions in tunnel linings should provide a sufficient basis of data, creating the first step for a proper description of thermal actions and the efficient and safe design of tunnel linings in the end. Therefore, further measurements and system control will follow.

Acknowledgements

This work has been supported by the Faculty of Civil Engineering of Czech Technical University under the student project SGS23/053/OHK1/1T/11 and by the Technology Agency of the Czech Republic under the project TK05020099.

References

[1] Ďuriš, L., Aldorf, J. and Geryk, J., "Thermal loading action on final linings of underground structures," *Tunnel*, **22**(2), pp. 44-52 (2013).

[2] Šourek, P., Vítek, L., Aldorf, J. and Ďuriš, L. "Measurement of deformations and temperatures on final tunnel liners." *Tunnel*, **17**(4), pp. 70-76 (2008).

[3] Technické kvalitativní podmínky pro dokumentaci staveb, Kapitola 24 - TUNELY, Ministerstvo dopravy, Praha, 12, 2006.

[4] Jonáš, M. and Zatloukal J., "Permanent tunnel lining monitoring system for the purpose of further design optimization," *Current Perspectives and New Directions in Mechanics, Modelling and Design of Structural Systems*. (CRC Press, 2022).

[5] ČHMÚ. Měsíční a roční data dle zákona 123/1998 Sb. *Available from* http://portal.chmi.cz/historicka-data/pocasi/mesicni-data/mesicni-data-dle-z.-123-1998-Sb# *Accessed:* 2023-05-12.

Determination of the Initial Anisotropy of Inconel 718 manufactured by Additive Manufacturing

Michal Kořínek[1,a], Radim Halama[1,b], Jiří Hajnyš[1,c] and Dušan Střílka[2]

[1]*VSB – Technical university of Ostrava; 17. Listopadu 2172/15, Ostrava-Poruba, Czech Republic;*

[2]*LabControl s.r.o.; Přerovecká 65/36, 747 95 Opava, Czech Republic;*

[a]*michal.korinek@vsb.cz*, [b]*radim.halama@vsb.cz*, [c]*jiri.hajnys@vsb.cz*, [d]*strilka@labcontrol.cz*

Abstract: This article deals with the determination of the initial plasticity area for the Inconel 718 nickel alloy produced by selective laser melting technology. A non-destructive method based on linearity deviation is used, which allows multiaxial loading with any combination of stress components. The printed material is tested in its as-built state and virgin powder is used. The results of a total of 4 mechanical tests show that the material produced in this way does not correspond to the plasticity criterion according to von Misses, and the yield strength in pure tension and the effective stress obtained in pure torsion do not have the same stress values, but the shear component of the stress is significantly dominant.

Keywords: Yield surface tracing; Inherent anisotropy; Inconel 718; Additive manufacturing; Selective laser melting

1 Introduction

Inconel 718 is a nickel-based steel alloy that has suitable mechanical properties, even at higher temperatures. Therefore, this alloy is mainly used in the aerospace industry. The disadvantage of this alloy is its poor machinability. Especially in the aerospace industry, there are components with complex shapes that need to be machined in a complicated way [1]. Additive manufacturing solves this problem to some extent. Using 3D printing technology such as Selective Laser Melting (SLM), it can produce elements with complex shapes. However, this technology has a disadvantage in build. The orientation of the building in 3D printing has an effect on the mechanical properties of the material.

This paper is focused on the investigation of a nickel-based alloy, specifically Inconel 718, which is made using additive technology, specifically SLM technology. This paper deals with the experimental determination of the initial yield surface under combined loading using a non-destructive method.

2 Materials and methods

The method of tracing the yield surfaces with a centre point is used. This method is suitable for combined loading with any ratio of axial and torsional stress components [2]. Tubular samples were used for the experiments, so it was possible to simultaneously control the components of the axial and torsional stress. The samples were produced using the SLM method with vertical orientation of the building, using virgin powder, and were tested in the as-printed state. The geometry of the sample is based on the standard for low-cycle fatigue [3], and the specific geometry used in these experiments is shown in Fig. 1. However, due to the manufacturing method, the real dimensions of the sample are slightly different. The real dimensions of the sample were 11.3 mm on the inner diameter and 13.9 mm on the outer diameter of the test part (length 28 mm in Fig. 1).

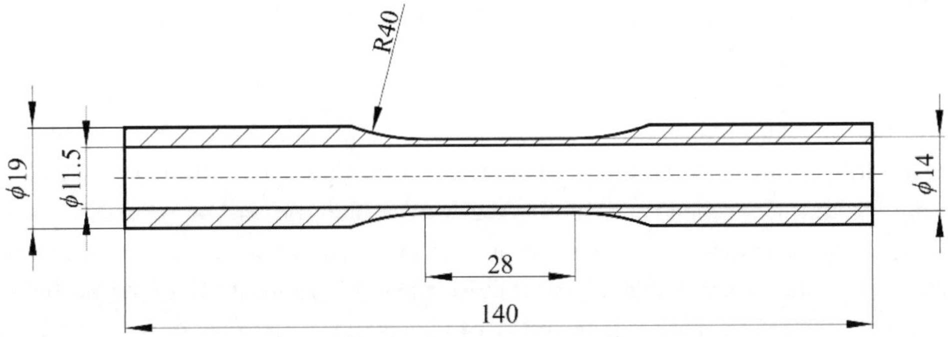

Fig. 1 Sample geometry for elastic region tracing.

All experiments performed were stress controlled. The ratio of axial and torsional stress rates was varied using a specific angle ψ_i. Thus, this angle defines a specific probing path. Using this angle, we can determine the effective stress rate, which depends on the increase of the stress components. These components of stress can be written in the following form

$$\dot{\sigma} = \dot{\sigma}_{eqv} \cdot cos\psi_i \,, \dot{\tau} = \dot{\sigma}_{eqv} \frac{cos\psi_i}{\sqrt{3}}, \tag{1}$$

where $\dot{\sigma}_{eqv}$ is the required equivalent stress rate. The same equivalent stress rate $\dot{\sigma}_{eqv}$ =1 MPa/s is used for all probing paths. On the contrary, relief was performed with a constant equivalent stress rate of 10 MPa per second to shorten the time of the entire test. The sequence and rate ratio of each probing path can be seen in Fig. 2.

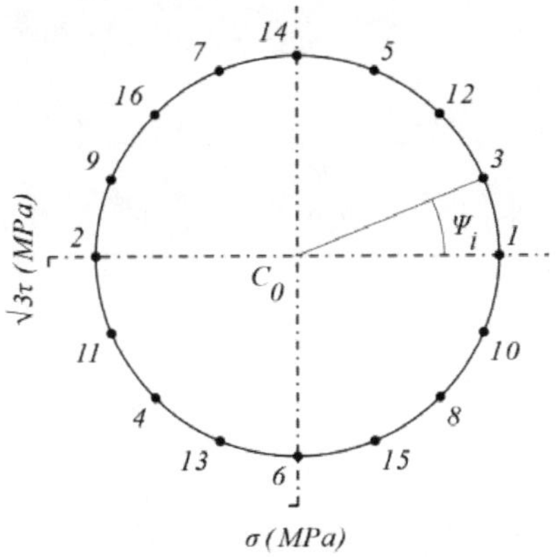

Fig. 2 Probing paths scheme for the experiment.

The LabControl hydraulic biaxial machine with maximum load values of 100 kN/1000 Nm was used for the tests. The criterion of the effective increment of the plastic strain, according to eq. (2), is used to terminate the given probing path and evaluate the point that corresponds to the specific yield point. This equation is evaluated online during measurement. As soon as this condition meets the prescribed value of 50 μS, the probe is terminated and the next in the sequence continues.

$$\Delta\varepsilon_{eff} = \sqrt{\Delta\varepsilon_{pl}^2 + \frac{1}{3}\Delta\gamma_{pl}^2}, \tag{2}$$

To determine the plastic components of strain in eq. (2), an additive rule is used in the following form

$$\Delta\varepsilon_{pl} = \varepsilon - \varepsilon_{el} - \varepsilon_R, \quad \Delta\gamma_{pl} = \gamma - \gamma_{el} - \gamma_R. \tag{3}$$

The total axial strain value ε is read by the extensometer during the test, as the total strain in shear γ. Epsilon 3550 extensometer with a gauge length of 25 mm was used for the tests. This type of extensometer allows us to measure both axial and torsional deformation simultaneously. The elastic components of strain ε_{el} and γ_{el} in relation (3) are determined as the ratio of the current stress value to the modulus of elasticity (for tensile component σ/E and for shear then τ/G). The moduli of elasticity (E and G) are determined as the slope of a straight line from the predetermined $\Delta\sigma$ region defined by eq. (4). For the following experiments, this value was set at 100 MPa. The modulus of tensile and shear elasticity determined from this region is then invariable for the evaluation of the given probe. Eq. (4) is repeated for each probing path. The result is 14 values of the modulus of elasticity of tensile and shear (for pure tension and pure torsion, the opposite component is neglected).

$$\Delta\sigma = \sqrt{\sigma^2 + 3\tau^2}, \tag{4}$$

The residual components of the strain in (3) ε_R and γ_R are then evaluated as a constant from the equation of a straight line that describes the modulus of elasticity. This value characterises, for example, the landing of the sample in the testing machine and is also evaluated for each probing path separately.

Conclusions

The total tracing of the yield surface, consisting of 16 partial probes, was repeated 4 times in total. All measurements were performed on the same sample and all specific yield stress values are listed in Table 1.

Table 1. Specific yield stress values for all tests and probing paths.

	Test 1		Test 2		Test 3		Test 4	
n. of path	σ (MPa)	$\sqrt{3}\tau$ (MPa)	σ (MPa)	$\sqrt{3}\tau$ (MPa)	σ (MPa)	$\sqrt{3}\tau$ (MPa)	σ (MPa)	$\sqrt{3}\tau$ (MPa)
1	225	0	198	0	206	0	207	0
2	268	111	256	118	291	120	281	116
3	158	378	167	403	157	378	165	394
4	-190	459	-205	493	-210	504	-208	501
5	-265	109	-265	109	-276	114	-276	115
6	-279	-116.4	-270	-112	-271	-114	-263	-111
7	-194	-471	-195	-473	-187	-450	-206	-499
8	171	-413	174	-421	182	-441	179	-429
9	-300	0	-311	0	-308	0	-277	0
10	-240	-241	-252	-252	-249	-250	-254	-254
11	0	-566	0	-574	0	-589	0	-623
12	234	-234	238	-239	232	-233	239	-241
13	263	-110	274	-114	275	-114	266	-111
14	235	235	240	240	240	240	241	243
15	0	591	0	594	0	625	0	633
16	-254	254	-259	257	-263	263	-256	258

Similar values as those given in this chapter were also obtained by control measurements on another sample of the same geometry. Fig. 3 shows the moduli of elasticity for individual load speed ratios using a box graph.

The first column, marked 0, indicates either pure tension or pure torsion. From the individual load speed ratios, it can be seen that the highest deviations for the tensile component give the probing path, when the shear component of the stress prevails over the tensile one. This is a probing path 5, 7, 13 and 15 from Fig. 2. The average value of the Young modulus is 137.7 GPa, which is lower compared to conventionally produced material [4]. However, lower values of the modulus of elasticity in Inconel 718 manufactured unconventionally were pointed out by [5]. On the contrary, the highest dispersion for the modulus of elasticity in shear is seen in all experiments by combinations where the tensile component dominates with respect to the shear component, in Fig. 2 it is probes 3, 9, 11 and 10. The other three ratios are very close to the average value of 92.8 GPa. This may to some extent be because the sample produced in this way shows significantly higher values of stress in torsion than in tension.

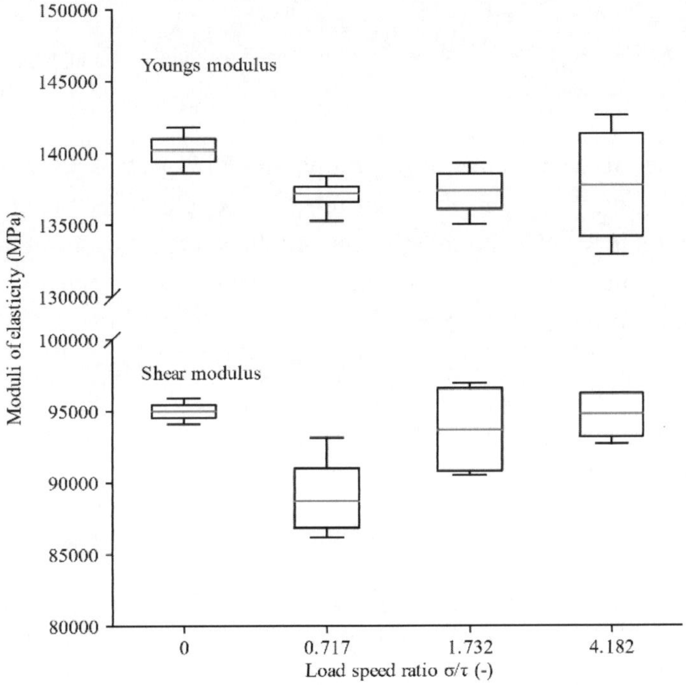

Fig. 3 Moduli of elasticity for Inconel 718 produced by additive manufacturing.

Fig. 4 shows the initial yield surface of the nickel-based alloy Inconel 718, produced by additive technology, specifically by SLM in the stress space $\sigma - \sqrt{3}\tau$. It can be seen from the results that the shear stress component is significantly dominant. The shear stress has an average value for all measurements of 610 MPa and in the opposite direction of 588 MPa. The average value for normal stress is 209 MPa, respectively, 298 MPa in pressure. Therefore, the data obtained cannot be fitted using the von Misses condition, as is the case with conventionally manufactured samples [6]. For a simple description, we can fit the points to the function of an ellipse with a main semiaxis size of 599 MPa and a minor semiaxis. size of 253.5 MPa, as can also be seen in Fig. 4.

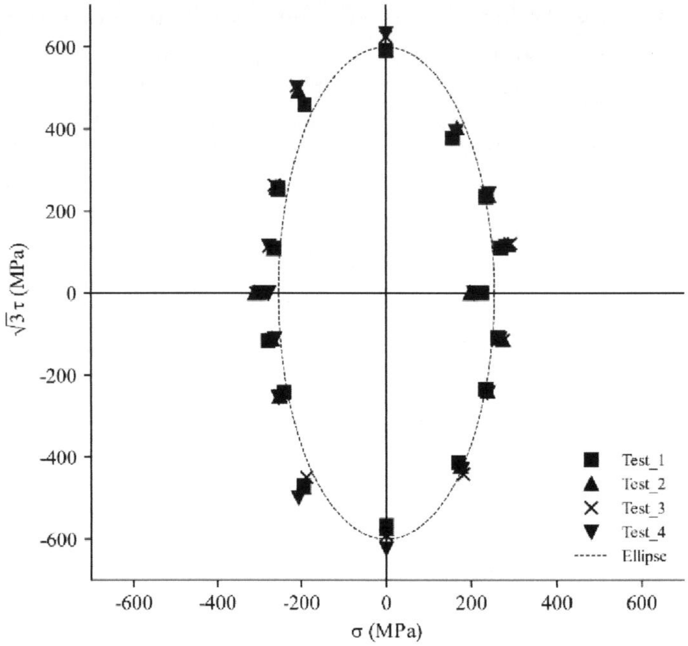

Fig. 4 Initial yield locus for Inconel 718 produced by additive manufacturing.

Acknowledgements

This work is an output of the project Computational modelling of ductile fracture of identical wrought and printed metallic materials under ultra-low-cycle fatigue created with financial support from the Czech Science Foundation under the registration no. 23-04724S and has been done is connection with the Specific Research Project (SP2023/27) supported by the Ministry of Education, Youth and Sports.

References

[1] Rahman, M., Seah, W.K.H. and Teo, T.T. The machinability of inconel 718. *J Mater Process Tech*, **63**(1-3), pp. 199-204 (1997).

[2] Štefan, J., Parma, S., Marek, R., Plešek, J., Ciocanel, C. and Feigenbaum, H. "Overview of an Experimental Program for Development of Yield Surfaces Tracing Method," *Applied Sciences*, **11**(16), 7606 (2021).

[3] STM E2207 - 15 Standard Practice for Strain-Controlled Axial-Torsional Fatigue Testing with Thin-Walled Tubular Specimens. West Conshoshocken, USA: ASTM International, 2015.

[4] Tayon, W.A., Shenoy, R.N., Redding, M.R., Keith Bird, R. and Hafley, R.A. "Correlation Between Microstructure and Mechanical Properties in an Inconel 718 Deposit Produced via Electron Beam Freeform Fabrication," *J Manuf Sci E-T ASME*, **136**(6), 061005 (2014).

[5] Kelley, P. F., Saigal, A., Vlahakis, J.K. and Carter, A. „Tensile and Fatigue Behavior of Direct Metal Laser Sintered (DMLS) Inconel 718“ in *Proceedings of the ASME 2015 International Mechanical Engineering Congress and Exposition, Volume 2A: Advanced Manufacturing*, Houston, Texas, USA, 2015.

[6] Gil C.M., Lissenden C.J. and Lerch B.A., "Determination of Yield in Inconel 718 for Axial-Torsional Loading at Temperatures up to 649 C," *NTRS – NASA/TM*, 208658, pp. 1-33 (1998).

Modal Analysis of Structural Element using Optical Experimental Methods of Mechanics

Ján Kostka[1,a], Peter Frankovský[1,b], Ingrid Delyová[1,c], Peter Sivák[1,d] and Michal Kicko[1,e]

[1]*Technical University of Kosice; Faculty of Mechanical Engineering; Letná 9, 042 00 Košice, Slovak Republic*

[a]*jan.kostka@tuke.sk*, [b]*peter.frankovsky@tuke.sk*, [c]*ingrid.delyova@tuke.sk*, [d]*peter.sivak@tuke.sk*, [e]*michal.kicko@tuke.sk*

Abstract: The content of the paper deals with a relatively new approach to the experimental determination of modal parameters (resonant frequencies, mode shapes and damping ratios) based on the measured displacements by the non-contact, full-field optical method of Digital Image Correlation (DIC). The output is an application module which, based on a three-dimensional displacement matrix from the measurement results, can compile a frequency response function (FRF) for the needs of experimental and operational modal analysis and determine the required modal parameters from it. The application module was designed in Scilab 6.1.0 software, while its script interferes directly with the ISTRA4D high-speed camera software from Dantec Dynamics company. The created application was based on the measurements of a simple steel sheet excited by an impact hammer for the simulation of experimental modal analysis. Verification of the calculation algorithm or the obtained modal parameters of the excited sheet steel plate was performed by means of a simulation in the numerical software Abaqus, whose mode shapes and resonant frequencies showed a high similarity with the results of the application.

Keywords: Digital Image Correlation; Experimental Modal Analysis; Operational Modal Analysis; Frequency Response Function; Resonant frequency; Mode shapes; Damping ratios

1 Introduction

Non-contact optical methods with the possibility of all-field analyses are among the current trends in the field of experimental mechanics. Laser Doppler vibrometry (LDV), electronic speckle interferometry, digital speckle shearography and digital image correlation fall into this field [1]. In vibration analyses, laser vibrometry is a frequently used method for practical reasons. The advantages of the laser vibrometry method are high sensitivity, frequency range, high accuracy, measurement speed and the ability to measure over large distances. The main disadvantage is the high cost of the measuring equipment. LDVs have a wide range of applications in science and research, as evidenced by the number of published papers devoted to experimental modal analysis, operational modal analysis, operational vibration waveform analysis, and vibrodiagnostics. Šároši et al. [2] used the LDV method for a comprehensive analysis of the disc during its rotation. Specifically, they dealt with experimental modal analysis of the disc in a stationary state, oscillation shape analysis at operating speed 5000 revolutions per minute and run-up analysis, which aimed to investigate the effect of the rotation speed on the change of the natural frequencies of the disc. A laser Doppler vibrometer and an optical derotator were used as measuring equipment. Zucca et al. [3] used the LDV technique to determine the frequency response functions and operating responses of 24 turbine blades at a specific angular velocity. Stanbridge et al. [4] streamlined the sensing approach which allowed modal shapes to be obtained from a relatively small set of measured data. Trebuňa et al. [5] used a laser vibrometer to identify the source of excessive

vibration of a gas compressor discharge line. The use of a laser vibrometer was necessary in this case as the vibration level limited the use of accelerometers. The operational modal analysis is also addressed using LDV in the following papers [6,7].
Digital image correlation (DIC) is another method that has potential for use in vibration analysis [8]. However, for these purposes, a measurement system with high-speed cameras must be used. Although the DIC method is primarily used for component testing and determination of material properties [14-16], various publications report its use for vibration analysis and experimental modal analysis as well. There is also known publication where DIC is used for in-plane or in-space motion analysis [17].

2 Application module for modal analysis

The ISTRA4D software is capable of recording and evaluating measurement data using the DIC optical method. It offers various options for the needs of laboratories and industrial practices. It allows further extensions in the form of post-process applications. These applications, also called application modules, are programmed in Scilab software and interfaced with ISTRA4D software, where the modules transform the measurement results according to defined mathematical functions. In this paper, the authors discuss the development and description of an application module for the determination of modal parameters based on experimental modal analysis (EMA) and operational modal analysis (OMA) from data obtained by digital image correlation (DIC) measurements [8]. The relationship between the ISTRA4D software and Scilab is described in Fig. 1. Based on the visual representation of the measured object, it is possible to identify a point, curve or polygon on its surface that defines our region of interest on the analyzed object. From the ISTRA4D software, it is then possible to export displacements, relative deformations, point coordinates, and time characteristics. It is also possible to obtain the outputs of the analog channel, which represents, for example, the excitation signal. The export of this data is then processed by Scilab.

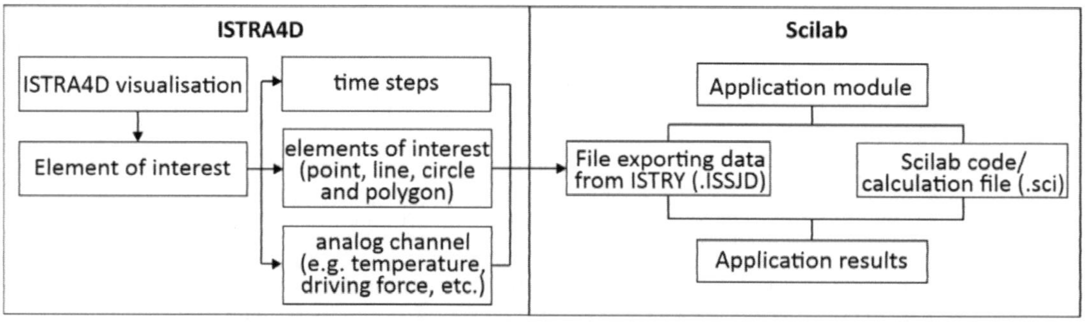

Fig. 1 ISTRA4D—Scilab relation.

The application module consists of two main files. They are text documents in the format:
a) .ISSJD, which defines the export of the required ISTRA4D measurement data to Scilab.
b) .SCE or .SCI, which works with the exported measurement data and transforms them on the basis of mathematical-physical relations into the required resulting parameters. It is the script itself created in Scilab.
The first interface of the modal analysis application module focuses on EMA, i.e., the estimation of modal parameters based on data from the frequency response function (FRF), which is influenced by the excitation frequency spectrum [8].

3 Measurement of input data using DIC for EMA solution

The application module was built based on a series of real sample measurements using the DIC method. For this purpose, a simple rectangular steel plate was chosen as the analyzed object, which had an analyzed area with the dimensions 156 × 152 mm and a thickness of 0.8 mm. The analyzed sample was made of DC03 steel. It is a deep-drawn steel, suitable for automotive body interior parts and other mouldings. The stochastic pattern was applied to the analyzed area in the form of pre-printed spots on a vinyl film. The steel sheet was woven at the bottom edge using two L-sections connected by screws (Fig. 2). Fig. 2 also shows the excitation location (EMA) where the impact hammer blow was located on the back side of the steel sheet [8].

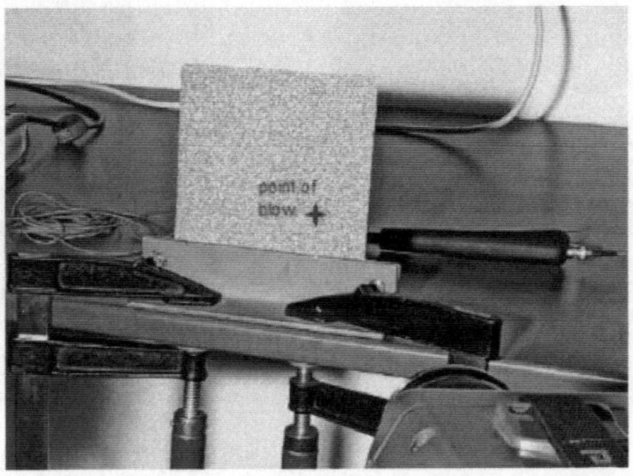

Fig. 2 Laying of the analyzed steel plate and the excitation point.

The measurement was carried out with the Q-450 system from Dantec Dynamics using high-speed cameras. Calibration and correlation of the images were carried out using Istra4D software [8].

The measurement and evaluation conditions were as follows:

- Image acquisition time = 1 s;
- Frame rate = 5000 fps;
- Facet size = 21 pixels;
- Application grid spacing = 17 pixels.

In the first step of the solution, all measurement images were exported from the ISTRA4D software into hierarchically arranged. HDF5 files containing temporal information, frame counts and displacement matrices (Fig. 3) [9]. The images were exported in the form of HDF5 files, with 2004 frames out of 5000 exported in total to reduce the computation time required. For the needs of the application module we developed, only the displacement matrices in the directions of all three axes are sufficient. The displacement matrices are loaded directly by a command from Scilab (.SCE file), which operates on the data of one of them as needed [8].

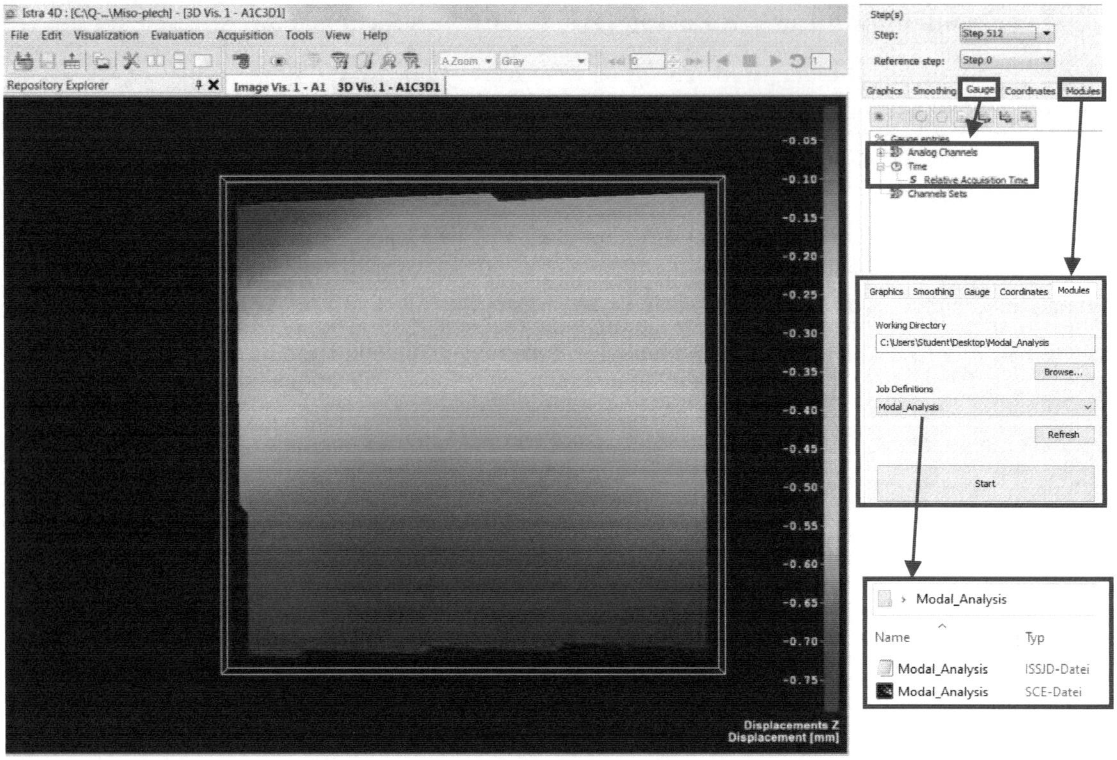

Fig. 3 ISTRA 4D working environment

The methodology for processing the measured and exported data using the Scilab mathematic-physical relationships is defined in Fig. 4. The .SCE file, which represents a computational script written in the Scilab programming language, must be placed in the same folder as the .ISSJD file that defines the export of the time increments.

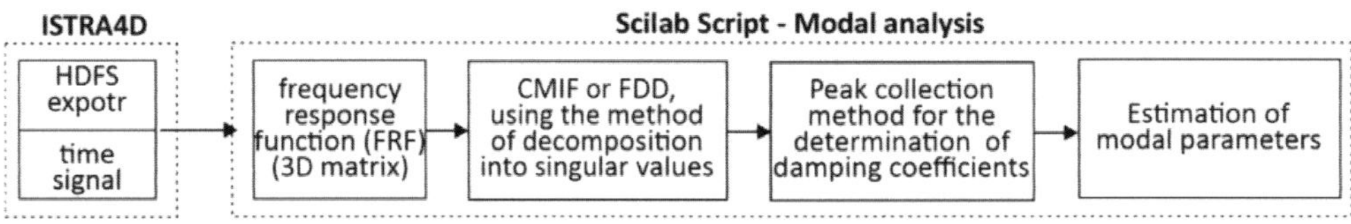

Fig. 4 Processing of exported measurement data in Scilab.

Starting the application displays a window (Fig. 5) that automatically defines the acquisition time based on the number of selected .HDF5 files.

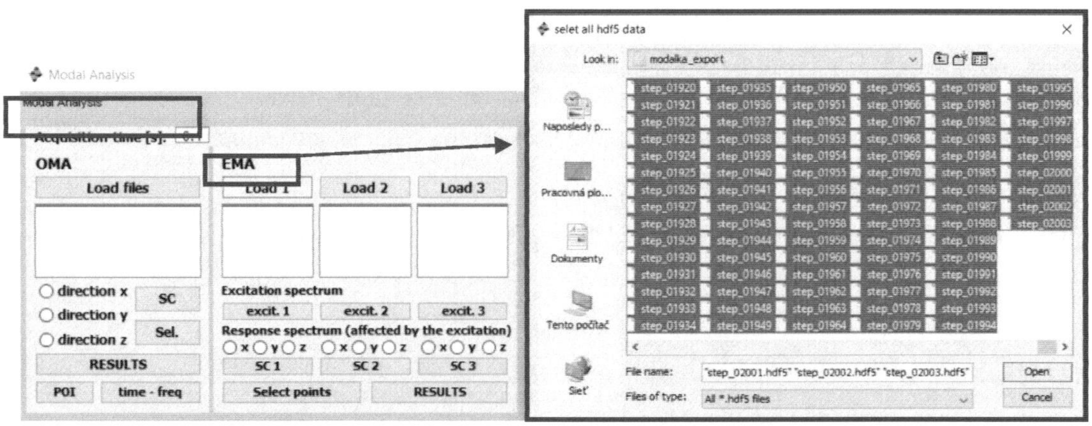

Fig. 5 Initial GUI window for determining modal parameters.

For the demonstration of the application module, only one measurement file was used, i.e., the system was built by an impact hammer strike at only one location; hence, we will only look for one singular curve.
The ISTRA4D software stores the temporal data from the measurements in terms of a column vector and orientates the spatial data into row vectors. Scilab command lines define the input parameter time, which is automatically calculated as the difference between the second and first time steps multiplied by the number of selected .HDF5 files. Since the acquisition time was 1 s at 5000 fps, when considering 2000 frames, the time is adjusted to 0.4 s.
The first function requires the user to select the path to the exported .HDF5 files from, the measurement to determine the number of time steps and the naming of the files to work with next. The quantities that are also required between functions are always defined as a global parameter.
For the analysis, the excitation signal from the analog output was adjusted to a timedomain waveform based on signal processing theory. Subsequently, it was transformed into a frequency spectrum by using the fast Fourier transform (FFT) (Fig. 6).

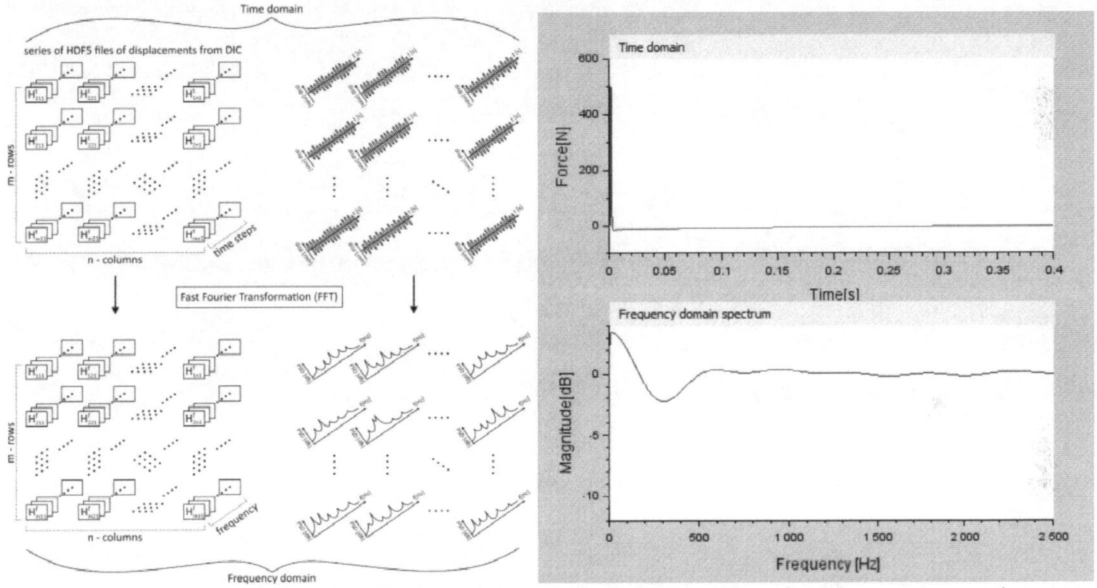

Fig. 6 Transformation of 3D displacement matrix from time domain to frequency spectrum and excitation signal in time domain and frequency spectrum.

A three-dimensional displacement matrix of the oscillating steel plate in the Z-axis direction, where the third dimension of the matrix represents time, was then used to determine the modal shapes. To describe each dimension of the matrix, a representation model was created, shown in Fig. 6. The product of the number of rows and columns gives the number of points (facets) at which displacements were measured. Each of these points has a specific characteristic function dependent on time t. These displacement functions of all points were transformed from the time domain to the frequency spectrum by a fast Fourier transform (FFT) (Fig. 6) [8,9].
Subsequently, the FRF matrix was determined based on the response in the form of the frequency spectrum of the displacement matrix, which is influenced by the frequency spectrum of the excitation signal and transformed into a vector notation (Fig. 7).

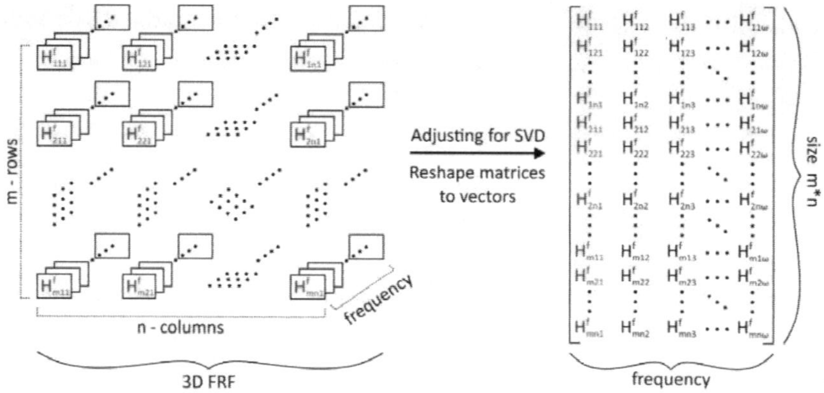

Fig. 7 Transformation of the FRF matrix into vector notation.

By plotting the singular values on the frequency axis, a singular waveform was obtained whose peaks indicated the presence of modes of oscillation (Fig. 8). The respective frequencies of the individual peaks are the natural or resonant frequencies. For each natural frequency, there is an intrinsic mode of oscillation. The eigenmodes of oscillation are exported for the selected frequency from the FRF matrix.

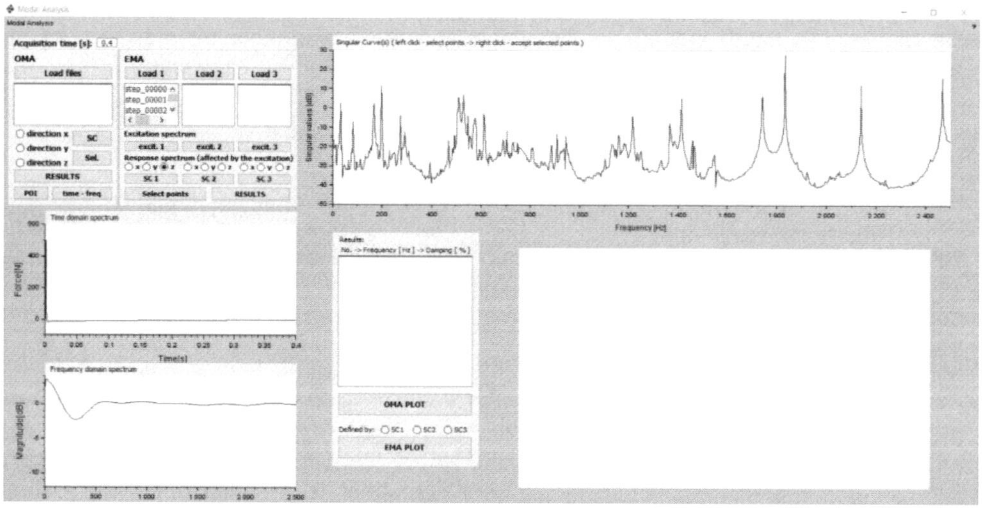

Fig. 8 Singular curve.

Another modal parameter that needs to be determined is the damping coefficient of the mode of oscillation. Since the Scilab software does not provide the ability to automatically select the half-power points based on the peaks, the nearby frequencies need to be manually selected.

Fig. 9 shows the natural frequencies and the corresponding mode damping coefficients determined from the Z-axis displacement matrix obtained by the DIC method (points 1–6). It is also possible to see peaks in the singular waveform that are not natural frequencies of the oscillating steel airfoil, but represent modes resulting from the fit (points X). In the same way, the modal parameters in the X and Y axis directions can be analyzed in the proposed application [8].

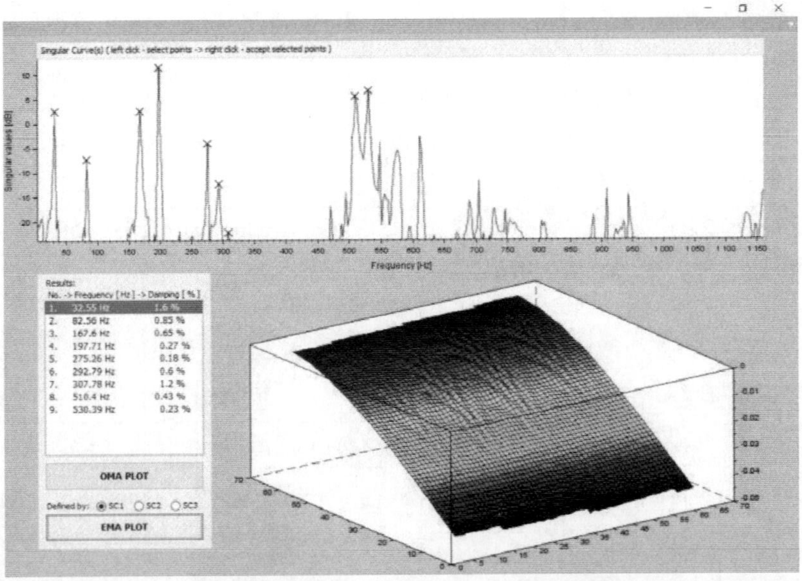

Fig. 9 Resulting modal parameters obtained with the application module.

Each natural frequency has its own specific mode of oscillation. The individual mode shapes with their respective frequencies are shown in Fig. 10.

Mode 1 = 32.55 Hz	Mode 2 = 82.56 Hz
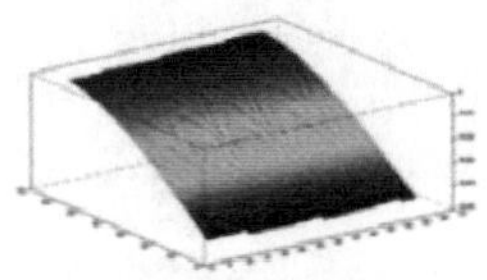	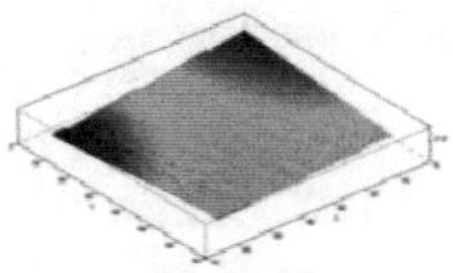
Mode 3 = 197.71 Hz	Mode 4 = 275.26 Hz
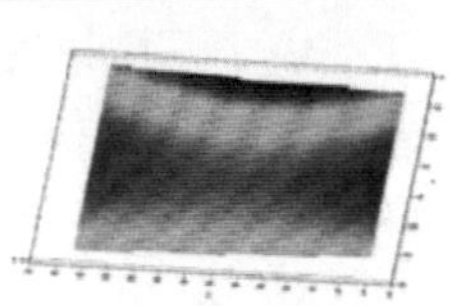	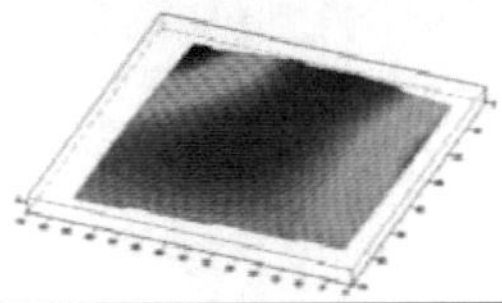
Mode 5 = 307.78 Hz	Mode 6 = 530.39 Hz
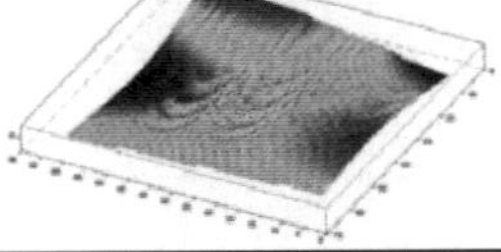	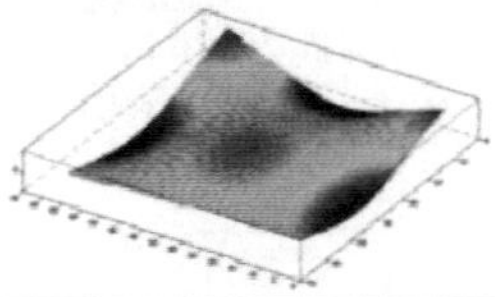

Fig. 10 Modal shapes and corresponding EMA eigenfrequencies.

4 FEM modal analysis

The correctness of the results obtained by experimental modal analysis based on DIC measurements was verified numerically. There are many numerical methods such as peridynamics [10], discrete element method [11], damage mechanics [12] and others, but due to the availability of software, we used the finite element method (FEM) to verify our solution. The numerical model was constructed in the form of a shell element to which the material properties of the steel and the dimensions of the real measured steel plate were assigned. The boundary condition of the fit was simulated by taking all degrees of freedom of motion at the bottom edge of the numerical model. The mesh was meshed from quadrilateral elements.

The resulting frequencies obtained by the FEM method are shown in Fig. 11a. To compare the results of the natural frequencies and modal shapes, the first six modes were selected whose natural shapes are shown in Fig. 11b [8,13].

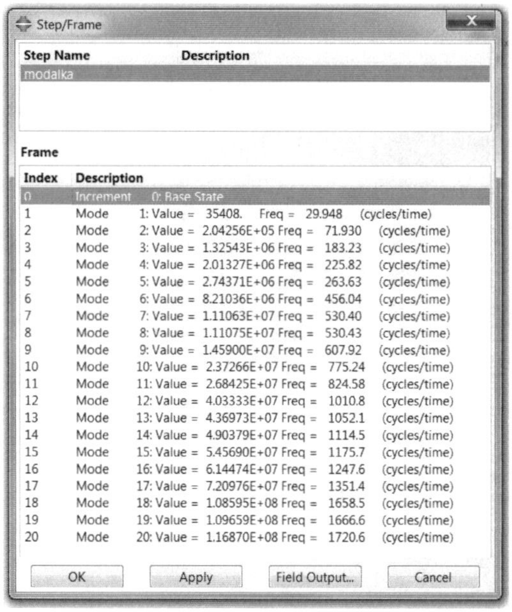

(a)

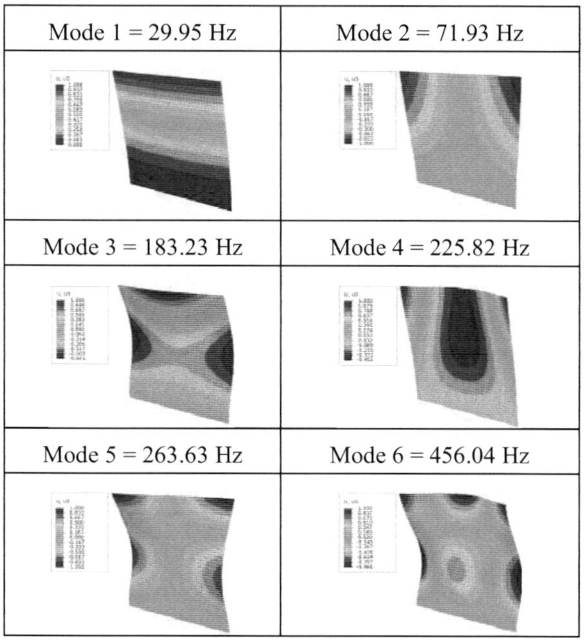

(b)

Fig. 11 (a) Eigenfrequencies of the modes determined by the FEM method, (b) Modal shapes (Z direction) and corresponding natural frequencies obtained by the FEM method.

5 Discussion

The resulting eigenfrequency values obtained experimentally by modal analysis using the DIC method and numerically by the FEM method are shown in Table 1. The differences of the natural frequencies in Hertz and percentage are also given in Table 1.

Table 1. Resulting eigenfrequency values obtained by EMA analysis and FEM analysis.

Mode	EMA Analysis [Hz]	FEM Analysis [Hz]	Difference [Hz]	Difference [%]
1	32.55	29.95	2.6	7.99
2	82.56	71.93	10.63	12.88
3	197.71	183.23	14.48	7.32
4	275.26	225.82	49.44	17.96
5	307.78	263.63	44.15	14.34
6	530.39	456.04	74.35	14.01

In Fig. 12, the deviations of the resulting natural frequencies of the different modes of oscillation obtained by numerical and experimental analysis are plotted [8].

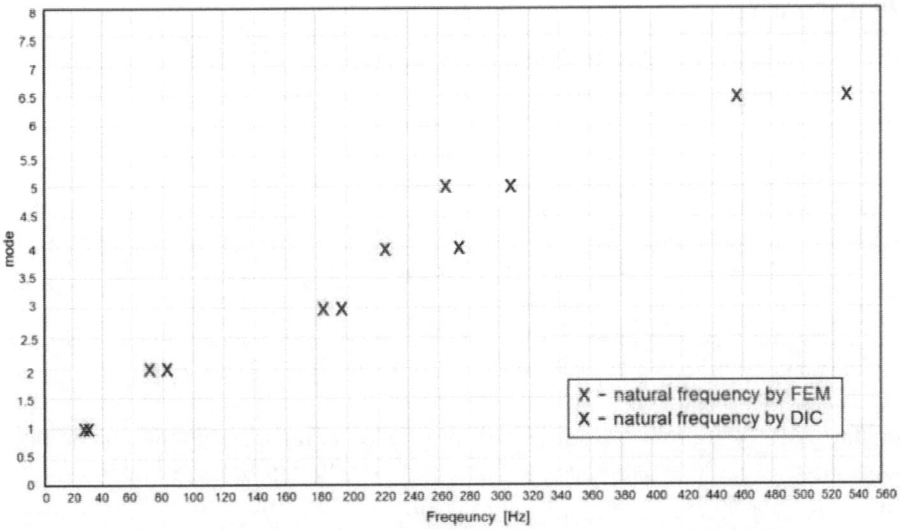

Fig. 12 Comparison of eigenfrequency results obtained by EMA analysis and FEM analysis.

The differences arising in the results are attributed to the choice of boundary conditions in the numerical FEM method. In the experiment, it is difficult to achieve a perfect fit of the analyzed sample.

Conclusion

Currently, either of the numerical modal analyses are used to estimate the modal parameters, or if a more accurate estimation corresponding to the real material properties and boundary conditions of the fit is required, an experimental solution is approached. The commonly used experimental modal analysis procedures are currently conditioned by the application of acceleration sensors on the object under analysis, by means of which the resulting frequency response function is obtained.

This paper describes a new approach to estimate the natural (resonant) frequencies, as well as the corresponding modal oscillation shapes and damping coefficients, based on the results of measurements obtained by the optical digital image correlation method. The resulting modal shapes are obtained based on a displacement matrix.

The main advantage of this new methodology is the fact that, by using high-speed measurements, a full-field analysis of the imaged area of the object is possible, which implies that responses at all points of the object can be obtained on the basis of a single measurement. Conventional methods, with their acceleration sensors, can negatively affect the modal parameters of the system by their mass for sensitive analyzed objects. The full-field analysis by the DIC method makes it possible to determine the responses of any point on the analyzed object without the additional mass of the sensors.

The application module which has been designed in Scilab is able to determine the modal parameters based on the measured displacement matrix by the DIC method.

The aforementioned application is able to determine natural frequencies, modal shapes, and damping coefficients based on experimental modal analysis when the FRF function takes into account the excitation of the system. It is also possible to estimate the modal parameters based on the principle of operational modal analysis when the excitation of the system is unknown.

Acknowledgements

The authors would like to thank the Slovak Grant Agency VEGA for financial support for this study under grant VEGA 1/0500/20.

References

[1] Trebuňa, F., Šimčák, F., Frankovský, P., Huňady, R. and Pástor, M. *Využitie optických metód v experimentálnej mechanike I.* (Technická univerzita v Košiciach, Slovensko, 2014).

[2] Šároši, P., Harčarík, T. and Huňady, R. "Vibrational study of the spinning disc using LDV technique," *Applied Mechanics and Materials*, **816**, pp. 469-473 (2015). ISSN 1660-9336.

[3] Zucca, S., Di Maio, D. and Ewins, D.J. "Measuring the performance of underplatform dampers for turbine blades by rotating laser Doppler vibrometer," *Mech Syst Signal Proc*, **32**, pp. 269-281 (2012). ISSN 1096-1216.

[4] Stanbridge, A.B., Martarelli, M. and Ewins, D.J. "Measuring area vibration mode shapes with a continuous-scan LDV," *Measurement*, **35**(2), pp. 181-189 (2004). ISSN 0263-2241

[5] Trebuňa, F., Šimčák, F., Huňady, R. and Pástor, M. "Identification of pipes damages on gas compressor stations by modal analysis methods". *Eng Fail Anal,* **27**, pp. 213-224 (2013).

[6] Marwitz, S. and Zabel, V. "Operational modal analysis with a 3d laser vibrometer without external referenc" in Proceedings of the 35th IMAC, A conference and Exposition on Structural Dynamics 2017: Rotating Machinery, Hybrid Test Methods, Vibro-Acoustics & Laser Vibrometry; Springer: Cham, Switzerland, 2016.

[7] Xu, Y.F. and Zhu, W.D. "Operational modal analysis of a rectangular plate using non-contact excitation and measurement". *J. Sound Vib*, **332**, pp. 4927-4939 (2013).

[8] Kicko, M. Analýza napätí dynamicky zaťažených konštrukčných prvkov využitím optických experimentálnych metód mechaniky". Ph.D. Thesis, Technical University of Košice, Košice, Slovakia, 2021; p. 73.

[9] Rainieri, C., Fabbrocino, G., Manfredi, G. and Dolce, M. "Robust output-only modal identification and monitoring of buildings in the presence of dynamic interactions for rapid post-earthquake emergency management". *Eng Struct,* **34**, pp. 436-446 (2012).

[10] Sanfilippo, D., Ghiassi, B., Alexiadis, A. and Hernandez, A.G. "Combined peridynamics and discrete multiphysics to study the effects of air voids and freeze-thaw on the mechanical properties of asphalt". *Materials*, **14**, 1579 (2021).

[11] Zhou, S. and Ju, J.W. "A chemo-micromechanical damage model of concrete under sulfate attack". *Int J Damage Mech*, **30**, pp. 1213-1237 (2021).

[12] Zhou, S. and Zhuang, X. "Micromechanical study of loading rate effects between a hole and a crack". *Undergr Space*, **4**, pp. 22-30 (2019).

[13] Xiong, C., Lu, H. and Zhu, J. "Operational modal analysis of bridge structures with data from GNSS/accelerometer measurements". *Sensors*, **17**, 436 (2017).

[14] Genovese, K., Cortese, L., Rossi, M. and Amodio, D. "A 360-deg digital image correlation system for materials testing". *Opt Lasers Eng,* **82**, pp. 127–134 (2016).

[15] Pan, B., Yu, L., Yang, Y., Song, W. and Guo, L. "Full-field transient 3D deformation measurement of 3D braided composite panels during ballistic impact using single-camera high-speed stereo-digital image correlation". *Compos Struct*, **157**, pp. 25-32 (2016).

[16] Gariboldi, E., Naumenko, K., Ozhoga-Maslovskaja, O. and Zappa, E. "Analysis of anisotropic damage in forged Al–Cu–Mg–Si alloy based on creep tests, micrographs of fractured specimen and digital image correlations". *Mater Sci Eng A*, **652**, pp. 175-185 (2016).

[17] Hagara, M., Huňady, R., Trebuňa, F. "Stress Analysis Performed in the Near Surrounding of Small Hole by a Digital Image Correlation Method". *Acta Mechanica Slovaca*, **18**, pp. 74-81 (2015).

Studying the Distribution of Stress in Two-dimensional Compression Piezoelectric Cells under Load-truck Conditions

Béla Kovács[1,a], Mohammed Alaa Alwafaie[1,b] and Jozef Bocko[2,c]

[1]*István Sályi Doctoral School of Mechanical Engineering Sciences, Miskolc, University of Miskolc;*

[2]*Technical University of Košice, Letná 1/9, 042 00 Košice-Sever, Slovakia;*

[a]*matmn@uni-miskolc.hu,* [b]*starala45@gmail.com,* [c]*jozef.bocko@tuke.sk*

Abstract: This research focuses on investigating the impact of stress on piezoelectric cells located on roads with electric voltage. We aim to optimize the design and placement of piezoelectric cells on roads to improve their accuracy and effectiveness in generating electricity from passing vehicles. Our study focuses on three types of loading: one-point loading, uniform distributed loading, and triangle-distributed loading. By analyzing the stress and shear exerted by the pressing truck on the piezoelectric cells, we aim to gain insights into the behavior of these cells under different loading conditions. To conduct this research, we use finite element analysis (FEA) simulations to model the compression cells and simulate the load-truck conditions. The piezoelectric cells are used to measure the electric voltage generated by the stress on the road surface, which is correlated with the stress and shear values obtained from the FEA simulations. The results of this research provide valuable insights into the impact of stress on piezoelectric cells and can be used to optimize the design and placement of these cells on roads, improving their accuracy and effectiveness in generating electricity from passing vehicles. Additionally, this research can inform the development of new materials and technologies for road surfaces that can better withstand the stress and shear exerted by heavy vehicles. Overall, this research has the potential to make significant contributions to the fields of transportation, energy, and materials science.

Keywords: Shear and stress research; Piezoelectric cells; Load-truck conditions; Finite element analysis; Electric voltage

1 Introduction

Shear and stress research has significant applications in various fields, including aerospace, construction, transportation, and materials science. This research is essential for developing new materials and designs that can withstand the stresses and strains of heavy use, as well as for testing the safety and reliability of critical infrastructure. One such application of shear and stress research is in the development of piezoelectric cells for generating electricity from passing vehicles on roads.

1.1 Piezoelectric cells

Piezoelectric materials are considered to be smart materials because of their unique ability to convert mechanical stress or strain into electrical potential and vice versa through the direct and reverse piezoelectric effects, respectively. This conversion process is the basis of piezoelectricity and has numerous practical applications. The piezoelectric material used in a piezoelectric system can operate in different modes, which are defined by the piezoelectric strain constant dij. This constant represents the relationship between the mechanical strain or deformation applied to the material and the resulting electrical voltage output. The subscript "i" indicates the direction of electrical voltage output, while "j" indicates the direction of applied mechanical stress.

There are primarily two constants, d33 and d31, which are commonly used in piezoelectric systems. These constants are characterized by the poling direction being along the 3-axis. The d33 constant represents the piezoelectric strain coefficient in the direction of the electric field, while the d31 constant represents the piezoelectric strain coefficient perpendicular to the direction of the electric field. Understanding these constants is essential for designing effective and efficient piezoelectric systems. In summary, piezoelectric materials are unique and valuable because of their ability to convert mechanical stress or strain into electrical potential, which has many practical applications. The different modes of operation in a piezoelectric system are defined by the piezoelectric strain constant dij, with d33 and d31 being the two most commonly used constants as see in Fig.1.

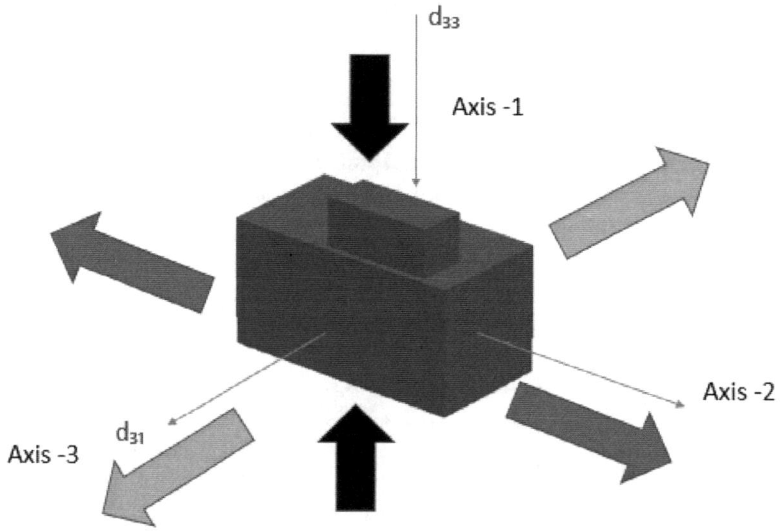

Fig 1. Reference axis generating for piezoelectric material.

1.2 Piezoelectric material

PZT-5J is a type of lead zirconate titanate (PZT) material that is commonly used in piezoelectric systems. It has a high electromechanical coupling coefficient, which makes it an efficient material for converting mechanical stress or strain into electrical energy. It is also more readily available and cost-effective than other PZT materials, making it a popular choice for many applications. The thickness of the PZT module is also an important factor to consider in the design of piezoelectric systems. The thickness of the module affects the sensitivity and responsiveness of the system to mechanical stress and strain. The availability and cost of the PZT-5J is more feasible than the other PZT materials. Therefore, PZT-5J material was chosen here and the module chosen is of thickness 2 mm.

2 Investigating the impact of stress on piezoelectric cells

2.1 Types of impact

In order to understand the impact of stress on piezoelectric cells located on roads with electric voltage, we will be investigating the behavior of these cells under different loading conditions. We will be conducting experiments to analyse the effect of one-point loading, uniform distributed loading, and triangle-distributed loading on the piezoelectric cells. These different loading conditions will be simulated to obtain a better understanding of how the cells react to the stress generated by passing vehicles on roads.

Fig.2 provides an illustration of the different loading trucks used in our research and their corresponding loading conditions on the piezoelectric cells on roads. By studying the behavior of the cells under these different loading conditions, we aim to determine the most effective and efficient design and placement of piezoelectric cells on roads to generate electricity from passing vehicles.

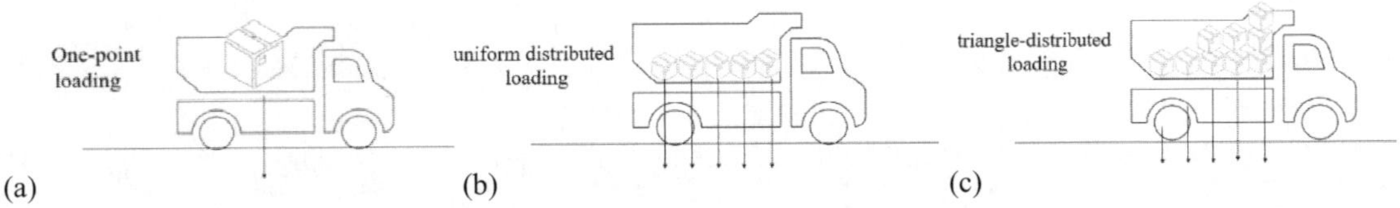

Fig. 2. Different loading truck on piezoelectric cells inside the roads. (a) one point loading, (b) uniform distibuted loading, (c) triangle distibuted loading.

2.2 Analysis for impact of stress on piezoelectric cells

The goal of this analysis is to optimize the design and placement of piezoelectric cells on roads to maximize their energy generation potential. We will use FEA simulations to model the compression cells and simulate different load-truck conditions. The FEA simulations by MATLAB will help us to understand how the cells react to different types and levels of stress generated by passing vehicles on roads so we have finite and infinite elements as shown in Fig. 3.

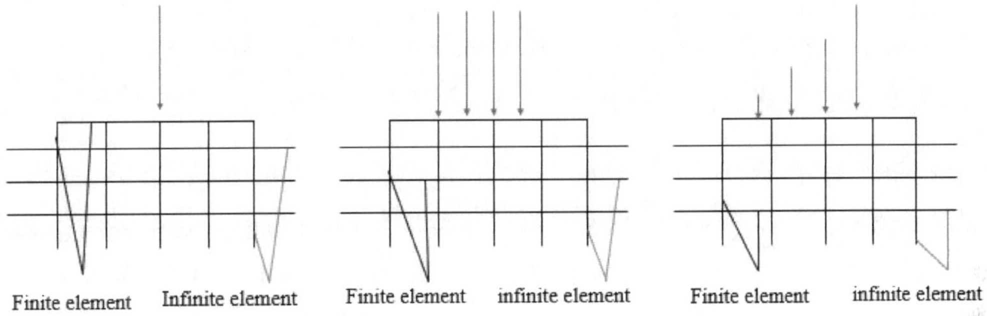

Fig. 3 Finite and infinite drawing under different loading.

Each stress vector (t) in indicial notation tx, ty in the coordinate system (x, y) can be resolved into components along the coordinate lines x, y:

$$\sigma_X\, i_X + \tau_{XY}\, i_Y = t_X \tag{1}$$

$$\tau_{XY}\, i_X + \sigma_Y\, i_Y = t_Y \tag{2}$$

Resolving stress vectors into their normal and shear components can be a valuable approach. This involves breaking down the stress vector (t) into two separate components: one that is parallel to the normal of the surface element (n), which is known as the normal stress (σ), and another component that is perpendicular to the normal (n), which is known as the shear stress (τ). This method is illustrated in Fig. 4, which demonstrates how the normal and shear stresses can be separated and analysed independently. By using this approach, we can gain a better understanding of how the stress is distributed across the surface and how it may impact the behavior of the piezoelectric cells.

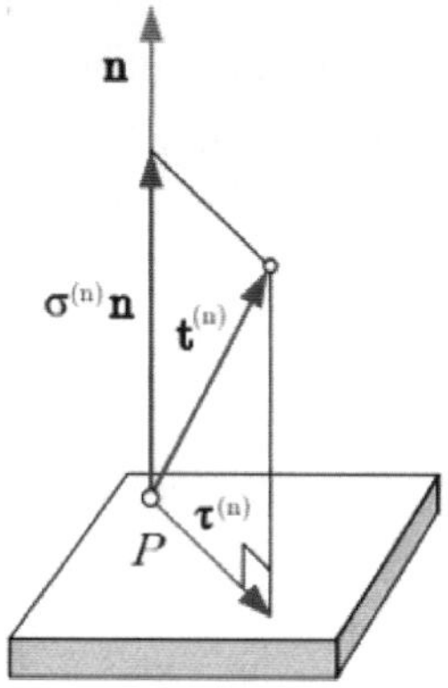

Fig. 4 Graphical representation of the resolution of the stress vector t according to normal element.

The equation of motion in coordinate system of x, y can give as:

$$\frac{\partial t_X}{\partial x} + \frac{\partial t_Y}{\partial y} + \rho b = 0 \tag{3}$$

ρb: it referred to forces which act on piezoelectric body in different loading truck.

When we substitute equations (1) and (2) into equation (3), we can derive scalar equations (4), (5) for the stress vectors. This involves using the equations to express the stress vectors in terms of their normal and shear components. By doing so, we can simplify the equations and make them easier to analyse. The resulting scalar equations provide a way to calculate the magnitude of the stress vectors based on their normal and shear components.

$$\frac{\partial \sigma_X}{\partial x} + \frac{\partial \tau_{XY}}{\partial y} + f_X = 0 \tag{4}$$

$$\frac{\partial \tau_{XY}}{\partial x} + \frac{\partial \sigma_Y}{\partial y} + f_Y = 0 \tag{5}$$

The model depicted in Fig. 5 shows an infinite medium that is subjected to various types of loading, shear and stress forces. The purpose of the model is to analyse the behavior of piezoelectric cells under different types and levels of stress. The arrows in the model represent various types and magnitudes of loading, which are applied to the medium to simulate different stress conditions. By analysing how the medium responds to these loadings, we can gain valuable insights into how the piezoelectric cells will behave under

different stress conditions, and how they will deform and generate electrical energy. This analysis is important for optimizing the design and placement of the cells, as it allows us to identify areas of high stress concentration and determine how the cells will perform under different stress conditions. The model may vary depending on the specific application and type of piezoelectric cell being used, but the basic principle is to subject the cell to different types of loading and analyse its behavior under these conditions. By optimizing the design and placement of the cells based on this analysis, we can maximize their energy generation potential and improve their overall performance.

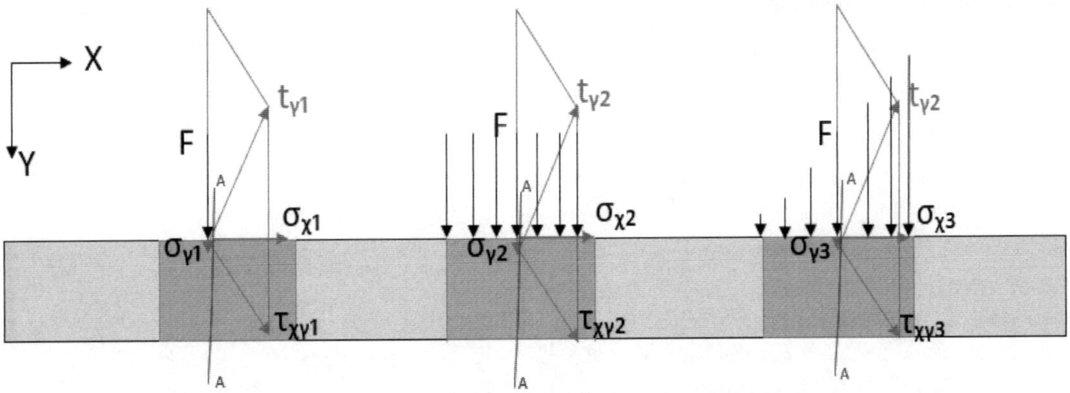

Fig. 5 Infinite medium model that is subjected to various types of loading, shear and stress forces.

Fig. 6 provides a visual representation of the stress distribution in the piezoelectric cell, which is non-uniform and varies across the cell. This stress distribution is observed in equations (4) and (5), as well as in Fig. 5. The figure shows the stress distribution across the cell on the A-A section, with different areas experiencing varying degrees of stress concentration. By analysing this distribution, we can gain a better understanding of how the cell will respond to different types and levels of pressure, and how we can optimize the design and placement of the cells for maximum energy generation potential. The figure models three types of loading on the piezoelectric cell dimension, with the length of the cell along the x-axis and the width of the cell along the y-axis both equal to 20 meters. The loadings include shear forces, compressive forces, and tensile forces, with a magnitude of 1000 kN/m. By modelling these different loadings, we can gain insight into how the cell will respond to different types and levels of stress. This information is important for optimizing the design and placement of the cells to ensure maximum energy generation potential.

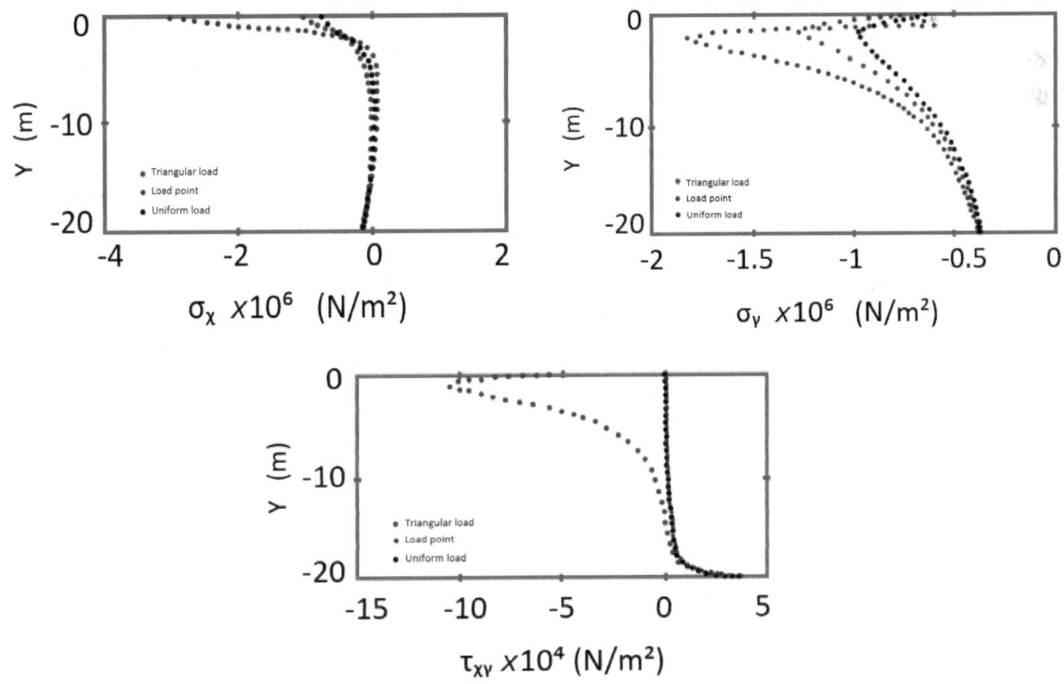

Fig. 6 A visual representation of the stress distribution σ_X, σ_Y, and τ_{XY}, in the piezoelectric cell along A-A section under three loads of point, triangular, and uniform.

The Open Circuit Voltage (OCV) is a measure of the voltage generated by the piezoelectric material when it is subjected to mechanical stress or strain. It is an important parameter to consider when designing piezoelectric devices for energy harvesting, as it determines the amount of electrical energy that can be harvested from the mechanical energy.

The OCV can be calculated using the following equation:

$$(OCV) = g_{33} . \sigma . T \tag{6}$$

where g_{33} is the voltage constant and it is a material property that relates the electrical and mechanical properties of the piezoelectric material. In our case of PZE the value of g_{33} is typically around: $g_{33} = 21.3\times10^{-3}$ Vm/N.

T: Thickness of piezoelectric cell and here its equal to 2.1 mm.

Similarly, the charge density (CD) is a measure of the amount of electric charge that can be stored in the piezoelectric material. The charge density can be calculated using the following equation:

$$(CD) = d_{33} . \sigma \tag{7}$$

where d_{33} is the electric charge constant, which is another material property that relates the electrical and mechanical properties of the piezoelectric material. In our cell the value of: $d_{33} = 485\times10^{-12}$ C/m².

By carefully selecting and designing the piezoelectric material and the device geometry, we can maximize the OCV and charge density, leading to more efficient and effective piezoelectric devices for energy harvesting applications by maximize the amount of stress or strain that it is exposed to, as well as selecting materials with high voltage and charge constants.

3 Results

The results of the stress analysis provide valuable information for the design and analysis of PZE materials, and can help ensure the safety, reliability, and performance of these systems.

According to the stress distribution plots in Fig. 6, it appears that the maximum stress occurs near the cell of the piezoelectric material. This is likely due to the localized deformation and strain that occurs in this region when the material is subjected to mechanical loading.

The maximum stress is observed under point load conditions and is equal to about 1.78×10^{6} N/m² for normal stress σ_Y. The minimum stress is observed under uniform distribution load conditions and is equal to 0.96×10^{6} N/m² for normal stress σ_Y. For normal stress σ_X, the maximum stress under point load conditions is found to be 0.12×10^{6} N/m², while the stress distribution under uniform and triangular loads appears to be relatively close together.

In terms of shear stress τ_{XY}, the maximum stress is observed under triangular load conditions and is equal to 10.3×10^{4} N/m². The stress distribution under point and uniform load conditions for shear stress τ_{XY} appears to be relatively close together. Using the OCV equation with a normal stress of 1.8×10^{6} N/m², the Open Circuit Voltage generated by the piezoelectric material was found to be 76.7 Volts. Using the formula for charge density (CD) with the area of the piezoelectric cell (400 m²), the charge density was calculated to be 0.34 A. Using the current and voltage values, the Power output was calculated to be 26.078 Watt. It's important to note that these calculations assume ideal conditions and may not be representative of the actual performance of the piezoelectric material in practical applications. Further research and testing may be needed to fully understand and optimize the performance of the material for specific applications.

Conclusion

This research focuses on investigating the impact of stress on piezoelectric cells located on roads with electric voltage and aims to optimize their design and placement to improve their accuracy and effectiveness in generating electricity from passing vehicles. Three types of loading (one-point loading, uniform distributed loading, and triangle-distributed loading) were analysed using finite element analysis (FEA) simulations to model the compression cells and simulate load-truck conditions. The results provide valuable insights into the behavior of piezoelectric cells under different loading conditions and can be used to inform the development of new materials and technologies for road surfaces that can better withstand the stress and shear exerted by heavy vehicles. we found that the normal stress of 1.8×10^{6} N/m^2, the charge density and power output were also calculated to be 0.34 A and 26.078 Watt, respectively, based on the area of the piezoelectric cell (400 m^2) and the current and voltage values. This research has the potential to contribute significantly to the fields of transportation, energy, and materials science, as it can inform the development of more efficient and sustainable road systems. By generating electricity from passing vehicles, piezoelectric cells can help reduce the reliance on non-renewable energy sources and promote the use of clean energy. Overall, this study highlights the importance of understanding the impact of stress on piezoelectric cells and provides a foundation for further research and development in this field.

Acknowledgements

We would like to acknowledge the contributions of several individuals and organizations who have made this research possible. First and foremost, we would like to thank Košice university to let us to provide this project. We would also like to thank the research participants like Miskolc university professors who generously gave their time and effort to participate in this study. Finally, we would like to acknowledge the broader scientific community whose work has informed and inspired our research. Their contributions have been instrumental in advancing our understanding of piezoelectric materials and their potential applications. This research was funded by Scientific Grant Agency of the Ministry of Education and Science of the Slovak Republic and the Slovak Academy of Sciences, VEGA 1/0500/20.

References

[1] El-Esnawy N. A., Akl A. Y. and Bazaraa A. S., "A new parametric infinite domain element," *Finite Elements in Analysis and Design*, **19**(1-2), pp. 103-114 (1995). ISSN 0168-874X.

[2] Zienkiewicz O. C., Emson C. and Bettess P., "A novel boundary infinite element," *International Journal for Numerical Methods in Engineering*, **19**(3), pp. 393- 404 (1983). ISSN 0029-5981.

[3] Erkal A., Laefer D. F. and Tezcan S. S., "Advantages of infinite elements over prespecified boundary conditions in unbounded problems," *Journal of Computing in Civil Engineering*, **29**(6), p. 1-28 (2015). ISSN 8873801.

[4] Løkke A. and Chopra A. K., "Direct finite element method for nonlinear analysis of semi-unbounded dam– water–foundation rock systems," *Earthquake Engineering & Structural Dynamics*, **46**, pp. 1267-1285 (2017). ISSN 0098-8847.

[5] Yerli H. R., Temel B. and Kiral E., "Transient Infinite Elements for 2D Soil-Structure Interaction Analysis," *Journal of Geotechnical and Geoenvironmental Engineering*, **124**(10), pp. 976-988 (1998). ISSN 1090-0241.

[6] Liu P., Wang D. and Oeser M., "Application of SemiAnalytical Finite Element Method Coupled with Infinite Element for Analysis of Asphalt Pavement Structural Response," *Journal of Traffic and Transportation Engineering (English Edition)*, **2**(1), pp. 48-58 (2015). ISSN 2095-7564.

[7] Wen P. H., Yang J. J., Huang T., Zheng J. L. and Deng Y. J., "Infinite Element in Meshless Approaches," *European Journal of Mechanics-A/Solids*, **72**, pp. 175-185 (2018).

[8] Lynn P. P. and Hadid H. A., "Infinite Elements with 1/rn Type Decay," *International Journal for Numerical Methods in Engineering*, **17**(3), pp. 347-355 (1981). ISSN 0029-5981.

[9] MATLAB 2020b Academic Version from MathWorks (https://www.mathworks.com/).

[10] Lindeburg M. R., *Mechanical Engineering Reference Manual for the PE Exam, 13th Ed.* (PPI, A Kaplan Company, 2013). ISBN 978-1591264149.

[11] Budynas R. and Nisbett J.K., Shigley's Mechanical Engineering Design, 8th Ed. (2006).

Mechanical Properties of Collagen-Polylactide/Caprolactone Composite for Pulmonary Artery Banding

Jakub Kronek[1,a], Lukáš Horný[1,b], Hynek Chlup[1,c], Zbyněk Sobotka[1,d], Zdeněk Petřivý[1,e], Tomáš Suchý[2,f], Lucie Vištejnová[3,g] and Eva Kuželová Košťáková[4,h]

[1]*Czech Technical University in Prague, Faculty of Mechanical Engineering; Technická 4 Prague 160 00, Czech Republic;*
[2]*Institute of Rock Structure and Mechanics of the Czech Academy of Sciences, V Holešovičkách 94/41 Prague 182 09, Czech Republic;*
[3]*Charles University, Medical Faculty in Pilsen, Alej Svobody 76, Pilsen 323 00, Czech Republic;*
[4]*Technical University of Liberec, Faculty of Science, Humanities and Education, Studentská 1402/2, Liberec 46117, Czech Republic;*

[a]*jakub.kronek@fs.cvut.cz*, [b]*lukas.horny@fs.cvut.cz*, [c]*hynek.chlup@fs.cvut.cz*, [d]*zbynek.sobotka@fs.cvut.cz*, [e]*zdenek.petrivy@cvut.cz*, [f]*suchyt@isrm.cas.cz*, [g]*lucie.vistejnova@lfp.cuni.cz*, [h]*eva.kostakova@tul.cz*

Abstract: Pulmonary artery banding is a surgical procedure used in neonates to induce a mechanobiological response, leading to natural repair of certain congenital heart defects. Current banding methods involve a two-step procedure that requires the use of a non-absorbable material, necessitating surgical removal after a specific period of time. The objective of our project is to develop a banding material for pulmonary artery banding that eliminates the need for surgical removal in the second step by enabling resorption. In this study, we compare the mechanical properties of a composite banding material consisting of a copolymer of caprolactone (PCL) and lactide acid (PLA) used as reinforcement in a collagen matrix (COLL) with a commercially available banding material from W. L. GORE. Through uniaxial tensile testing, we demonstrate that our COLL-PLCL banding material exhibits greater compliance compared to GORE materials while maintaining sufficient strain at failure.

Keywords: Collagenous nanofiber matrix; Polycaprolactone; Polylactide; Uniaxial tensile test; Young modulus of elasticity

1 Introduction

Cardiovascular diseases (CVD) remain a leading cause of death worldwide, contributing to significant morbidity and prompting the ongoing development of both non-invasive and invasive treatment approaches [1-3]. Our research group focuses on the development of new materials for cardiovascular surgery, with the ultimate aim of creating a resorbable fabric composed of a composite material comprising a polylactide and polycaprolactone (PLA/PCL) copolymer nanofibrous reinforcement combined with a collagen matrix. One potential application for this material, among others, is in pulmonary artery banding (PAB) procedures [4,5]. PAB is a palliative surgical technique employed to reduce pulmonary overcirculation in neonates with various types of congenital heart disease [6-8]. It is primarily performed in cases where a left-to-right shunt is present. According to various databases, the total number of PAB procedures accounts for approximately 2% of all surgeries for congenital heart disease [9]. The key advantage of utilizing our COLL-PLCL material for PAB is its resorbability, eliminating the need for debanding surgery and minimizing the burden on patients due to repeated surgical interventions.

2 Methods

Composite COLL-PLCL (Fig. 1A) bandings were fabricated using nanofibrous layers based on a L-lactide/caprolactone copolymer (PLCL; 70/30 mol/mol; Purasorb PLC 7015, Corbion, The Netherlands). The nanofibrous layers were prepared using a needleless electrospinning technique (Production line NS 1S500U with precisely controlled air conditioning unit NSAC150, Elmarco, Czech Republic) with a direct current applied to a 10 wt% (PLCL) solution of chloroform and ethanol (Penta, Czech Republic) in an 8:2 weight ratio. The surface density of the layers was 30 g/m^2, and the fiber diameter ranged from 50 to 1000 nm. The matrix of the composite was composed of collagen type I (calf skin, VUP Medical, Czech Republic). Five separate layers of nanofibrous reinforcement were impregnated with a 5 wt% aqueous dispersion, placed in a mold, and left at room temperature for 36 hours. The volume fraction of all fibrous reinforcement in the composite was 70 ± 10 vol%. The stability of the collagen matrix was further enhanced by cross-linking using a 95 wt% ethanol solution containing EDC (N-(3-dimethylaminopropyl)-N-ethylcarbodiimide hydrochloride) and NHS (N-hydroxysuccinimide) at a 4:1 w/w ratio (Sigma Aldrich, USA).

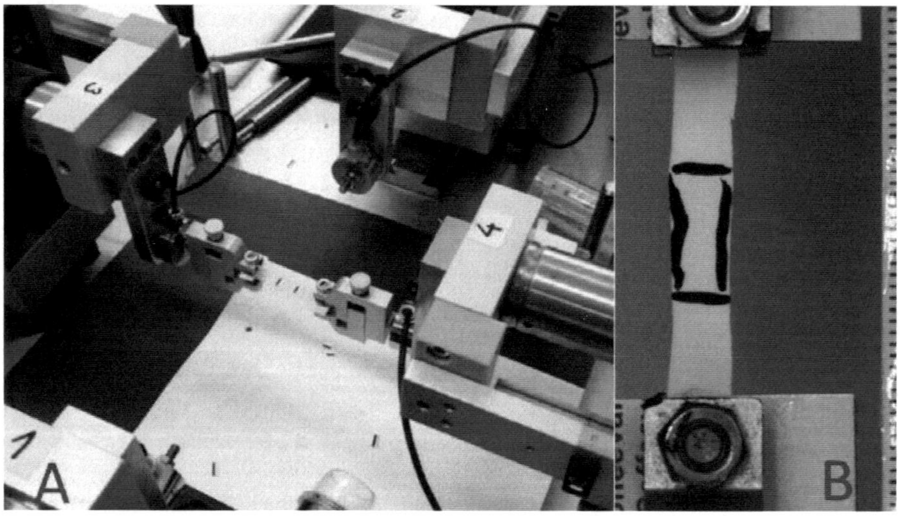

Fig. 1 Experimental setup with mounted COLL-PLCL specimen (panel A), GORE Cardiovascular patch strip (panel B).

For comparison with our artery banding material, we acquired commercially available materials from GORE (W. L. Gore & Associates, Inc., USA) that are currently used for banding purposes. These were strips prepared from GORE Preclude Pericardial Membrane (3 pieces) and GORE Cardiovascular Patch (3 pieces, Fig. 1B). Further details can be found at https://www.goremedical.com/products.

The mechanical behavior of all materials was assessed through uniaxial tensile testing using a Zwick/Roell universal testing machine (Zwick/Roell, Germany, Fig. 1A) specifically designed for testing soft tissues (blood vessels) and polymeric materials. The testing machine was equipped with electromechanical actuators with a working load range of -200 N to 200 N in compression-tension mode. In our experiments, HBM U9C +/- 25N load cells were utilized. The displacement range of the machine is from 0 mm to 140 mm, with a position resolution of 1 μm. The deformations of the specimens were measured using a built-in video extensometer consisting of a 5 Mpx uEye 3.0 CMOS camera. The applied loading rate was set to 0.5 mm/s. All data were recorded and stored on a control PC at a sampling rate of 20 Hz for subsequent post-processing.

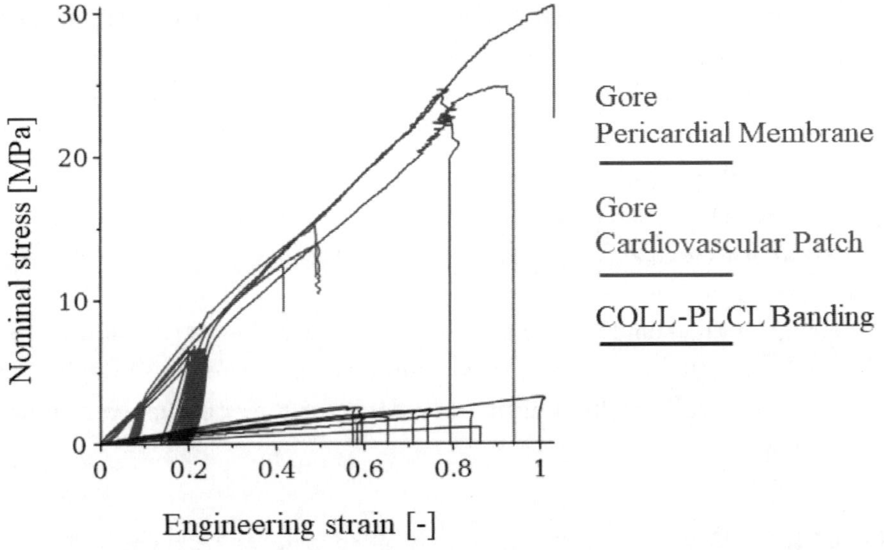

Fig. 2 Stress-strain relationships determined in uniaxial tensile tests within the primary loading (pristine material).

The mechanical response of the measured specimens was described by the nominal stress $\sigma = F/S_0$, where F is the recorded loading force and S_0 is the reference cross-sectional area of the specimens. Deformation was expressed as the engineering strain $\varepsilon = l/L - 1$. Here, L is the distance between the marks made on the surface of a specimen prior to the test, and l is this distance in the deformed state.

The mechanical properties of the COLL-PLCL material were determined in its hydrated state, with the experiment conducted in air immediately after exposing the material to a phosphate buffer solution at 37°C for 2 hours. As it was already obvious during manual (sensory) testing that the elongation at intermediate strength for the GORE material was far beyond what was expected under physiological loading (several hundred percent of the original length), and on the other hand we knew nothing about COLL-PLCL at the beginning, it was decided to use a slightly different loading scheme for each type of material. The COLL-PLCL material was tested monotonically, while the GORE materials underwent 5 loading-unloading cycles, followed by a 6th loading branch that continued until failure (refer to Fig. 2).

3 Results and discussion

Fig. 2 presents the results obtained from our experiments, which involved a total of 15 uniaxial tests (9 using COLL-PLCL strips and 6 using GORE materials). All three materials displayed an approximately linear mechanical response under primary loading. For the GORE materials, Fig. 2 illustrates the complete cyclic loading pattern. In contrast, Fig. 3 provides a detailed view of only the 1st and 5th segments of the loading branch. The cyclic loading induced a preconditioned state in the GORE materials, resulting in a slightly non-linear response that can still be effectively modeled using a linear model. Furthermore, the preconditioned GORE-Tex materials were considerably stiffer compared to their pristine state (Fig. 3).

The GORE materials exhibited higher stress at failure than the COLL-PLCL material, while the strain at failure was approximately the same (Fig. 2). Notably, the COLL-PLCL material demonstrated significantly greater compliance than the GORE materials (Figs. 2 and 4). Based on our findings, we hypothesize that the COLL-PLCL material will induce a reduced mechanobiological response to compliance-mismatch during banding procedures.

If a linear constitutive model is used to describe the material behavior, the mechanical properties can be expressed in terms of the Young's modulus of elasticity (E). Our experiments revealed that the Gore Pericardial Membrane and Gore Cardiovascular Patch exhibited a Young's modulus of E = 35.7 MPa ± 2.12 MPa and E = 33.7 MPa ± 1.76 MPa, respectively, under primary loading. After preconditioning, their response stiffened, resulting in a Young's modulus of E = 167.4 MPa ± 6.88 MPa (Pericardial membrane) and E = 119.9 MPa ± 10.9 MPa (Cardiovascular patch). On the other hand, the COLL-PLCL material demonstrated a Young's modulus of E = 3.78 MPa ± 1.02 MPa, indicating significantly greater compliance. While some may express concerns regarding this result, it is important to note that the COLL-PLCL material falls within the range of physiological properties observed in arteries, thereby facilitating the avoidance of the compliance mismatch effect [10].

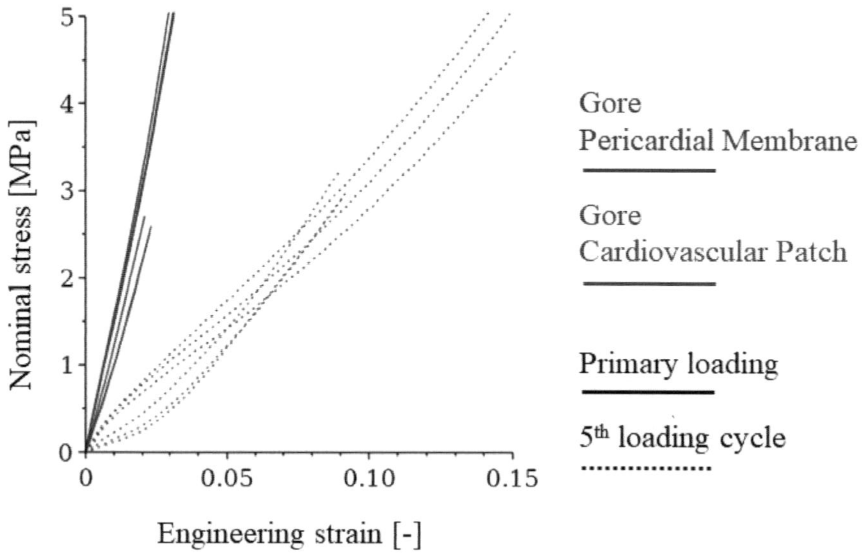

Fig. 3 Comparison of the primary loading response and 5th cycle (preconditioned).

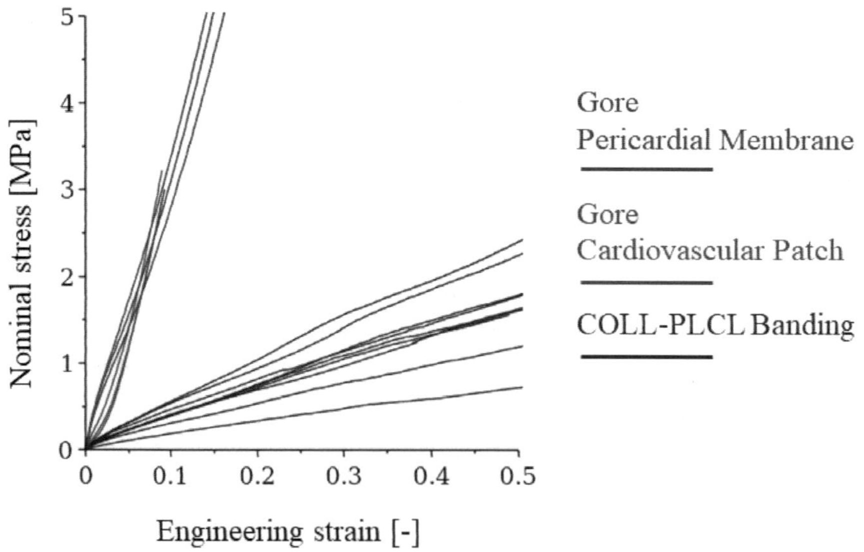

Fig. 4 Stress-strain relationships determined in uniaxial tensile tests within the primary loading of commercially available banding materials (GORE) and our COLL-PLCL composite.

Conclusions

In this study, we presented the results of our comparison of the mechanical properties between our COLL-PLCL material and two commercially available Gore materials suitable for pulmonary artery banding: Gore Pericardial Membrane and Gore Cardiovascular Patch. Uniaxial tensile tests were conducted on strips made of all three materials. The COLL-PLCL material underwent monotonic tensile testing, while the Gore materials were subjected to cyclic loading. The COLL-PLCL material was tested in its hydrated state. The results of the uniaxial tensile tests demonstrated that our COLL-PLCL banding material exhibited greater compliance compared to the GORE materials while still exhibiting sufficient strain at failure.

Acknowledgements

This study has been supported by the Czech Ministry of Health in project NU20-02-00368 Biomechanically defined absorbable materials for cardiovascular surgery.

References

[1] McAloon, C.J., Boylan, L.M., Hamborg, T., Stallard, N., Osman, F., Lim, P.B. and Hayat, S.A. "The changing face of cardiovascular disease 2000-2012: An analysis of the world health organisation global health estimates data," *Int J Cardiol*, **224**, pp. 256-264 (2016).

[2] Virani S.S, et al., Heart Disease and Stroke Statistics - 2021 Update: A Report From the American Heart Association, *Circulation*, **143**(8), pp. e254-e743 (2021).

[3] Im, S.H., Jung, Y. and Kim, S.H. "Current status and future direction of biodegradable metallic and polymeric vascular scaffolds for next-generation stents," *Acta Biomater*, **60**, pp. 3-22 (2017).

[4] Moláček, J., Vištejnová, L., Klein, P., Suchý, T., Horný, L., Kuželová Košťáková, E. and Třeška, V. "Experimental surgery as part of the development of degradable biomaterials for cardiovascular surgery," *Rozhledy v Chirurgii: Měsíčník Československé Chirurgické Společnosti*, **101**(12), pp. 599-606 (2022).

[5] Daley, M., Brizard, C.P., Konstantinov, I.E., Brink, J., Jones, B. and d'Udekem Y. "Absorbable pulmonary artery banding: a strategy for reducing reoperations," *Eur J Cardio-Thorac*, **51**(4), pp. 735-739 (2017).

[6] Juan, S.M. and Jing, H. "Palliative procedures for congenital heart defects," *Arch Cardiovasc Dis*, **102**, pp. 549-557 (2009).

[7] Hoseinikhah, H., Moeinipour, A., Zarifian, A., Andalibi, M. S. S., Moeinipour, Y., Teshnisi, M. A. and Bahreini, A. "Indications, results and mortality of pulmonary artery banding procedure: A brief review and five-year experiences," *International Journal of Pediatrics*, **4**(5), pp. 1733-1744 (2016).

[8] Yoshimoto, A., Miyamoto, T., Ozaki, S., Kobayashi, T. and Kobayashi, T. "Flow-adjustable bilateral pulmonary artery banding in the neonatal period for severe congenital heart diseases," *General Thoracic and Cardiovascular Surgery*, **61**(6), pp. 340-344 (2013).

[9] Brown, S., Boshoff, D., Rega, F., Eyskens, B., Meyns, B. and Gewillig, M. "Dilatable pulmonary artery banding in infants with low birth weight or complex congenital heart disease allows avoidance or postponement of subsequent surgery," *Eur J Cardio-Thorac*, **37**(2), pp. 296-301 (2010).

[10] Kassab, G. S. "Biomechanics of the cardiovascular system: The aorta as an illustratory example," *J Roy Soc Interface*, **3**(11), pp. 719-740 (2006).

Identification of the Bending Moment on the Paddle of a Kayaker

Bohumil Kropík[1,a], Karel Doubrava[1,b] and Tomáš Mareš[1,c]

[1]*Czech Technical University in Prague – Faculty of Mechanical Engineering; Technická 4, Praha 6, Czech Republic;*

[a]*bohumil.kropik@fs.cvut.cz*, [b]*karel.doubrava@fs.cvut.cz*, [c]*tomas.mares@fs.cvut.cz*

Abstract: This contribution deals with the experimental identification of the loading acting on the paddle of a kayaker. To identify loading, strain gauges were installed at three positions of the paddle shaft, always in two directions. The paddle was then loaded with known force in the laboratory to identify the relation between the bending moment and the bending strain. Operational loading followed to identify the level of bending strain during use. With knowledge of the relation between the bending moment and the bending strain, the bending moment during operational use was identified in each measured position and direction and also the global bending moments were identified for each position combining both directions in position.

Keywords: Paddle; Shaft; Composite; Kayak; Loading identification

1 Introduction

The aim of the work is to identify the loading that acts on the paddle of the kayaker during sprint kayaking. Knowledge of loading is important for the development of an optimised design and will be used as one of the entries for multicriteria optimisation. In general, sports equipment has been manufactured and improved for decades. Our long-term aim is to propose a design that will help improve the performance of an athlete. Thus, the first step in this long-term objective is to identify the loading. Usually, the experimental identification of the loading is done using one measurement position on each side of the paddle as in [1,2]. Based on [3], the loading during stroke is quite complex, as the paddler moves with the paddle at a specific angle from the boat to the side. Also, the paddle blade is not just a flat plane shape, but it has curvature using the same idea as the aeroplane wing profile to help the athlete move forward.

2 Method

Positions of interest on the paddle are shown in Fig. 1 and described in Table 1. An approach of identification of the bending moment over the bending strain was chosen. Three positions were chosen at the paddle shaft for the strain measurement and, at each of these positions, strain gauges (SGs) were installed. Linear SGs were used, half-bridge connected for bending strain measurement. At the positions on the left and right sides of the paddle, the SGs were orientated in the expected maximum strain position and a perpendicular position close to the blades to be able to identify the global bending of the shaft. The last measured place was between the hands to obtain two measurement places for each (left and right) stroke. As the measured paddle was feathered, the SGs between the hands were rotated against the SGs close to the blades by ± half the feathering angle. With coefficient k we can get the relation between bending strain and bending moment as the formula:

$$M_o = k \cdot \varepsilon_o. \tag{1}$$

The SGs were wired to the small amplifier that was attached to the paddle shaft together with a small battery and controlled remotely via Bluetooth. Data were recorded in the memory card inside the amplifier.

Fig. 1 Positions of interest of the paddle.

Table 1. Description of positions of interest.

Position	Description
PI_1	Position of the left blade
PI_2	Position of the measurement, left close to the throat
PI_3	Left hand grip position
PI_4	Position of the measurement, between hands
PI_5	Right hand grip position
PI_6	Position of the measurement, right close to the throat
PI_7	Position of the right blade

3 Laboratory loading

To identify the relation k between bending moment M_o and bending strain ε_o, it was necessary to load the paddle with known force. This was done separately for the left and right sides of the paddle to have the maximal bending moment close to the measured place. The paddle was loaded at first only in the expected maximal orientation as the perpendicular one is unstable. In both cases, the paddle was placed inside the loading machine and loaded over the strap as shown in Fig. 2. The loading scheme with PI_2, PI_4, and PI_6 representing the places with SGs is shown in Fig. 3. The position of SGs between the hands was not loaded separately as it was loaded during the loading of the left blade.

Fig. 2 Paddle during the laboratory loading.

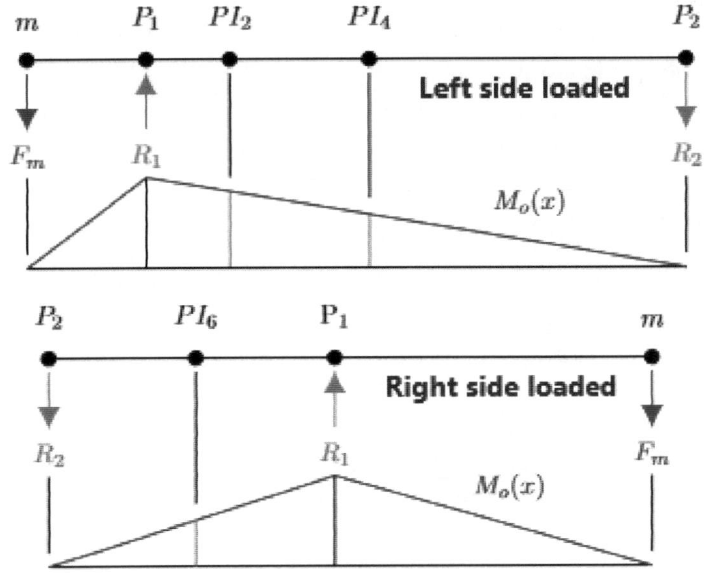

Fig. 3 Scheme of the laboratory loading with the bending moment.

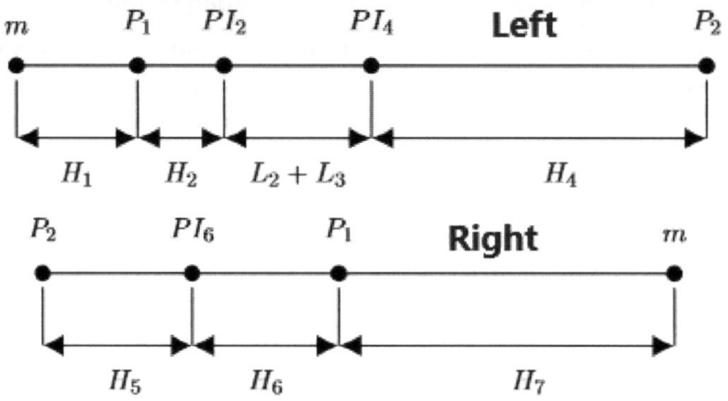

Fig. 4 Distances between positions during the laboratory loading.

3.1 Left side

For the left side, weight m = 10 946 g was used during laboratory loading. Distances according to the schemes in Fig. 3 and Fig. 4 are in Table 2. We can construct the equilibrium equations and identify the bending moments at the PI_2 and PI_4 positions:

$$M_{o(PI_2)} = R_2 \cdot (L_3 + L_4 + H_4), \tag{2}$$

$$M_{o(PI_4)} = R_2 \cdot H_4. \tag{3}$$

By expressing the k in formula (1), we can obtain the coefficient for the left position directly using the measured strain at this position $\varepsilon_{o(PI_2)}$ = 568 μm/m. The coefficient is:

$$k_{(PI_2)} = \frac{M_{o(PI_2)}}{\varepsilon_{o(PI_2)}} = \frac{36\ 370}{568} = 64.03\ \frac{\text{Nmm}}{\mu\text{m/m}}. \tag{4}$$

For the central section, the loading is not orientated directly on any of the SGs. Considering the approach of finding the resultant of generally acting moment, in case of perpendicular positions, it is:

$$M_o = \sqrt{M_{o(M)}^2 + M_{o(R)}^2}, \tag{5}$$

where $M_{o(M)}$ and $M_{o(R)}$ are representing two perpendicularly oriented moments. Using formula (1) in (5), we get:

$$M_{o(PI_4)} = \sqrt{\left(k_{(PI_4)} \cdot \varepsilon_{o(PI_4-M)}\right)^2 + \left(k_{(PI_4)} \cdot \varepsilon_{o(PI_4-R)}\right)^2} =$$

$$= k_{(PI_4)} \cdot \sqrt{\varepsilon_{o(PI_4-M)}^2 + \varepsilon_{o(PI_4-R)}^2}, \tag{6}$$

and thus, we can find the global strain for the PI_4 position as:

$$\varepsilon_{o(PI_4)} = \sqrt{\varepsilon_{o(PI_4-M)}^2 + \varepsilon_{o(PI_4-R)}^2}. \tag{7}$$

The coefficient for the PI_4 position is then:

$$k_{(PI_4)} = \frac{M_{o(PI_4)}}{\varepsilon_{o(PI_4)}} = \frac{15\ 008}{253} = 59.32\ \frac{\text{Nmm}}{\mu\text{m/m}}. \tag{8}$$

Table 2. Distances during laboratory loading of the left side of the paddle.

Position	Distance [mm]
H_1	368
H_2	45
$L_2 + L_3$	306
H_4	215

3.2 Right side

For the right side, weight m = 14 891 g was used during laboratory loading. Distances according to the schemes in Fig. 3 and Fig. 4 are in Table 3. Again, we can construct the equilibrium equations and identify the bending moments at the PI_6 position:

$$M_{o(PI_6)} = R_2 \cdot H_5, \tag{9}$$

Again, by expressing the k in formula (1), we can obtain the coefficient for the right position directly using the measured strain at this position $\varepsilon_{o(PI_6)}$ = 1014 μm/m. The coefficient is:

$$k_{(PI_6)} = \frac{M_{O(PI_6)}}{\varepsilon_{O(PI_6)}} = \frac{65\ 155}{1014} = 64.2\ \frac{\text{Nmm}}{\mu\text{m/m}}. \tag{10}$$

Table 3. Distances during laboratory loading of the right side of the paddle.

Position	Distance [mm]
H_5	530
H_6	32
H_7	473

3.3 Result

As the $k_{(PI_2)}$ value is almost the same as the $k_{(PI_6)}$, average of these two values was used for both, and for the perpendicular positions on the left and right. The $k_{(PI_4)}$ value is slightly different from them as it is affected by the joint between the hands.

4 Operational loading

The paddle was loaded on a short track by a kayaker simulating sprint kayaking. The strains were meassured, section of recorded signals is shown in Fig. 5. With the use of (1), we can identify the bending moments from them as shown in Fig. 6. With the use of formula (5) we obtained the global bending moments at PI_2, PI_4, and PI_6 positions, which is shown in Fig. 7. Maximal values of the bending moments during measurement were identified as $PI_{2\text{-}max}$ = 101 011 Nmm, $PI_{4\text{-}max}$ = 89 352 Nmm and $PI_{6\text{-}max}$ = 99 575 Nmm. Unfortunately, during this first measurement, the $PI_6 - R$ channel did not measure any signal, which means that the absolute value for the right side is affected and the real value will probably be slightly higher.

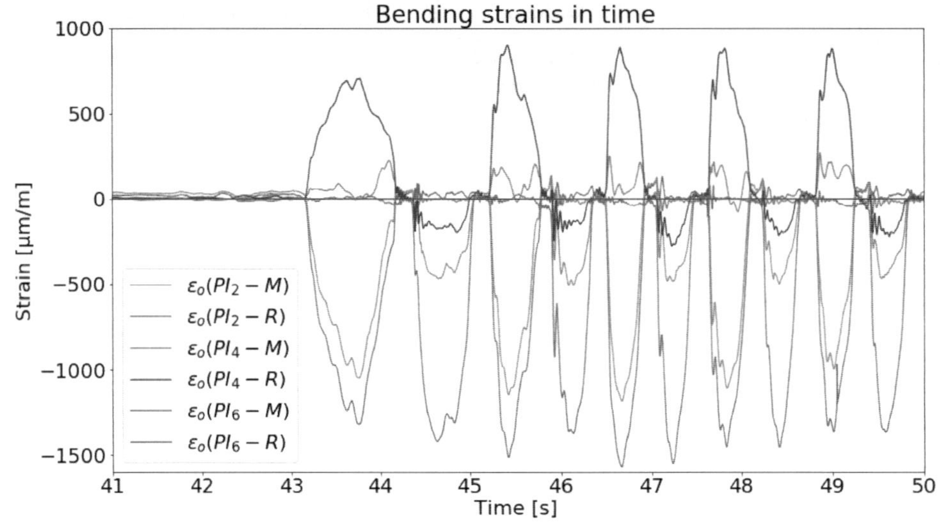

Fig. 5 Section of measured bending strains during the operational loading.

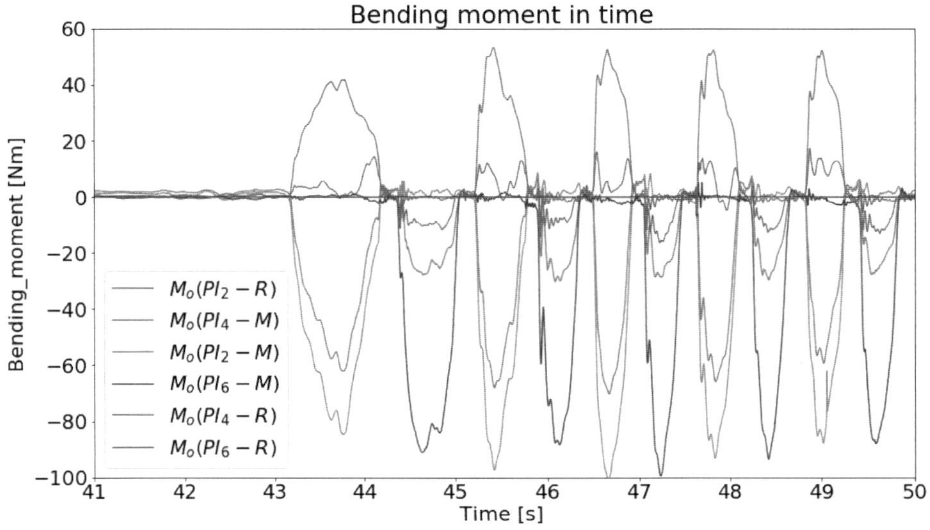

Fig. 6 Section of calculated bending moments during the operational loading.

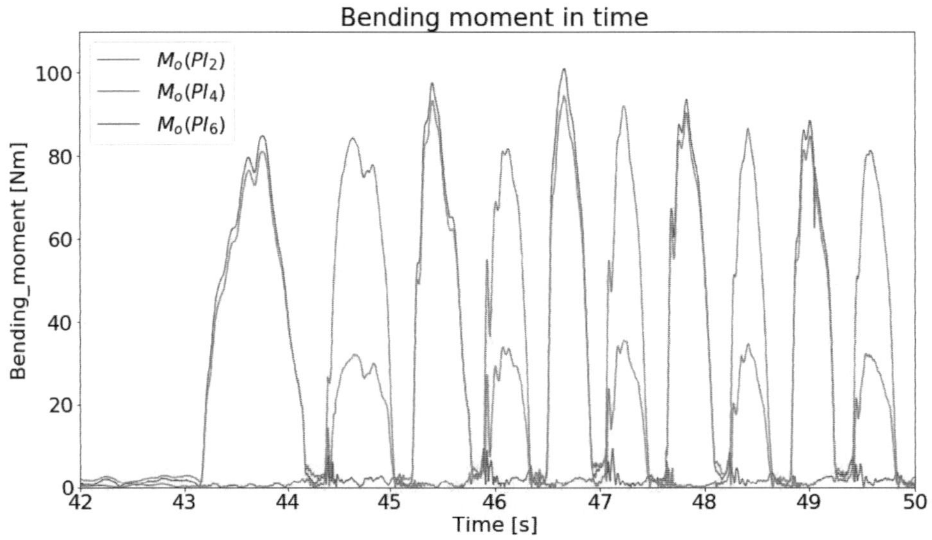

Fig. 7 Section of global bending moments during the operational loading.

Conclusion/Summary

The bending moment identification approach proved to be useful. The choice of two perpendicular positions of SGs in each *PI* position helped to improve the precision of the identified values. It was possible to identify the relation between the bending moment and the bending strain and use it for identification of bending strains/moments during operational loading. With a malfunction of the $PI_6 - R$ channel during operational loading, we probably got slightly underestimated values of global bending moment at PI_6 position which we can estimate from the behaviour of PI_2 position. During each stroke, two places were actively measuring, which is important for further identification of forces acting on the paddle shaft. Thus, this setup will be used for future measurements to identify the forces acting on the paddle, which will serve as input for the optimisation of the paddle shaft.

Acknowledgements

This work has been supported by the project No. SGS21/151/OHK2/3T/12 of the Grant Agency of the Czech Technical University in Prague.

References

[1] Prétot C., et al., "On the Physics of Kayaking," *Available from* https://doi.org/10.3390/app12188925 *Accessed:* 2023-11-30.

[2] Gomez B., et al., "Analysis of the on-water paddling force profile of an elite kayaker," *Available from* https://ojs.ub.uni-konstanz.de/cpa/article/view/4822 *Accessed:* 2023-11-29.

[3] Social Nature Movement, "How does a kayak wing paddle blade work?" *Available from* https://socialnaturemovement.nz/wing-blades-explained/ *Accessed:* 2023-06-05.

On a Possibility of High-Resolution Neutron Diffraction Observation of Inconel 718 Substructure Phases

Pavol Mikula[1,a] and Vasyl Ryukhtin[1,b]

[1]*Nuclear Physics Institute CAS, 250 68 Řež, Czech Republic;*

[a]*mikula@ujf.cas.cz*, [b]*ryukhtin@ujf.cas.cz*

Abstract: A feasibility of using an unconventional high-resolution neutron diffraction three-axis setting for studies of substructure phases of some alloys is presented. The usefulness of the diffraction setting is documented on measurements of powder diffraction profiles of several samples of Inconel 718 alloy. Even though the measurement at the medium power research reactor is rather time consuming, together with the γ phase matrix, it was possible to distinguish portions of other phases γ' and γ'' corresponding to created precipitates which depending on percentage content could have a significant influence on material properties.

Keywords: Neutron diffraction; Inconel 718 alloy; Substructure phases; Experimental Analysis

1 Introduction

Polycrystalline Ni-based superalloys are one of the very frequently used materials for high temperature load-bearing applications due to their high mechanical strength and chemical resistance. When looking for good mechanical properties and corrosion resistance of the material at elevated temperatures, a Ni-based superalloy 718 (also known as Inconel 718) is widely used e.g. in gas turbines in jet engines. The excellent properties of the Inconel 718, namely, strength and ductility, result from a good combination of the phases [1-3]: γ phase matrix with the fcc crystal structure, the coherent γ'' phase precipitates with Ni3Nb composition and a tetragonal bct ($D0_{22}$) crystal superstructure. Then, less often a fcc crystal structure of γ' precipitates could be present. The precipitation of either γ' or γ'' is dependent on the concentration of Ti and Nb. Typically, Inconel 718 is used in applications carried out at medium temperature, because above 650 °C the thermodynamically stable δ-phase forms, with Ni3Nb composition and an orthorhombic crystal structure (at the expense of the γ''phase) which degrades the mechanical properties of the material. The δ phase precipitates either at the grain boundaries or intragranular.

Thanks to their excellent properties, fabricating components of the Alloy 718 using additive manufacturing (AM) methods are very attractive for different applications because they provide large possibilities of design flexibility. The most usually used AM processes are the so-called powder bed fusion ones, which include the following methods: direct metal laser sintering (DMLS), electron beam melting (EBM), selective heat sintering (SHS), laser powder bed fusion (LPBF usually referred as selective laser melting, SLM), and selective laser sintering (SLS). In the LPBF method a laser source is utilized for melting and selectively fusing together metallic powders, which are spread on a moving plate when the designed component is built. In the LPBF method, processing parameters as the laser power, scanning speed, hatch spacing, and layer thickness, are the most influential ones when producing microstructure for particular metals/alloys. For the studies of the microstructure and determination of the individual phases in the material, conventional X-ray diffraction (XRD) and neutron diffraction are usually used. Namely, synchrotron XRD, providing the high flux, is very important for indexing and quantifying different phases in low fractions in the material. However, it is well known that due to a low penetration of X-rays into the material (for low energies X-rays), XRD provides required information related to the surface of the studied samples. On the other hand, neutron diffraction can use much larger probed volumes when fighting with a relatively much lower flux and thus also with much longer measurement time. As a feasibility study, in this paper we introduce neutron powder diffraction results of several samples of Inconel 718 as obtained by unconventional high-resolution neutron diffraction method in order to identify some phases having very close lattice spacings to the matrix one.

2 Experimental procedure

The neutron diffraction experiment was carried out by a high-resolution three-axis setting as schematically shown in Fig. 1. Following the sketch shown in Fig. 1 (for small widths of the samples), a maximum resolution from this arrangement can be achieved for minimal dispersion of the whole system [4,5]. When treated in momentum space, it means that the orientation of the Δk momentum domains related to the monochromator and analyzer should be matched to that of the sample, and this can be achieved by proper radii of curvature of the bent perfect crystal (BPC) monochromator and analyzer. Thus, focusing in real and momentum space can be realized. Contrary to the conventional two axis diffractometer, the diffraction profiles with the three-axis set-up are obtained by rocking (step-by-step) the (BPC) analyzer situated on the third axis of the diffractometer and the neutron signal is registered by a point detector (see Fig. 1).

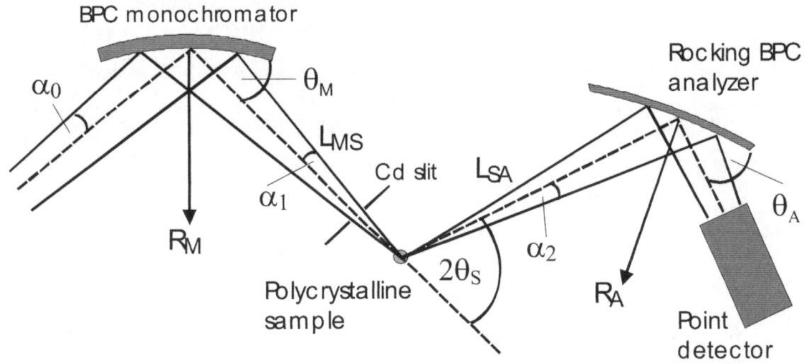

Fig. 1 Three-axis diffractometer setting employing BPC monochromator and analyzer as used in the experiment (R_M, R_A - radii of curvature, θ_M, θ_A - Bragg angles).

For $L_{MS}/(R_M \cdot \sin \theta_M) \neq 1$ and $L_{SA}/(R_A \cdot \sin \theta_A) \neq 1$, a general form for focusing in momentum space (not dependent on α_1 and α_2) which minimizes the dispersion between all elements can be derived as [6]

$$2 \tan \theta_S = \tan \theta_M /(1 - L_{MS}/(R_M \cdot \sin \theta_M)) + \tan \theta_A /(1 - L_{SA}/(R_A \cdot \sin \theta_A)). \quad (1)$$

When fulfilling the condition (1), a maximum peak intensity and a minimum *FWHM* of the analyzer rocking curve can be expected. The experimental tests were carried out on the 3-axis neutron diffractometer installed at the Řež research reactor LVR-15. The diffractometer employs the Si(111)-monochromator and Si(311)-analyzer single crystals which have the dimensions of 200x40x4 mm^3 and 20x40x1.3 mm^3 (length x width x thickness), respectively. The monochromator Si(111) provides the neutron wavelength of 0.162 nm and has a fixed curvature with a radius R_M of about 12 m. The radius of curvature of the analyzer was changeable in the range from 3.6 m to 36 m. It permitted us to determine an optimum analyzer curvature minimizing the dispersion and thus optimizing the resolution of the setting.

3 Experimental results

Examples of an excellent resolution of the 3-axis setting as documented by *FWHMs* of the analyzer rocking curves obtained on well annealed α-Fe(110) standard samples are shown in Fig. 2. Both samples were put on the second axis in the vertical position. High resolution of the setting points out an applicability of using it for studies of special material research tasks e.g. determination and volume content of different material phases, namely, when they have very close values of lattice spacing. Furthermore, microstructure analysis of plastically deformed samples, or samples under the external thermomechanical load during a deformation process, can be studied on the basis of diffraction profile analysis [7-10]. Recently, it has been found that the setting provides a sufficiently high resolution, though slightly relaxed, when wider slits (e.g. up to 20 mm) i.e. wider samples are used [11]. This phenomenon can be used e.g. for strain measurement on samples of large dimensions which is based on the measurement of the resulted peak shift $\Delta\theta_S$. When the sample of large dimensions is used, namely, its width introduces a too large resolution uncertainty in the case of the conventional two-axis strain/stress scanners.

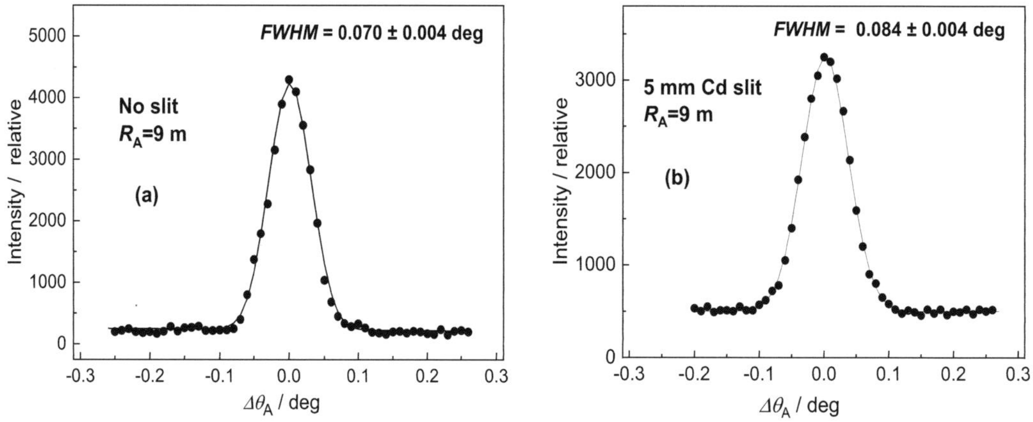

Fig. 2 α-Fe(110) diffraction profiles when using a well annealed 2 mm and a small steel plate of the dimensions of 2x7x15 mm³ – (b).

Recently, it has been found out that the high-resolution setting can be used to distinguish diffraction profiles corresponding to very close lattice constants e.g. of different phases, as it is in the case of Inconel 718. For the measurement test we used several samples of Inconel 718 of the dimensions of 2x7x18 mm³ which were put on the second axis vertically. The impinging beam on the sample was limited by a 5 mm Cd slit in order to be smaller than the width (7 mm) of the sample. Fig. 3 - Fig.5 show several results obtained on the wrought, SLM and wrought–aged samples which document the feasibility of the three-axis diffraction setting for such demanding experimental investigations. When using the values of the lattice spacings from ref. 3 and a simple formula $\Delta\theta_A \approx - \Delta(2\theta_S)$ valid for large values of R_A, our results well correspond to the ones obtained in ref. 1.

The inspection of Fig. 3 reveals that the main maximum (the red line) corresponds to the γ phase of the matrix while the small maxima at the shoulders correspond to the separated γ' (green line) and γ'' (blue line) phases, respectively. Then, Fig. 4 reveals that the main maximum (the red line) corresponds to the γ phase while the small maximum at the shoulder (green line) corresponds to the γ' phase. On the other hand, Fig. 5 shows that together with the main γ phase, there is a considerable contribution of a mixture of γ' and γ'' phases (green line). The difference of the lattice spacings of the individual phase in the particular sample can be determined from the relation $\Delta\theta_A = (\Delta d_S/d_{0,S})\cdot\tan\theta_S$.

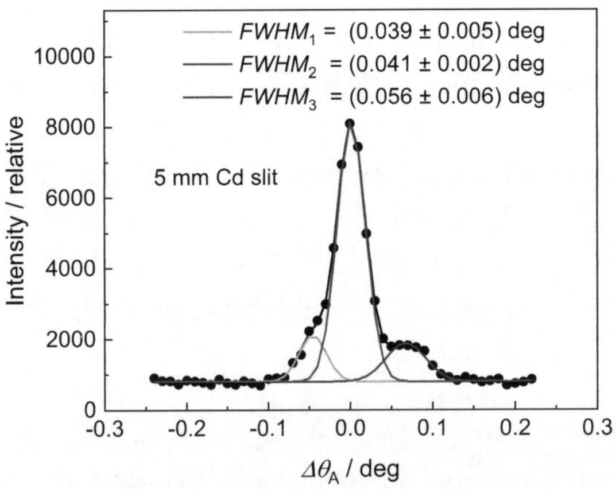

Fig. 3 Analyzer rocking curve related to the wrought Inconel 718 sample.

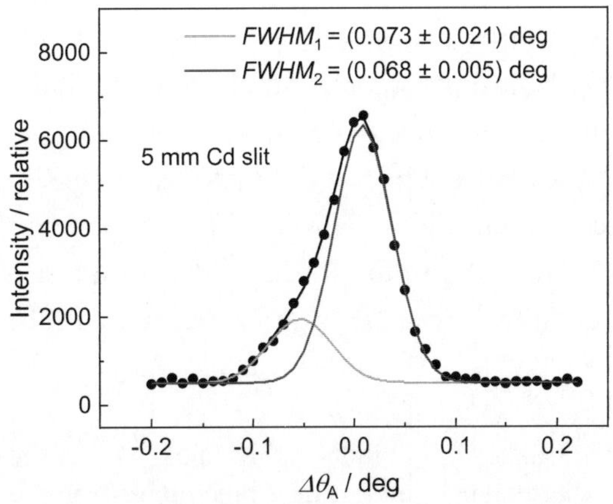

Fig. 4 Analyzer rocking curve related to the SLM Inconel 718 sample.

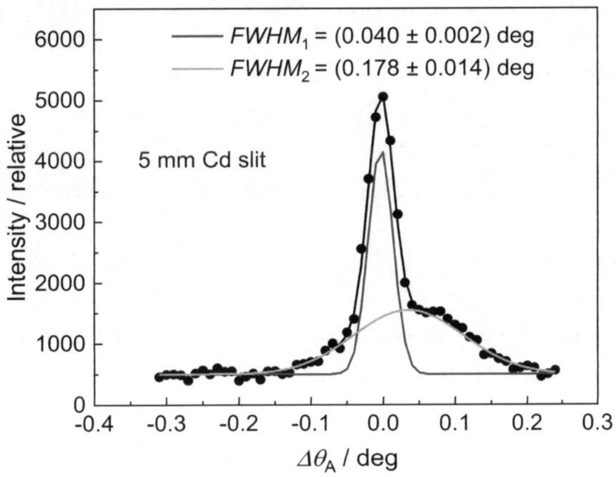

Fig. 5 Analyzer rocking curve related to the wrought-aged Inconel 718 sample.

Conclusion/Summary

High-resolution three-axis setting employing BPC Si(111) monochromator and BPC Ge(311) analyser with a polycrystalline sample between them was tested on the NPI neutron optics diffractometer and its feasibility of using it for investigation of special tasks of powder diffractometry was demonstrated. Namely, the excellent properties of the diffractometer setting permitted us to identify a substructure of Inconel 718 polycrystalline alloy containing several phases having close lattice spacings. As the mechanical properties considerably depend on the volume content of additional phases accompanying the main one their determination could be highly important. Naturally, the diffractometer setting could also be exploited for measurements of diffraction profiles of plastically deformed samples subjected e.g. to thermo-mechanical load and from their analysis to evaluate the dislocation density and a mean grain size in the polycrystalline sample. Of course, this setting can also be used for conventional macro-strain scanning based on the angular shifts $\Delta\theta_A$ of the obtained diffraction profiles, but due to the step-by-step analysis with the analyzer, rather time-consuming measurement could be impractical in comparison with the conventional strain/stress diffractometer. We hope that the presented neutron diffraction setting can offer an additional support to complement the information achieved by using the other conventional characterization methodologies.

Acknowledgements

Measurements were carried out at the CANAM instrument of NPI CAS Řež installed the CICRR infrastructure, which is financially supported by the Ministry of Education and Culture - project LM2023041. The authors acknowledge support from ESS participation of the Czech Republic – OP (CZ.02.1.01/0.0/0.0/16_013/0001794) and from the project ESS Scandinavia-CZ II (LM2018111), respectively. Furthermore, they acknowledge support from the CAS in the frame of the program "Strategie AV21, No. 23". The authors thank B. Michalcová from NPI CAS for significant help with measurements and data elaborations.

References

[1] Čapek, J., Polatidis, E., Knapek, M., Lyphout, C., Casati, N., Pederson, R. and Strobl, M. "The effect of γ' and δ phase precipitation on the mechanical properties of Inconel 718 manufactu-red by selective laser melting: An in-situ neutron diffraction and acoustic emission study," *JOM* **73**, pp. 223-232 (2021).

[2] Solís, C., Munke, J., Bergner, M., Kriele, A., Mülbauer, M.J., Cheptiakov, D.V., Gehrmann, B., Rosler, J. and Gilles, R. "In-situ characterization at elevated temperatures of a new Ni-based superalloy VDM-780 Premium," *Metall Mater Trans A*, **49**, pp. 4373-4381 (2018).

[3] Repper, J. "Einfluss mikroskopischer Eigenspannungen auf die makroskopische Eigenspan-nungsanalyse mittels Neutronenbeugung," *PhD Thesis Technische Universität München*, 2010.

[4] Mikula, P., Šaroun, J., Ryukhtin, V. and Stammers, J. "An alternative neutron diffractometer performance for strain/stress measurements," *Powder Diffr*, **35**, pp. 185–189 (2020).

[5] Mikula, P., Šaroun, J. and Ryukhtin, V. "High-resolution residual strain/stress measurements on three-axis neutron diffractometer," *Adv Mat Res*, **1166**, pp. 33-40 (2021).

[6] Vrána, M., Lukáš, P., Mikula, P. and Kulda, J. "Bragg diffraction optics in high resolution strain measurements," *Nucl Instrum Meth A*, **338**, pp. 125-131 (1994).

[7] Delhez, R., De Keijser, T.H. and Mittemeijer, E.J. "Determination of crystallite size and lattice distortions through X-ray diffraction line profile analysis," *Fresenius' Zeitschrift für analytische Chemie,* **312**(1), pp. 1-16 (1982).

[8] Davydov, V., Lukáš, P., Strunz, P. and Kužel, R. "Single-line diffraction profile analysis method used for evaluation of microstructural parameters in the plain ferritic steel upon tensile straining," *Mater Sci Forum*, **571-572**, pp. 181-188 (2008).

[9] Macek, K., Lukáš, P., Janovec, J., Mikula, P. Strunz, P., Vrána, M. and Zaffagnini, M. "Austenite content and dislocation density in electron beam welds of a stainless maraging steel," *Materials Science and Engineering: A,* **208**, pp. 131-138 (1996).

[10] Hirschi, K., Ceretti, M., Lukáš, P., Ji, N., Braham, C. and Lodini, A. "Microstrain measurement in plastically deformed austenitic steel," *Texture Microstruct*, **33**, pp. 219-230 (1999).

[11] Mikula, P., Ryukhtin, V., Šaroun, J. and Strunz, P. "On possible high-resolution residual strain/stress measurements by three-axis neutron diffractometer," *Materials – MDPI*, **13**, pp. 5449-5455 (2020).

Residual Stress Measurement Measured by Different Type of Method on the 3D Print-thermal Spray

Radka Musilová[1,a], Marek Vostřák[1,b] and Karel Trojan[2,c]

[1]*Dynamic Testing Laboratory, Research and Testing Institute Pilsen, Tylova 1581/46, 301 00 Plzeň, Czech Republic,*
[2]*Department of Solid State Engineering, Faculty of Nuclear Sciences and Physical Engineering, Czech Technical University in Prague, Trojanova 13, 120 00 Prague, Czech Republic*

[a]*musilova@vzuplzen.cz*, [b]*vostrak@vzuplzen.cz*, [c]*karel.trojan@fjfi.cvut.cz*

Abstract: The development of 3D print-thermal spray systems, resistant against wear and corrosion is the main objective of an extensive research project handled in our institute. Additive technologies have potential to produce light and complex-shaped components. However, required service life and reliability under different load modes has to be ensured. Components surface needs to be protected against wear and corrosion. Such combination enables to exploit advantages of 3D printing with benefits provided by thermal spraying. One of the goals of the project is to determine the effect of residual stresses on the quality of the thermal coating and at the same time to optimize the application of the coating in terms of the level of residual stresses. First, the spray optimization was carried out on standard steel. The goal of presented research was to compare residual stresses in thermal coatings for different coating materials and process conditions. To be able to choose the optimum measurement procedure, several measurements method were compared. Both semi-destructive hole-drilling and groove grinding methods and non-destructive X-ray diffraction were used.

Keywords: Residual stresses; Hole-drilling method; X-ray diffraction; 3D print-thermal spray; Additive technologies; Thermal spraying technology

1 Introduction

3D-printing is often the technology of choice for manufacturing intricate-shaped parts with cooling channels and a light-weight structures. However, one of the disadvantages is lower strength when subjected to dynamic and cyclic loading, others are the presence of residual stresses, higher surface roughness and poor wear resistance.

Thermal spraying technology is a modern technology that offers the protection of the substrate surface from external influences, such as wear, corrosion, high temperature, oxidation or thermal shock. As well as 3D printing technology, the results are highly dependent on the process parameter settings. The microstructure of the coatings, the adhesion to the substrate and the development of the residual stresses are strongly influenced by the material used, the deposition technology and the process parameters.

The 3D printed-thermally spray systems will be analyzed to study their mutual influence on adhesion strength and residual stress development in a more advanced stage of the project. The coatings will be deposited from powders and the spray parameter optimization will be done regarding coatings microstructure, phase composition and residual stress development from commercially available powders, WC- and CrC-based hard metals with variable amount of composition of matrix and new powder composition. Optimization of the spray parameters for coatings microstructure, phase composition and residual stress development will also be performed. Heat treatment will be used for further residual stress elimination.

At the beginning of the project, MS1 tool steel and following types of thermal sprays Amperit 588, Amperit 543 (42WC 42"CrC"- 16Ni) and WOKA 7502 (37WC-43"CrC"-18NiCo) were used. Thickness of the thermal spray layer was 0.5 mm. Main goal was to introduce compressive stresses into the surface layer.

2 Measurement of residual stresses with semi-destructive methods

The combination of 3D printing and thermal spray technology has not been studied yet, neither the residual stress of the layer, 3D printed material, nor the system.

Therefore, theoretical calculation of residual stress in the layers were performed at the beginning. Determination of the residual stresses was measured using three methods, two of them were performed by Research and Testing Institute Pilsen, namely semi-destructive hole-drilling method [1] and groove-grinding method and the last one X-ray diffraction, which was performed by Department of Solid State Engineering, in Prague. Both devices for semi-destructive measurement of residual stresses are presented in Fig. 1.

(a) (b)

Fig. 1 Measuring devices which were used for measurement of residual stresses: (a) hole-drilling method, (b) groove-grinding method.

In a first step, the residual stresses were measured on the opposite side of a 5 mm thick thin plate. Here, two strain gauges LWK-06-W250D-350 in the longitudinal and transversal direction and Pt-thermal resistance sensor were fixed to the bottom side using spot welder, see Fig. 2. The layer was deposited on the front side. The goal was to measure the residual stresses through the change of the curvature. However, the sensitivity of such a measurement was very low, only approx. 12% of residual stresses of the layer were detected on the bottom. On the other hand, the strain gauges withstood a relatively high temperature without destruction.

Fig. 2 Plate with the strain gauges and Pt-thermal sensor.

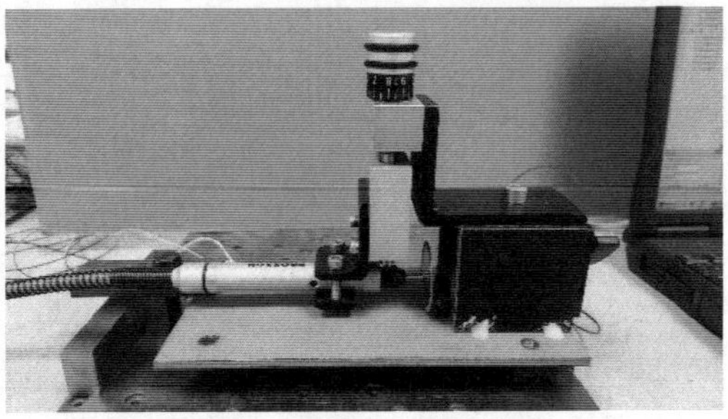

Fig. 3 Groove grinding method on the plate of 5mm thickness.

The groove-grinding method was used at the beginning of the investigations on a 5 mm thick plate, see Fig. 3 and the results in Fig. 4. The HBM 0,6/120LY11 strain gauge was used for the measurement and the groove was performed 2 mm from the center of the grid of strain gauge, a diamond wheel with a diameter of 38 mm was used as a tool. Depth of the groove was 2 mm and the step of grooving was 0.05 mm. Grinding was used to achieve the base material, where stress relaxation could be observed.

Based on FEM calculations, the size of the steel cube for applying the layer was optimized to minimize the bending stresses. The result was the cuboid with dimensions of 100×100×150 mm, with which further tests were performed.

In the first step, the residual stresses were measured in the CrC-NiCr layer by the groove-grinding method on the optimized cuboid, see Fig. 1b). The goal was to compare residual stresses in the layer of the 5 mm plate and the layer on the cuboid. For both material thicknesses the depth courses of residual stresses were comparable, see Fig. 4.

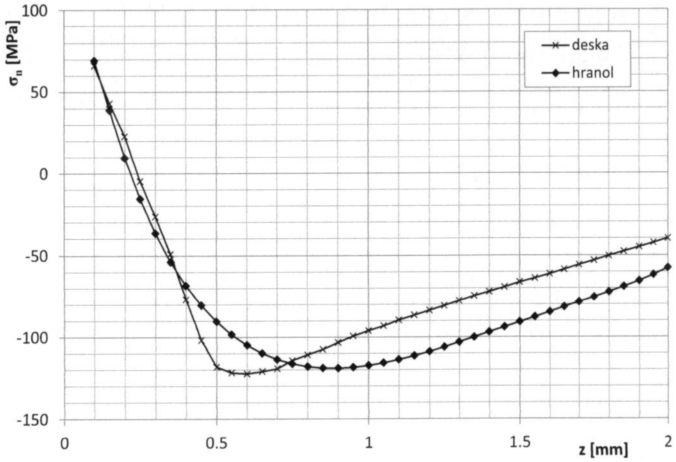

Fig. 4 Comparison of the depth courses of residual stresses measured on the plate – "deska" and on the cuboid – "hranol".

Next step was the comparison of different types of Amperit 543 (42WC42"CrC" 16Ni) and WOKA 7502 layers. Residual stress measurement was performed by hole-drilling method using Vishay RS-200 device with air turbine with the speed of over 200 000 rpm. A TML FRS-2-11 strain gauge rosette and a SINT D2/1 diamond cutter (ø 1.6 mm) were used. An 8-chanell HBM SPIDER8 amplifier and Catman Professional software were applied for the data acquisition. Evaluation of residual stresses was performed by the SINT EVAIL7.2 software from SINT and also using user developed software in spreadsheet.

Three specimens with different coating parameters were prepared using Amperit 543 and WOKA 7502, corresponding to the main optimalization, see Table 1.

Table 1. Deposition parameters.

Sample				**Layer param.**	
Ser. 1 #	**Material**	**Ser. 2 #**	**Material**	**Φ**	**Press. [Pa]**
1	Amperit 543	1	WOKA	0,66	6,8
2	Amperit 543	2	WOKA	0,75	7,8
3	Amperit 543	3	WOKA	0,85	8,5

Sample No. 3 was measured both with hole-drilling and X-ray diffraction.

3 Measurement of residual stresses using X-ray diffraction

The X-ray diffraction measurements were performed using the X´Pert Pro MPD diffractometer in Bragg-Brentano geometry. Evaluation was focused for two major phases in the layer, namely WC and Cr_3C_2, and for the base material. The {*111*} diffraction lines of WC phase, {*361*} of Cr_3C_2 phase and {*211*} of α-Fe phase were analyzed using chromium radiation. The residual stresses values were calculated from lattice deformations determined on the basis of experimental dependencies of $2\theta(\sin^2\psi)$ assuming a biaxial state of residual stress (θ is the diffraction angle, ψ the angle between the sample surface and the diffracting lattice planes) [2].

In order to determine the depth courses of macroscopic residual stresses, gradual electrolytic removal of the surface layers of the material was chosen using electrolytic polishing realized using the PROTO Electro polisher with A2 electrolyte. The thickness of the removed layer was measured using a digital micrometer.

Based on the results of strain analysis, it can be said that majority WC phase shows homogeneous tensile residual stresses (RS) at the surface, whereas RS values of minority Cr_3C_2 phase shows moderate compressive values. The depth courses of the RS of both analyzed phases have a decreasing trend. RS are tensile in the case of the WC phase and compressive in the case of Cr_3C_2. At depths greater than 0.5 mm, residual stresses were determined for MS1 tool steel as the major phase. Only compressive residual stresses of about 130 MPa were described (see Fig. 4).

4 Evaluated residual stresses

Evaluated results are presented in Fig. 5. Residual stresses in the Amperit 543 layer which have decreasing trend from tensile residual stresses to the compressive for each sample. A very high residual stress gradient was found at the interface between the layer and the base material, and rapid relaxation accompanied by an audible crack was observed when the hole was drilled. This behavior was able to observe only in the Amperit 543 layer.

The evaluated residual stresses in this paper are the mean values of the measured stresses in the principal directions (the both stresses in the principal directions are nearly the same). This is described as a hydrostatic stress state with zero shear stresses.

It is evident that samples using Amperit 543 have higher tensile residual stresses than the others. Samples with WOKA 7502 have compressive residual stresses.

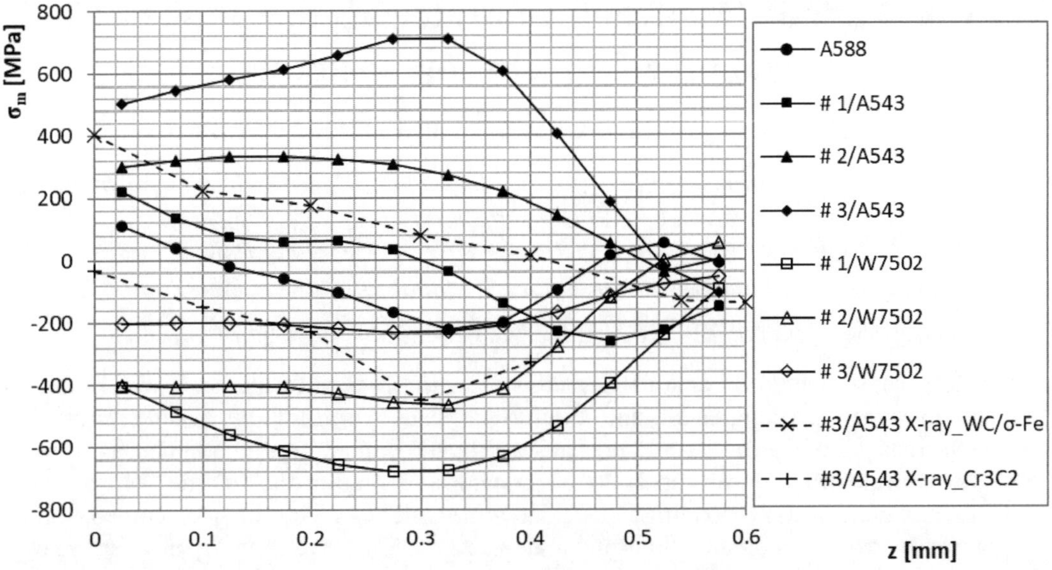

Fig. 5 Comparison of the mean component of depth courses of the residual stresses measurement including measurements by the X-ray diffraction.

Conclusion

The compressive residual stresses were proven after some optimization procedures in WOKA 7502 layer using hole-drilling method. The main goal was achieved, compressive residual stresses were obtained for all selected deposition parameters. Further, it was also found that X-ray diffraction and hole drilling method give approximately the same values of macroscopic residual stresses.

Acknowledgements

This work was supported by M-ERA.Net 2 Call 2020 with the name „Development of "3D print-thermal spray" systems for applications with dynamic and impact loading" Epsilon under the name TH75020003.

References

[1] ASTM Determining residual stresses by the hole-drilling strain-gage method. Standard test method E837-20. American Society for Testing and Materials, West Conshohocken, PA (2013).

[2] Kraus, I. and Ganev, N. *Technické aplikace difrakční analýzy* (ČVUT, Praha, 2004).

From Fatigue Behaviour of Materials to Fatigue Life of Machine Elements or Structures

Adam Niesłony[1,a], Karol Czekaj[2,b] and Bartosz Mazurek[3,c]

[1]*Opole University of Technology; Faculty of Mechanical Engineering, Department of Mechanics and Machine Design, Prószkowska, Opole, Poland;*
[2]*ZF CV Systems Poland; Ostrowskiego 34, Wrocław, Poland;*
[3]*Science and Technology Park in Opole; Technologiczna 2, Opole, Poland;*

[a]*a.nieslony@po.edu.pl*, [b]*karol.czekaj@zf.com*, [c]*b.mazurek@pnt.opole.pl*

Abstract: Fatigue failure remains a major concern in the design and performance evaluation of machine components and structures as it accounts for a significant proportion of mechanical failures. This article presents a fatigue evaluation methodology based on SN (stress-cycles to failure) curves to understand and predict the fatigue behavior of components under various loading conditions with widely varying device geometry and dynamics. In order to accurately interpret and utilize the SN curves, the paper outlines key factors influencing material fatigue, including stress amplitude, mean stress, stress concentration, environmental effects, and surface finish. The integration of these factors into the SN curve-based assessment is discussed to tailor fatigue evaluations to specific machine components and structures. To demonstrate the practical application of SN curves in fatigue assessment two case studies of machine components and structures are presented. The paper ends with a summary and conclusions, the most important of which is that the greatest impact on the design fatigue life has accurately estimated stresses resulting from the load conditions and the dynamics of the structure.

Keywords: Fatigue assessment; Machine components; Structural fatigue; SN curves

1 Introduction

Fatigue analysis is an integral part of the design of machine elements and structures subjected to variable load. It is carried out at the design stage and during the improvement of the structure with the use of basic fatigue information of materials. These are mainly fatigue stress or strain characteristics, so called fatigue curves, that are performed in the laboratory under simple loading conditions such as axial push-pull or reverse torsion [1–4]. In order to use information about materials to predict fatigue life, an algorithm should be developed that takes into account all aspects affecting the fatigue life of machine and structure components. Typically, the following elements are considered [5–8]:

<u>Load Analysis</u>: The algorithm should incorporate an analysis of the applied loads, including both static and dynamic loads. This involves understanding the magnitude, frequency, and distribution of the loads that the machine elements or structures will experience during their service life.

<u>Stress and Strain Analysis</u>: The algorithm should account for the stresses and strains generated in the machine elements or structures due to the applied loads. This involves performing stress analysis and calculating the stress and strain amplitudes at critical locations. It is crucial to account for the response of the material to variable loads by employing suitable hypotheses for the accumulation of fatigue damage such as Miner's rule or other more advanced models based on the specific application. Additionally, the algorithm should address the multiaxial nature of the stress state.

<u>Material Properties</u>: It is crucial to consider the material properties, such as yield strength, ultimate tensile strength, fatigue strength limit and finite fatigue strength. These properties

determine the response of the material to cyclic loading and are essential for accurate fatigue life prediction. This data is typically represented by an S-N curve, which relates stress amplitudes to the number of cycles to failure.

Environmental Factors: The algorithm should account for environmental factors that can affect fatigue life, such as temperature, corrosive conditions and exposure to other potentially damaging agents.

Scale and Stress Concentration Effects: If the machine elements or structures contain notches, fillets, or other stress concentration features, the algorithm should incorporate methods to assess their impact on fatigue life. This may involve applying appropriate stress concentration factors and combine them with finite element analysis (FEA) to model local stress concentrations. The scale effect should be considered if the volume of material subjected to fatigue stresses is not comparable with laboratory samples.

Validation and Calibration: It is important to validate and calibrate the algorithm against experimental data or real-life case studies to ensure its accuracy and reliability. This may involve performing fatigue tests on representative samples or comparing the algorithm's predictions with known fatigue failures.

Bearing in mind the above factors affecting fatigue life, new and more accurate algorithms for determining fatigue life are being developed. These algorithms are built to take into account all factors, creating a universal way of solving the problem. Nevertheless, each case is different and even the most universal algorithm needs to be adapted to the current case. The reason for using the algorithm for current needs is primarily the speed of calculations and their accuracy. A decision should be made regarding:

- durability estimation methods i.e. the stress or strain approach,
- domain in which the calculations will be performed, i.e. in time or frequency,
- whether adjustments to the impact of notches on fatigue life will be necessary,
- whether "pure material" fatigue or fatigue of joints, such as welds or bolted components, determine durability,
- whether the influence of external factors such as temperature change or corrosive environment should be taken into account,
- from where they will be taken or by what method stresses or strains will be determined.

In order to present the complexity of this issue, two cases of determining the fatigue life of structures are presented below, which differ significantly from each other and the basis for their determination are classic stress fatigue characteristics, i.e. SN curves for basic material.

2 Case studies

Two case studies are presented. The first concerns the compressor mounting bracket in passenger cars and the second concerns the device for transporting motorcycles. In both cases, the SN curve is known, however, the degree of complexity of the shape of the element and load conditions complicate the process of determining the fatigue life and requires a different computational path. Therefore, appropriate algorithms for fatigue assessment were constructed for both cases and presented in subsections 2.1 and 2.2.

2.1 *Fatigue assessment of the compressor mounting bracket*

2.1.1 Background

This case involves suspension system with nonlinear dynamic behaviour caused by collisions among suspension subcomponents during operational loading and vibration tests. The structure consists of a vibrating mass, functioning as a compressor for car air suspension system, which is securely attached to compressor brackets. To isolate vibrations, the structure is suspended on vibration isolators that include springs, rubber, and a lower bumper. Additionally, the entire assembly is fastened to a rigid frame that serves as the upper bumper for the rubber (Fig. 1).

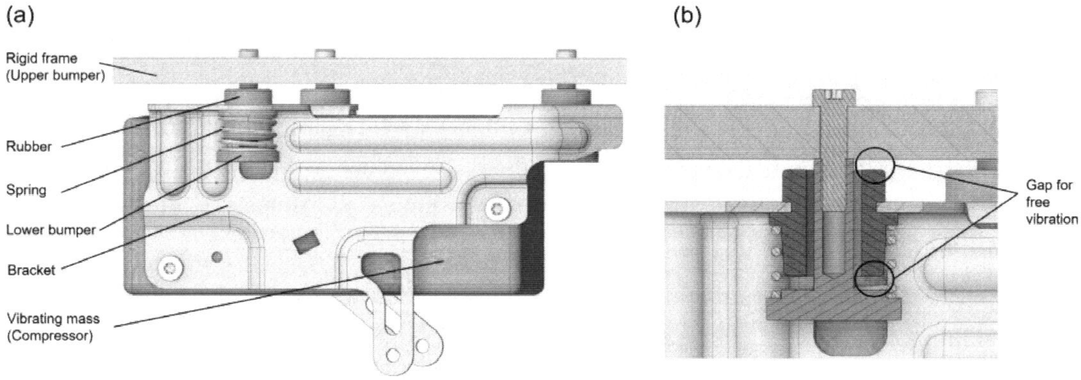

Fig. 1 (a) Simplified scheme of suspension system of compressor. (b) Vibration isolator cross section.

During working conditions, the structure operates close or at its natural frequency which leads to the rubber-bumpers collisions even under low level excitation. Collisions are realized double-sided, occurring both between the rubber and the upper bumper, and between the rubber and the lower bumper. As a consequence, the fatigue life of the compressor brackets is significantly affected.

Effective methods for vibration fatigue calculations called spectral methods are performed in frequency domain [9,10]. Frequency response analysis in Finite Element Method (e.g. Nastran SOL 111 or ANSYS Harmonic Response) are used as an input for the spectral methods. Obviously, this are linear solvers which do not consider structure nonlinearities like collisions in the described system and cannot be used directly for stress history prediction. To overcome this limitation, an adaptation was needed to make the linear response of the structure compatible with the real nonlinear dynamic caused by collisions.

2.1.2 Proposed solution

Due to the occurrence of collisions the main problem in the algorithm for determining the fatigue life is the correct determination of stresses in the points of interest. In order to use efficient linear FEM solvers for stress calculations the method is proposed where, vibration test was conducted to obtain dynamic response of the structure under collisions conditions. The idea is to modify in FEM simulation the excitation input to obtain response of vibrating mass which will be in line with registered in the test. The modified PSD input for FEM simulation can be obtained from Eq. (1), where PSD_{otput_test} is power spectral density based on accelerations recorded in the test, and $S(f)$ is the transfer function of the system.

$$PSD_{input\_modified} = \frac{PSD_{output\_test}}{S(f)^2} \tag{1}$$

The system was excited on electrodynamic shaker in vertical direction using PSD profile with RMS acceleration of 1g (see Table 1). Fig. 2 displays the graphical representation of the PSD input. The shape and test frequency range of the defined PSD input is based on automotive standards like DIN EN 60068-2-64 "Environmental testing; Part 2-64: Tests; Test Fh: Vibration, broadband random and guidance" or DIN EN 60068-2-47 "Environmental testing; Part 2-47:Tests; Mounting of specimens for vibration, impact and similar dynamic tests".

Table 1. Vibration test profile used in the test.

Point	Frequency [Hz]	PSD [g^2/Hz]
1	5	0.0250
2	10	0.0300
3	20	0.0300
4	200	0.0004
RMS acceleration	1.02g	

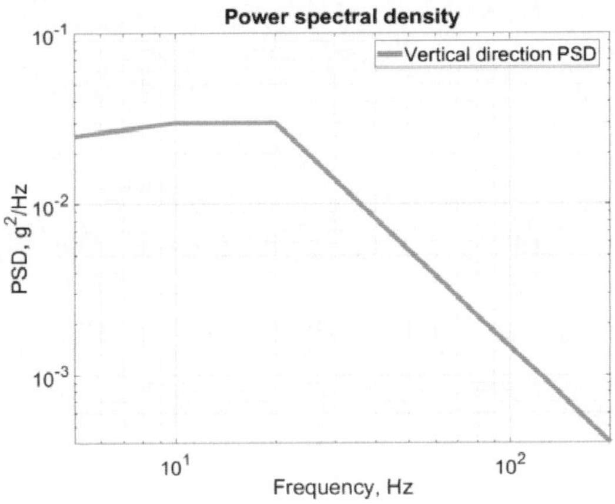

Fig. 2 Vibration test profile according to Table 1.

Two acceleration sensors were installed on the test rig. The 1st one close to shaker table to control the input PSD signal, and the 2nd sensor is mounted on the vibrating mass (compressor) to control the system PSD response required for further FEM simulation (Fig. 3). The structure was excited for 100 s using sampling rate 1 600 Hz.

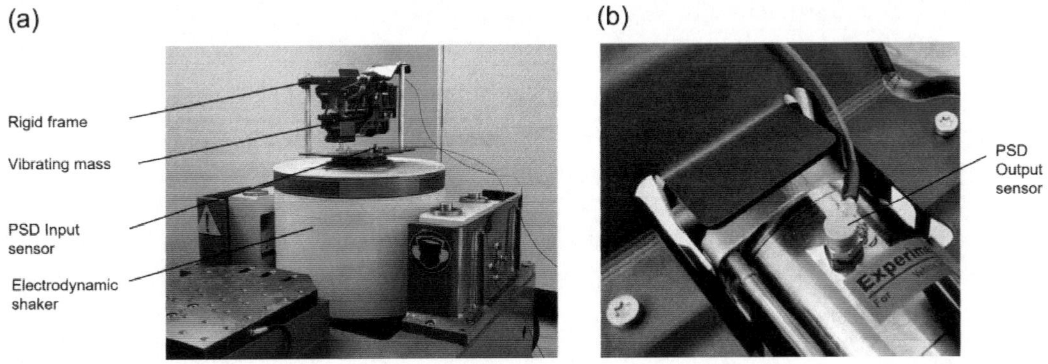

Fig. 3 (a) Test rig for compressor dynamic response measurement. (b) Response measurement point (compressor).

2.1.3 Results for loading PSD

The test aimed to measure the dynamic response of the vibrating mass (compressor) in the suspension system, following vibration test profile presented in Table 1. Registered results are presented in Fig. 4. To obtain the PSD function, the acceleration time history data on the compressor (Fig. 4a) was subjected to analysis through the Welch method [11], resulting in the presentation of the PSD function in Fig. 4b.

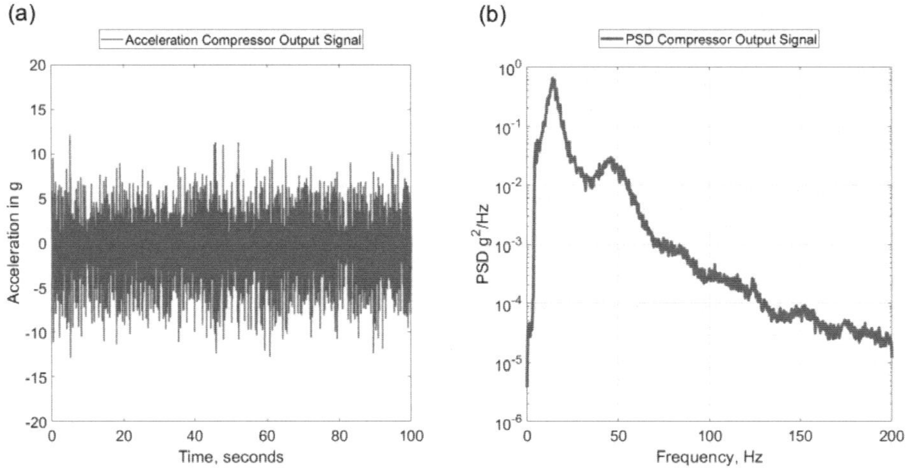

Fig. 4 Response of the compressor in accordance to the input shown in Table 1. (a) Response acceleration in time domain, (b) response in frequency domain.

The acquired results were utilized to adjust the fundamental PSD input for finite element method simulation. In place of the vibration profile detailed in Table 1 and Fig. 2, a modified PSD input was calculated using Eq. (1) and visualized in Fig. 5.

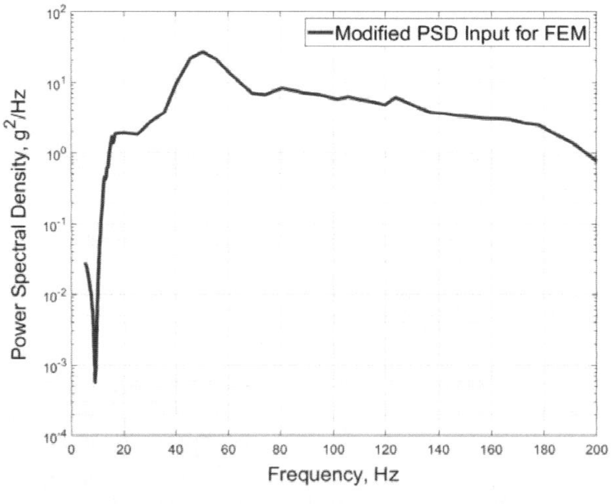

Fig. 5 Modified PSD input signal for FEM simulation.

Fig. 6 shows the difference between RMS stress results on the compressor bracket when collisions are disregarded (Fig. 6(a)) and when they are taken into consideration (Fig. 6(b)).

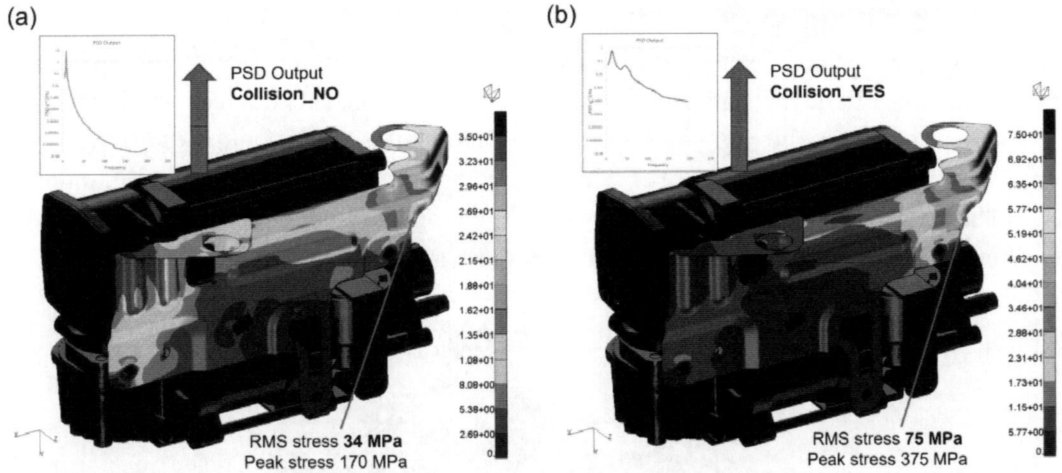

Fig. 6 Von Mises stress results during random vibration test: (a) results when collisions are not included, (b) Results with modified PSD Input – collisions included.

The PSD functions for both the model with collision consideration and the model without collision consideration were extracted from the bracket notch area, where fatigue cracking is anticipated. These functions are presented in Fig. 7.

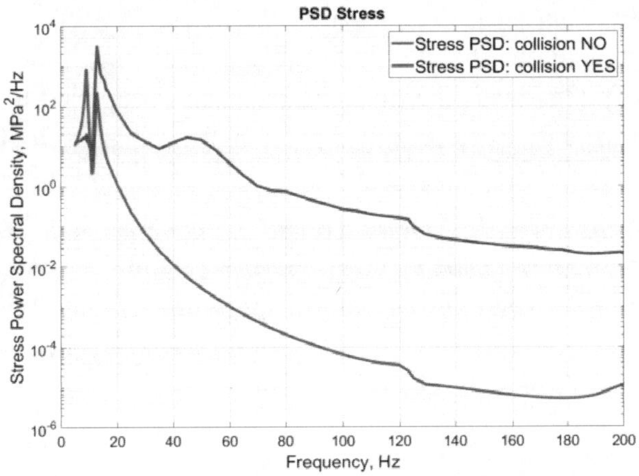

Fig. 7 Von Mises stress results during vibration test: (a) results when collisions are not included, (b) results with modified PSD Input – collisions included.

The bracket material is specified as fine-grained steel suitable for cold-working. S-N curve was derived based on the fatigue limit σ_{af} = 185 MPa and the corresponding number of cycles N_0 = 10^6 at which this limit was initially observed. The curve slope, represented by the parameter k, is determined to be 5. Following that, fatigue calculations were conducted employing the methodologies proposed by Dirlik and Tovo-Benasciutti [12–14]. The outcomes of these fatigue calculations are presented in Table 2.

Table 2. Bracket vibration fatigue calculation results.

Stress PSD	RMS, MPa	Peak stress, MPa	E[P], 1/s	According Dirlik, days	According Tovo-Benasciutti, days
Without collision	34	170	12.34	467.6	495.4
With collision	75	375	50.38	3.22	3.33

2.2 Fatigue assessment of the motorcycle transport device

The second case study focuses on analysing a transportation system consisting of a motorcycle transport device, an automobile, and a motorcycle. This device is mounted on the car's hitch, as shown in the Fig. 8. During transportation, it is subjected to random vibrations originating from the road surface over which the movement takes place [15]. While the S-N curves (fatigue strength) of the materials used in the device's construction are well-known, the exact loads and excitations acting on this device remain unknown.

Fig. 8 Transport device according to patent PL 224395 B1; 1. Mounting element (for tow hitch); 2. Wheel base; 3. Stabilizing strap

In the context of the discussed system, which is related to the field of automotive engineering, there is an opportunity to reference the characteristics of excitations – standardized Power Spectral Density (PSD) functions that are commonly used in fatigue calculations within the automotive industry [16]. However, the presented system is exposed to specific excitations resulting from interactions with the suspended mass on the car's hitch. To precisely analyse these excitations, real acceleration measurements were conducted directly on the device at the hitch mounting location (Fig. 9). These measurements were conducted on two distinct road surfaces: asphalt roads and unpaved roads. The device underwent these measurements while being subjected to a full load during the transportation of a motorcycle (Fig. 10)

Fig. 9 Measurement point for acceleration in three axes.

Fig. 10 The device with a motorcycle attached.

Based on the measured accelerations, utilizing MATLAB tools, particularly the PWELCH function, the power spectral density function was determined for both the asphalt (Fig. 11) and non-paved road (Fig. 12) routes. The obtained PSD characteristics constitute significant input data for further analysis. To comprehensively assess the impact of these excitations on the device's behaviour, a numerical simulation was employed using the FEM within ANSYS software.

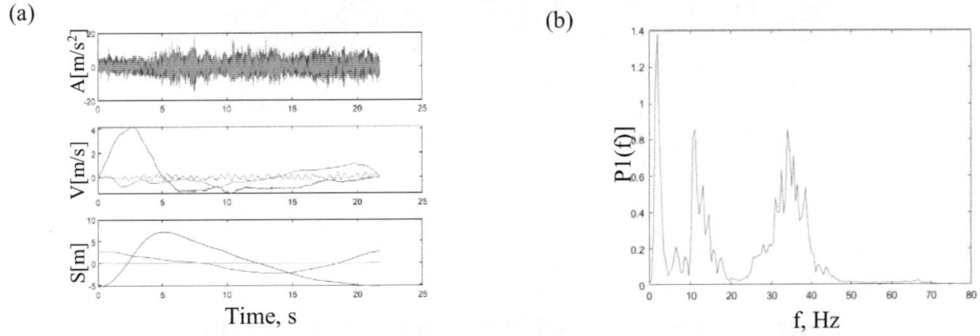

Fig. 11 Asphalt road: (a) raw recorded signal, (b) determined PSD function.

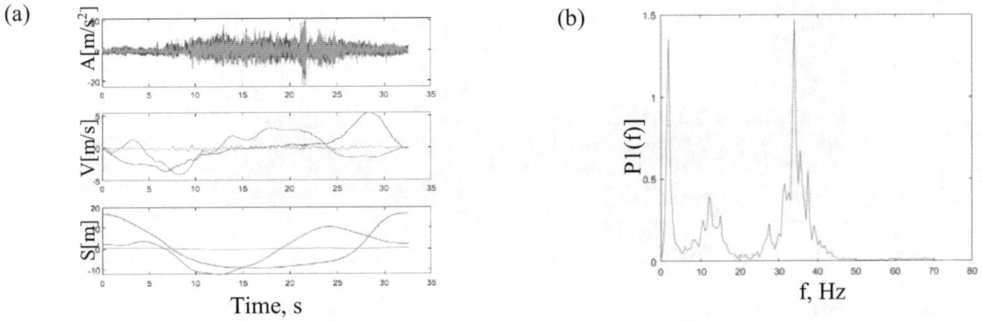

Fig. 12 Unpaved road: (a) raw recorded signal, (b) determined PSD function

The incorporation of PSD excitation characteristics in the computer simulation allowed for the replication of a realistic dynamic environment in which the analysed device operates. Such an approach facilitated a detailed assessment of stresses, deformations, and potential critical areas with regard to material fatigue. Comparing simulation results with previously known S-N curves of materials will enable an evaluation of safety and durability for the designed structure under real operating conditions.

After constructing a simplified computer model of the analysed system, a detailed simulation was initiated using the Ansys software (Fig. 13). As part of the basic strength analysis, a node most susceptible to potential damage was identified. This pivotal node is located within a triangular support where significant loads and stresses are present.

To conduct a more advanced analysis, Ansys Random Vibration tools were employed, enabling the modelling of interactions caused by random vibrations. These simulations were conducted while considering boundary conditions, which were selected in accordance with the research carried out in the publication by Mazurek and Mamala [17]. The device's mounting location was modelled as a fixed support, the rear car wheel as a cylindrical support, and the entire motorcycle system with a mass of 200 kg was appropriately considered. For the purpose of representing real-world conditions, an excitation in the form of PSD was applied to the fixed support point.

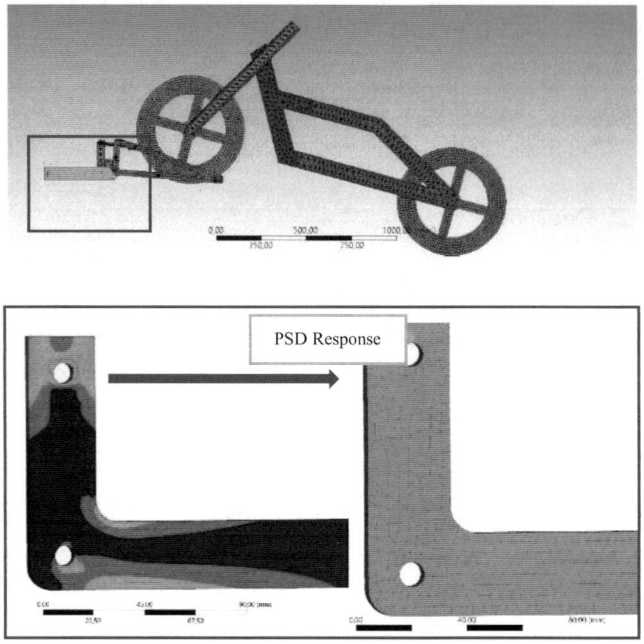

Fig. 13 Determination of the PSD response reading location.

Conclusion and Summary

Fatigue characteristics of materials in the form of SN curves are only the basis for fatigue calculations. Equally important here are all elements of the fatigue life determination algorithm, in particular stress determination and consideration of factors influencing fatigue, such as scale factor, stress concentration effect, average stress, amplitude distribution and sequence, and environmental factors. At the forefront of factors influencing the fatigue life is the correct determination of stresses, on the basis of which the analysis is carried out. In real elements subjected to operating loads, it is difficult to assess the actual state of the load. Therefore, this element is burdened with the greatest uncertainty.

For fatigue calculations in real conditions, the most suitable is the spectral method, which statistically reflects the load. This is a well-known method and is often referred to as vibration fatigue method.

The influence of the notch on the fatigue life is usually expressed by an increase in the stress value. In the spectral method, if the notches are not sharp, their influence is taken into account in the FEM calculation, also by increasing the stress near the notch. Of course, the scale factor for the volume of loaded material should not be neglected.

References

[1] Niesłony, A., el Dsoki, C., Kaufmann, H. and Krug P., "New method for evaluation of the Manson-Coffin-Basquin and Ramberg-Osgood equations with respect to compatibility," *Int J Fatigue*, **30**(10–11), pp. 1967-1977 (2008).

[2] Fouchereau, R., Celeux, G. and Pamphile, P. "Probabilistic modeling of S–N curves," *Int J Fatigue*, **68**, pp. 217-223 (2014).

[3] Ramberg, W. and Osgood, W.R. "Description of stress-strain curves by three parameters", Technical Notes, National Advisory Committee for Aeronautics No. 902 (1943).

[4] Bandara, C.S., Siriwardane, S.C., Dissanayake, U. I. and Dissanayake, R. "Developing a full range S–N curve and estimating cumulative fatigue damage of steel elements," *Comp Mater Sci*, **96**(A), pp. 96-101 (2015).

[5] Lagerblad, U., Wentzel, H. and Kulachenko, A. "A methodology for strain-based fatigue damage prediction by combining finite element modelling with vibration measurements," *Eng Fail Anal*, **121**, 105130 (2021).

[6] Niesłony, A. and Böhm, M. "Frequency-domain fatigue life estimation with mean stress correction," *Int J Fatigue*, **91**(2), pp. 373-381 (2016).

[7] Augustyniak, M., Gajewski, P. and Świątek, K. "Adaptation of engineering FEA-based algorithms to LCF failure and material data prediction in offshore design," *J Theor App Mech-Pol*, **54**(4), pp. 1345-1356 (2016).

[8] Fricke, W., Cui, W., Kierkegaard, H., Kihl, D., Koval, M., Mikkola, T., Parmentier, G., Toyosada, M. and Yoon, J.H. "Comparative fatigue strength assessment of a structural detail in a containership using various approaches of classification societies," *Mar Struct*, **15**(1), pp. 1-13 (2002).

[9] Zorman, A., Slavič, J. and Boltežar, M. "Vibration fatigue by spectral methods–A review with open-source support," *Mech Syst Signal Pr*, **190**, 110149 (2023).

[10] Benasciutti, D., Sherratt, F. and Cristofori, A. "Recent developments in frequency domain multi-axial fatigue analysis," *Int J Fatigue*, **91**, pp. 397-413 (2016).

[11] Same, M.H., Gandubert, G., Gleeton, G., Ivanov, P. and Landry, R. "Simplified Welch Algorithm for Spectrum Monitoring," *Applied Sciences*, **11**(1), 86 (2021).

[12] Dirlik, T. "Application of computers in fatigue analysis," PhD Thesis, University of Warwick, 1985. *Available from* http://webcat.warwick.ac.uk/record=b1445503~S9 *Accessed:* 2015-02-26.

[13] Benasciutti, D. and Tovo, R. "Spectral methods for lifetime prediction under wide-band stationary random processes," *Int J Fatigue*, **27**(8), pp. 867-877 (2005).

[14] Dirlik, T. and Benasciutti, D. "Dirlik and Tovo-Benasciutti Spectral Methods in Vibration Fatigue: A Review with a Historical Perspective," *Metals*, 11(9), 1333 (2021).

[15] Cao, D., Xu, X., Li, W., Hu, J. and Kong, Z. "Research on random vibration condition of driving motor system for new energy passenger vehicle," *Energy Reports*, **8**, pp. 988-996 (2022).

[16] Su, H. "Vibration Test Specification for Automotive Products Based on Measured Vehicle Load Data," SAE Technical Paper 2006-01–0729, 2006.

[17] Mazurek B. and Mamala, J. "Development of a design concept for a prototype transport device using the Design Thinking method," *MATEC Web Conf.*, **338**, 01016 (2021).

Podometric Assessment of Gait Parameters

Bibiána Ondrejová[1,a], Lucia Bednarčíková[1,b], Monika Michalíková[1,c] and Jozef Živčák[1,d]

[1]*Technical University of Košice; Letná 1/9, 042 00 Košice-Sever, Slovakia;*

[a]*bibiana.ondrejova@tuke.sk;* [b]*lucia.bednarcikova@tuke.sk;* [c]*monika.michalikova@tuke.sk;* [d]*jozef.zivcak@tuke.sk*

Abstract: This pilot study is aimed to compare the gait parameters of two healthy subjects when wearing barefoot shoes and classic shoes, using a pedogait device. The study was conducted in a laboratory setting, and each participant performed three walking trials (barefoot, barefoot shoes, and classic shoes). The pedogait device recorded various gait parameters, including gait line, foot pressure, step length, and gait speed. Results showed that when wearing classic shoes, participant that is not barefoot user had a significantly shortened support phase of the step, and the maximum forces at the initial contact of the heel and the rebound from the toe increased compared to barefoot. These preliminary findings suggest that barefoot shoes may alter walking parameters compared to normal shoes and it is necessary to emphasize the correct walking technique since there is no compensatory cushioning in the shoe itself. However, the small sample size of this pilot study limits the generalizability of the results, and further research with larger sample sizes and more diverse populations is needed to confirm these findings.

Keywords: Barefoot shoes; Gait analysis; Gate line

1 Introduction

The first interface between the body and the ground during the human gait are shoes. Their primary purpose is mainly ensuring to protect the feet against thermal, chemical, or mechanical influences and possible injuries [1]. During one step, the human foot is transformed from a flexible shock absorber, which occurs when the heel strikes, to a rigid lever, which occurs when stepping forward. The foot is defined as a structure formed by transverse and longitudinal arches. During the stance phase of the gait cycle, the human foot is an elastic and arched structure that actively works to maintain body stability through small mediolateral and anteroposterior movements [2]. Footwear significantly affects the distribution of pressure on the foot, which is often associated with changes in upright posture. Shoes also significantly affect postural and dynamic stability, it degrades somatosensory feedback to the foot through the tactile proprioceptive system, because of the design elements of the footwear itself [3]. The aim of the presented work is to compare and evaluate the gait lines recorded during the measurements on the sensory treadmill. The DIERS 4D Lab Motion - Pedogait device was used for the individual measurements. Two subjects took part in the podometric assessment, one of them is a user of barefoot shoes and the other wears classic shoes. The parameters such as: gait line, average pressure, and basic walking parameters such as: step length, step width, speed, cadence, and foot rotation were evaluated.

2 Materials and methods

During the pilot study aimed at assessing the biomechanical parameters of gait, the following methodology was determined, and the described equipment was used. Two female participants were measured. The subjects were familiarized with the entire measurement procedure, they filled in and confirmed the informed consent and a brief questionnaire before the measurement. The measurement was non-contact, painless and safe and took a few minutes. The participants were measured barefoot, in classic and barefoot shoes, between individual measurements there were breaks to avoid the fatigue effect. During the adaptation to walking on the surface of the sensory treadmill, the gait velocity gradually increased to the standard 4.5 km/h. Each measurement lasted 5 minutes, while a time of 20 seconds was recorded, representing a path of 25 meters.

The first measured participant was a 22-year-old woman with a weight of 59 kg and a height of 162 cm. The number of shoes used during the measurement was 36. Classic footwear was sport shoes with a sole thickness of 3 cm (Fig. 1a), and barefoot footwear was characterized by a wider and rounded shape of the front part of the shoe with a sole thickness of 0.6 cm (Fig. 1b). Subject 1 is not a user of barefoot shoes; in this study, it is her first contact with this type of shoes.

The second measured participant was a 42-year-old woman with a height of 169 cm and a weight of 61 kg. The number of shoes used when measuring the subject was 40. Classic footwear was sport shoes with a sole thickness of 3 cm, and a higher slope of the toe (Fig. 1c), and barefoot footwear was characterized by an elongated shape of the front part of the shoe, which provides enough space for the toes, but is not of a particularly wide type with a sole thickness of 0.65 cm (Fig. 1d). Subject 2 common uses barefoot shoes during her daily life.

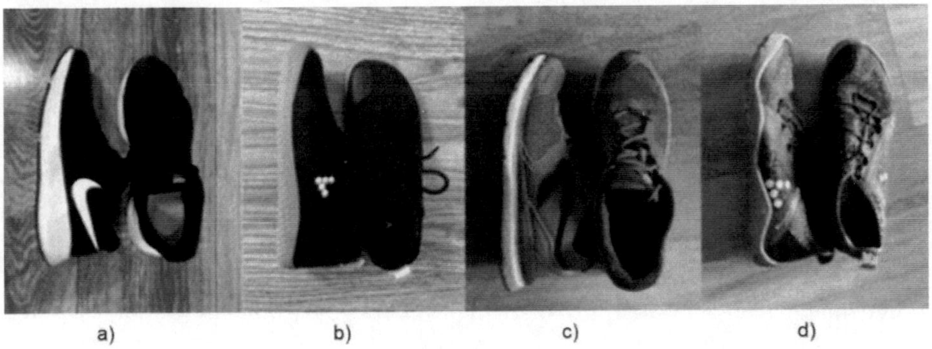

Fig. 1 Classic (a) and barefoot (b) shoes of the first subject. Classic (c) and barefoot (d) shoes of the second subject.

The measurement was carried out in cooperation with MUDr. Ján Sekáč in the Orthopedic Clinic and Movement Laboratory with the DIERS pedogait system (DIERS International GmbH, Germany) allows the functional representation of the foot pressure reaction forces while walking. The integrated measuring platform is 1.0 m long with 5.376 sensors for an exact capture of the pressure values. The admission frequency amounts to 100 Hz The standing treadmill can be used for static measurements of the foot pressure measurement as well as for the stabilometry.

2.1 Measured and analyzed biomechanical parameters of the gait

One of the main parameters evaluated during the dynamic test is the level and distribution of plantar pressure. Average and maximum pressure are mostly evaluated. In the graphical display, the pressure values are displayed as a line of points of the maximum load that was recorded during the dynamic analysis. The curve of this graph should rise for normal values in the middle part and slightly decrease at the end of the graph. The curve of the graph of the average value of the pressure shows the line of the average pressure recorded.

Another important parameter is impulse force over time. It can be expressed as the area under the pressure versus time curve. The greater the pressure and the longer it acts, the greater the impulse, so it is the total load.

An important aspect of plantar pressure measurement is the centre of pressure line. The physiological gait line should start in the rear third of the foot, continue towards the 5th metatarsal, then proceed centrally to the 4th metatarsal, then the 3rd to 2nd metatarsal, and end at the big toe, as it is the last in contact with a ground. This movement represents a shift in about 83% of the length and 18% of the width of the contact surface of the foot in middle-aged adults. Pressure line velocity is also an important indicator, which should be approximately 22-27 cm/s and 38 cm/s for middle-aged adults and young adults [4].

3 Results

The evaluated parameters are the distribution of plantar pressures, the course and length of the gait lines, the maximum pressures during contact of the heel with the mat and the toe off, mediolateral deviations of the centre of gravity, the percentage of the individual phases of the gait cycle, the step parameter (length, width, foot rotation, cadence) and also the total force of segment [N] in time on individual foot areas. Tables 1 and 2 represent evaluated biomechanical parameters of gait for different types of shoes and bare feet of subject 1 (Table 1) and subject 2 (Table 2).

Table 1. Biomechanical parameters of gait for subject 1.

Step parameters	Classic shoes		Barefoot shoes		Bare feet		Standard
	R	L	R	L	R	L	
Step length [cm]	64+/-1	63+/-2	64+/-1	63+/-2	63+/-21	62+/-2	
Step time [ms]	514+/-2	499+/-2	511+/-2	495+/-2	503+/-24	486+/-2	
Stride length [cm]	127+/-6		126+/-6		124+/-6		141
Stride time [ms]	1013		1005		990		1200
Track width [cm]	12+/-1		12+/-1		9+/-2		10
Foot rotation [°]	7	2	4	2	4	10	10-15
Cadence [step/min]	114		114		117		113

Table 2. Biomechanical parameters of gait for subject 2.

Step parameters	Classic shoes		Barefoot shoes		Bare feet		Standard
	R	L	R	L	R	L	
Step length [cm]	66+/-2	64+/-2	64+/-1	63+/-2	62+/-2	62+/-1	
Step time [ms]	512+/-2	518+/-2	504+/-1	512+/-2	498+/-3	496+/-2	
Stride length [cm]	129+/-6		127+/-6		124+/-6		141
Stride time [ms]	1029		1016		994		1200
Track width [cm]	8+/-1		8+/-1		9+/-1		10
Foot rotation [°]	3	5	3	7	4	8	10-15
Cadence [step/min]	114		114		117		113

The pressure distribution of subject 1 (Fig. 2) on individual feet is uneven in classic and barefoot shoes, in the case of barefoot, the pressure is adequately distributed, and subject 1 loads the foot equally on the heel and in the forefoot. In none of the measurements was pressure exerted over the 5th metatarsal, which also captures the shape of the rolling curve, which is flat (i.e. straight). In classic shoes, can be observed that the gait line did not go through 5., 4., 3. nor the 2nd metatarsal. It's beginning was not in a third of the heel, and the line of pressure went directly to the toe, which was the most loaded. The pressure was exerted through the centre of the foot and not through the individual metatarsals.

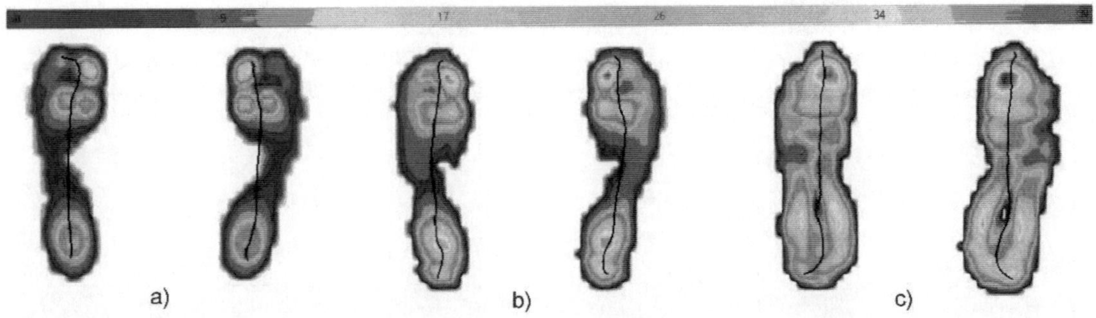

Fig. 2 Graphical representation of the average pressure distribution and gait line in subject 1: a) bare foot vs. mat, b) barefoot shoes vs mat, c) classic shoes vs mat.

On all prints of subject 2 (Fig. 3), the pressure lines are also unevenly distributed. In barefoot shoes, the medial part of the forefoot was the most loaded. The gait line was flat, i.e. straight throughout its entire course in barefoot shoes on the left foot. Barefoot was also the most loaded forefoot in the medial part. The most deformed gait line can be seen in classic footwear. Its beginning is not in a third of the heel and does not gradually pass through the metatarsals. In this case, the forefoot and the area of the toes, i.e. the thumb in particular, are more stressed. The pressure is not exerted through the metatarsals but through the middle of the foot and the roll curve ends at the 2nd metatarsal.

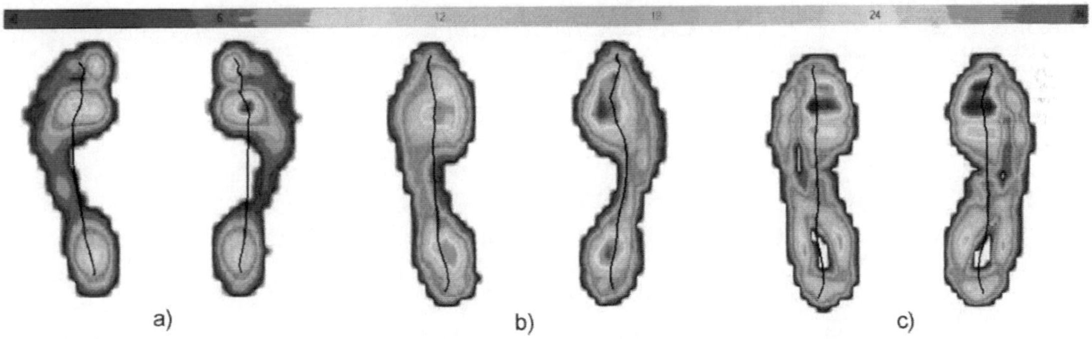

Fig. 3 Graphical representation of the average pressure distribution and gait line in subject 2: a) bare foot vs. mat, b) barefoot shoes vs mat, c) classic shoes vs mat.

The maximum forces at the heel strike of subject 1 were lower in each of the measurements than at the toe-off phase and their values ranged from 437-528 N (Fig. 4). The maximum forces at the toe-off phase were higher than at the heel strike in each of the measurements. Their values ranged from 530-592 N (Fig. 5). The biggest difference between impact and reflection was captured barefoot on the right foot. The smallest values were measured barefoot shoes and barefoot. The measured values of the maximum forces of subject 1 are shown in the Table 3.

Table 3. Maximum forces during heel strike and toe-off gait phase for subject 1.

Maximum forces	Classic shoes		Barefoot shoes		Bare feet	
	L	R	L	R	L	R
Heel strike [N]	528	470	486	485	480	437
Toe off [N]	530	531	570	530	592	540

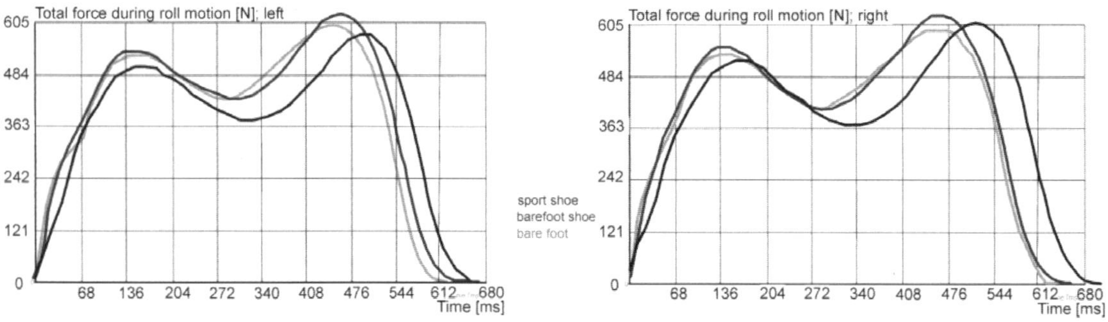

Fig. 4 Force-time diagram subject 1: a) left foot, b) right foot; the blue curve represents classic shoes, the red curve represents barefoot shoes and the green curve represents bare feet.

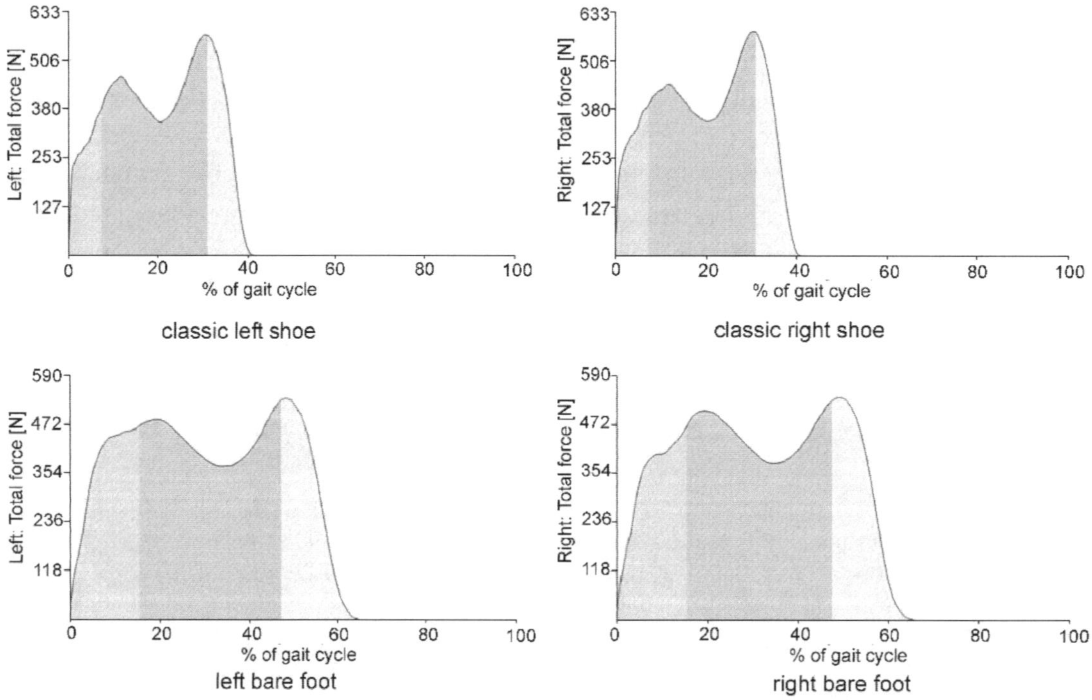

Fig. 5 Total force diagram - subject 1.

The maximum forces of subject 2 at heel strike were lower in each of the measurements and their values ranged from 506-565 N (Fig. 6). The maximum forces at the toe-off phase were higher than at impact in each of the measurements. Their values ranged from 588-630 N (Fig.7). The biggest difference between impact and reflection was in barefoot on the left foot. The smallest values were measured barefoot. We did not notice significant differences between the shoes. The measured values of the maximum forces of subject 2 are shown in the Table 4.

Table 4. Maximum forces during heel strike and toe-off gait phase for subject 2.

Maximum forces	Classic shoes		Barefoot shoes		Bare feet	
	L	R	L	R	L	R
Heel strike [N]	530	513	565	515	510	506
Toe off [N]	620	612	615	630	600	588

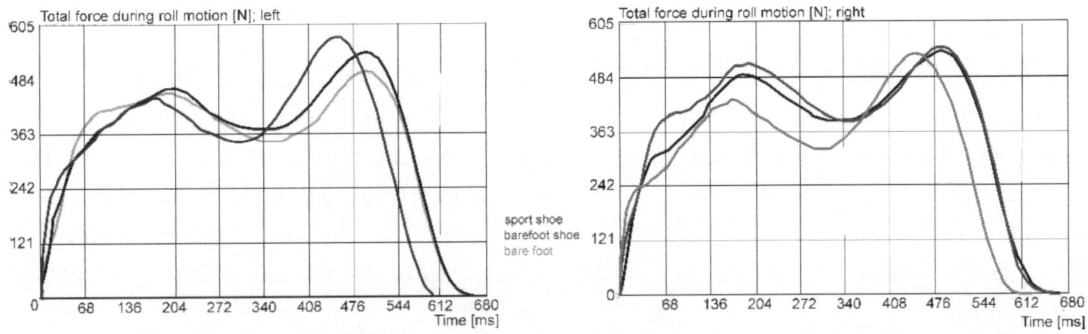

Fig. 6 Force-time diagram subject 2: a) left foot, b) right foot; the blue curve represents classic shoes, the red curve represents barefoot shoes and the green curve represents bare feet.

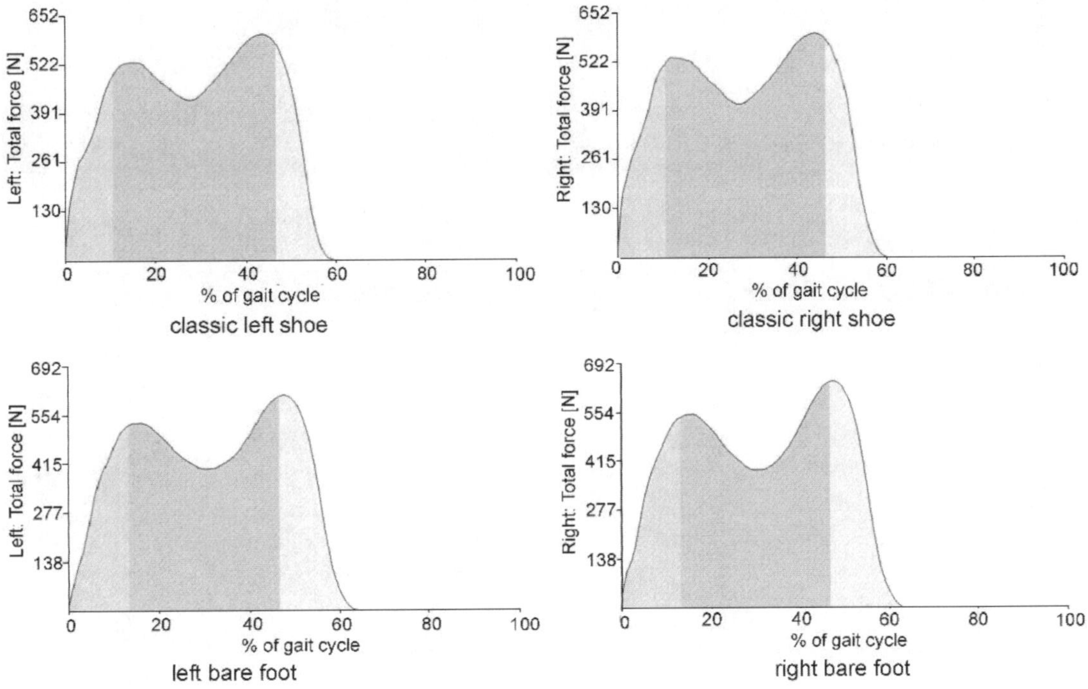

Fig. 7 Total force diagram - subject 2.

The length of the gait line of subject 1 (Fig. 8a) varied between 14.8-17.3 cm during the measurements. Significant differences were not captured between individual measurements. In the total length of the gait line, we noticed a significant difference between classic shoes and bare feet, and that's on the left foot.

Table 5. Gait line length and mediolateral deviations of COG for subject 1.

Shoe type	Classic shoes		Barefoot shoes		Bare feet	
	L	R	L	R	L	R
Length of gait line in midstance phase	16+/-4,2	15.6+/-1,2	17.3+/-1,1	14.8+/-4	16.3+/-1	15.8+/-3.9
Total length of gait line	26.4+/-6.6	27.5+/-0.2	25.4+/-0.2	23.5+/-5.9	21.6+/-0.3	20.6+/-5
Mediolateral deviations of COG	6.9+/-1.3	7.3+/-0.9	7.3+/-1.2	6.0+/-1.2	7.1+/-1.7	6.4+/-1.3

The length of the gait line of subject 2 (Fig. 8b) varied between 17.3-19.3 cm during the measurements. Significant differences between individual measurements were not captured. In the total length of the gait line, we detected slight differences between the classic and barefoot, and the significant difference was in the right foot. During the measurement, individual fluctuations of the centre of gravity were recorded, which are related to the course of the gait line.

Table 6. Gait line length and mediolateral deviations of COG for subject 2.

Shoe type	Classic shoes		Barefoot shoes		Bare feet	
	L	R	L	R	L	R
Length of gait line in midstance phase	19.3+/-0.8	17+/-4.4	17.4+/-4.4	18.2+/-0.5	17.3+/-4.2	18+/-0.4
Total length of gait line	28.9+/-0.4	27.5+/-6.9	25.5+/-6.4	27+/-0.3	22.1+/-5.4	23.1+/-0.4
Mediolateral deviations of COG	3.1+/-1.8	5.5+/-1.5	4.3+/-1.0	3.8+/-1.4	5.1+/-1.1	3.4+/-1.3

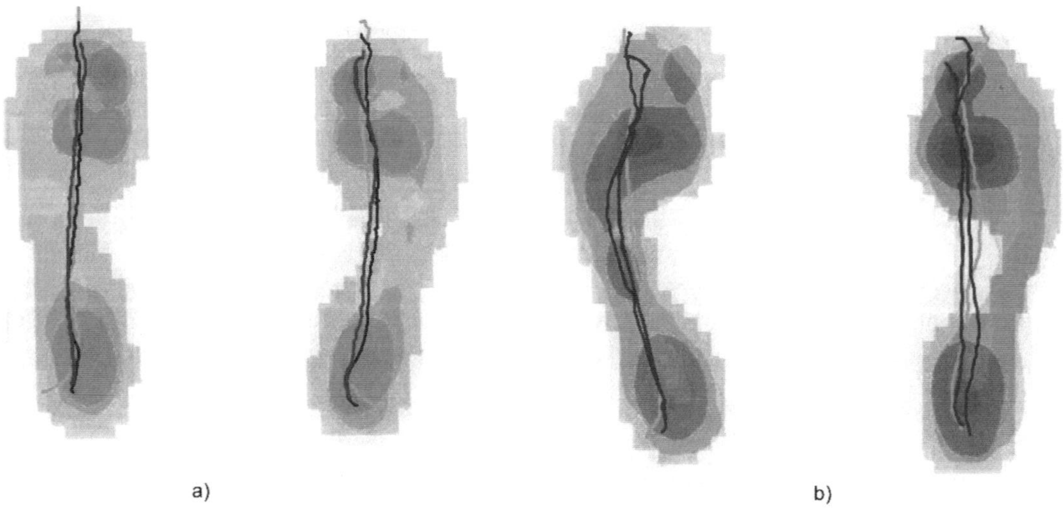

Fig. 8 Gait line during gait cycle: a) subject 1, b) subject 2, the blue curve represents classic shoes, the red curve represents barefoot shoes, and the green curve represents bare feet.

Conclusion

Significant differences in the movement pattern were observed in subject 1, who is not a regular user of barefoot footwear. The difference between barefoot shoes and bare feet walking was minimal, but significantly different from classic footwear. In classic shoes, the support phase of the step has been significantly shortened and the maximum forces at the initial contact of the heel and the rebound from the toe have increased. This may be due to the subject relying on the cushioning ability of the sole. When walking barefoot and walking in barefoot shoes, the subject focused more on the initial contact of the heel, which increased its duration and also reduced the forces acting in this phase. Since subject 2 is a regular barefoot shoes user, the walking pattern differed minimally.

The presented work is a pilot study from which the following hypotheses result, which need to be subsequently verified. In further studies, it would therefore be appropriate to verify different movement patterns in individual types of footwear, especially in subjects who are not regular users of barefoot footwear. Step time decreases significantly during barefoot walking in both measurement cases. In classic shoes, the impact on the heel is lower than in barefoot or bare feet. When walking bare feet, the length of the gait line is shortened. As it results from the biomechanics of gait, the area of the heel during contact with the mat and the toes during the toes off phase were the most heavily loaded. These values were mitigated by the influence of a thicker sole in classic shoes. Classic shoes also ensured a more optimal distribution of pressure in the area of the metatarsals. From the results, it can be concluded that when walking barefoot or in barefoot shoes, it is necessary to emphasize the correct walking technique, since there is no compensatory cushioning in the shoe itself.

Acknowledgements

This work was supported by the Slovak Research and Development Agency under contract No. APVV-19-0290 and by project KEGA 044TUKE-4/2022 Implementation of progressive technologies in prosthetics and orthotics education and support integration with practice.

References

[1] Mosai, G., Descarreaux, M. and Cantin V. "The influence of footwear on walking biomechanics in individuals with chronic ankle instability," *PLoS One*, **15**(9), e0239621 (2020).

[2] Sorrentino, R. et al. "Morphometric analysis of the hominin talus: Evolutionary and functional implications," *J Hum Evol*, **142**, 102747 (2020).

[3] Cudejko, T., Gardiner, J., Akpan, A. and D'Août K. "Minimal footwear improves stability and physical function in middle-aged and older people compared to conventional shoes," *Clin Biomech*, **71**, pp.139-145 (2020).

[4] Michalíková, M., Bednarčíková, L., Staško, R. and Živčák, J. "The comparison of the dynamic tests results from sensory platforms," *Acta Tecnología*, **7**(4), pp. 131-134. (2021).

Concrete Loaded in the Axial Direction with Strengthening Reinforcement in the Transverse Direction

Pavel Padevět[1,a], Petr Bittnar[1,b] and Zdeněk Bittnar[1,c]

[1]*CTU in Prague, Faculty of Civil Engineering, Thákurova 7, 166 29, Prague 6, Czech Republic;*

[a]*pavel.padevet@fsv.cvut.cz*, [b]*petr.bittnar@fsv.cvut.cz*, [c]*bittnar@fsv.cvut.cz*

Abstract: The attention of the article is focused on the use of additional concrete reinforcement to increase material parameters. Concrete structures are demanding in terms of material consumption, energy used in cement production and CO_2 production. The first part of the article describes the method of preparation and production of test specimens. Subsequently, the methodology for measuring material properties is discussed. The result of the measurement is a comparison of contact and non-contact methods of measuring deformations, ascertaining the material properties of the tested concrete.

Keywords: Concrete, Strain and stress, Digital Image Correlation, Strain measurement, Transverse direction reinforcement

1 Introduction

The social effort to reduce the energy demand of the construction industry leads to the search for economical solutions. One of the many possibilities for savings is the more economical use of building materials. Concrete, as the most widespread building material, can offer an opportunity to find savings. The production of concrete, and especially cement, is a very energy-intensive process, with a high production of CO_2. There are more options for optimizing concrete production. For example, more efficient use of production raw materials in the production of cement, more appropriate use of selected types of cement in concrete for a specific type of construction, or appropriate use of the mechanical properties of concrete in building construction.

This paper is focused on the last of the mentioned optimizations, i.e. influencing the mechanical properties of concrete in the structure. Specifically, attention is paid to increasing the load-bearing capacity of concrete by means of transversely oriented external reinforcement, which increases the compression strength of concrete in the axial direction.

The compressive strength of concrete is mainly influenced by the composition of the concrete. Reinforced concrete structures have increased compressive strength due to concrete reinforcement. Nevertheless, the basic component of the concrete strength of reinforced concrete remains the strength of the concrete and the strength of the steel reinforcement [1]. Often in construction applications there is no need for a rapid increase in concrete strength and the ability of concrete to increase compressive strength over time can be used.

Many concrete building structures are loaded with long-term loads after a few months of their construction. The increase in strength of concrete over time strongly depends on the type of cement used [2]. The Portland cements type I have the fastest increase in strength. In contrast, the use of blended Portland cement results in a lower evolution of heat of hydration during concrete maturation, as well as a significantly slower increase in compressive and tensile strength. Nevertheless, over a longer period of time, concrete made from mixed cement can achieve the same parameters as concrete in which Portland cement was used.

2 Tested material

The work aims to capture the change in strength over time, which is influenced by the composition of the concrete, as well as by the method of reinforcement. The question is whether reinforcement has an effect on increasing the bearing capacity of concrete?

Concrete from mixed Portland cement CEM II was chosen for testing. This cement has a slower increase in strength than CEM I cement. Reducing the amount of cement would result in low strength. A replacement for cement was chosen using ground limestone. The detailed composition of the concrete mixture is described in Table 1.

Table 1. Mixture of the tested concrete.

Part	Description	Content [kg/m³]
1	CEM II/A-LL 42,5	272
2	Limestone	148
3	Sand 0-4	980
4	Aggregate 4-8	216
5	Aggregate 8-16	380
6	Plasticizer	2.5
7	Stabilizer	1.5
8	Water	175

The plasticizer and stabilizer were added to the mixture for processing without vibrating the mixture and for preventing the separation of individual components from the mixture.

The production of concrete specimens consisted of concreting cylindrical test specimens with a base diameter of 100 mm and a cylinder height of 200 mm. The basic cylinders were concreted into steel forms without reinforcing rings. These specimens were used to determine the basic strength of the produced concrete. Other cylinders were made by inserting aluminum rings with an outer diameter of 100 mm into steel molds. Plastic spacer rings of the same diameter and thickness were inserted between the individual rings. Two ring thicknesses of 2 and 3 mm were chosen for reinforcement. The plastic intermediate rings were removed after removing the specimens from the steel molds.

Three specimens without reinforcing rings, three cylinders with 2 mm thick reinforcing rings and three bodies with 3 mm thick rings were prepared for testing. The production was prepared for the verification of properties at the age of 28 and 90 days. After concreting, the specimens were left in a laboratory environment, i.e. at 20°C and a relative humidity of 50%. Storing the specimens in an environment with 100% relative humidity was not feasible due to the need to exclude a chemical reaction between concrete and aluminum.

As can be seen from Fig. 1, cylindrical specimens with 12 reinforcing aluminum rings of two thicknesses of 2 mm and 3 mm were used for testing.

3 Measurement of the parameters

The specimens were tested in compression at the age of 28 and 96 days. Table 2 shows how the strength of the cylinders without reinforcement and with reinforcement increased in the given period. Simultaneously with the determination of the compressive strength, other mechanical parameters such as the modulus of elasticity and Poisson's ratio are possible determinate.

A compression test for concrete specimens was chosen for testing. The loading speed was chosen in accordance with standard regulations [3]. For the measurement, as can be seen from Fig. 1, two attached extensometers with a measuring length of 150 mm were used. The measured length of the extensometers guaranteed the measurement of deformations that could occur on the cylinder. The sensors were mounted in the direction of loading of the test specimens. The force signals and total deformation of the cylinder in the axial direction were also recorded from the test device.

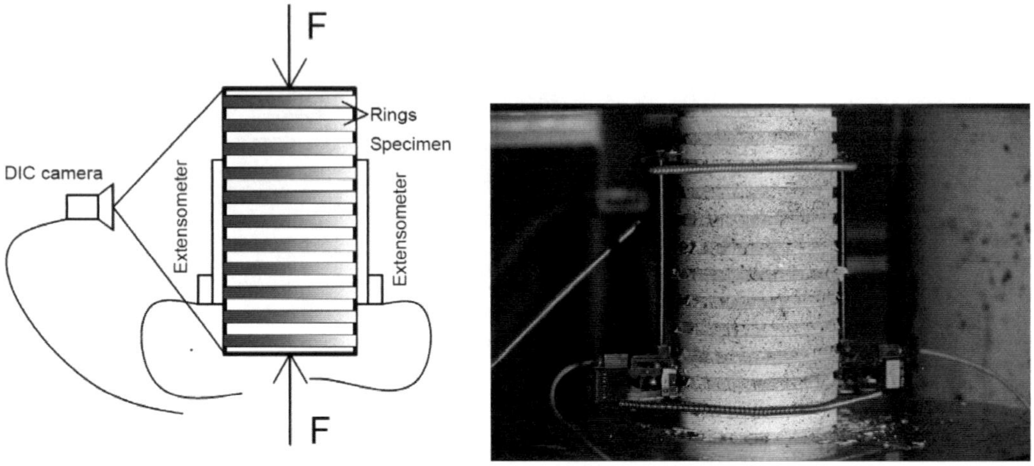

Fig. 1 Concrete specimen strengthened by external reinforcement in the shape of rings. The cylinder after completion of the compressive test (right) and principle of the measurement of deformation (left).

The measurement was carried out until the destruction of the surface reinforcement, which was made of an aluminum alloy. Axial deformations were measured using attachment extensometers. From the previous measurements, a DIC (Digital Image Correlation) with a stereo deformation recording device manufactured by Dantec Istra was chosen for a more detailed analysis.

To measure deformations using DIC, all specimens were provided with a pattern. The image recording from both cameras was stored in a storage device. The time step between frames was chosen to be 5 seconds. The test of the unreinforced specimen is quite fast and approximately 150 images could be recorded during the test. Testing cylinders with reinforcement to complete failure is more time-consuming. In this case, a record of 300 to 400 images was taken, which were used for processing by digital image correlation.

The processing of the measured signals enables the creation of working diagrams. Using extensometers, it is possible to determine the changing strain in the axial direction. From the total deformation and strength record, a working diagram can be established, including the strength behavior of the concrete. The overall working diagram cannot be determined from extensometer measurements and force recording. The result is strongly dependent on the fact whether the crack formation takes place in the area measured by the extensometer or not.

This aspect can be avoided by using non-contact measurement. With the help of DIC, suitable places on the body can be selected for the placement of virtual extensometers, both in the axial and transverse directions with respect to the loading axis.

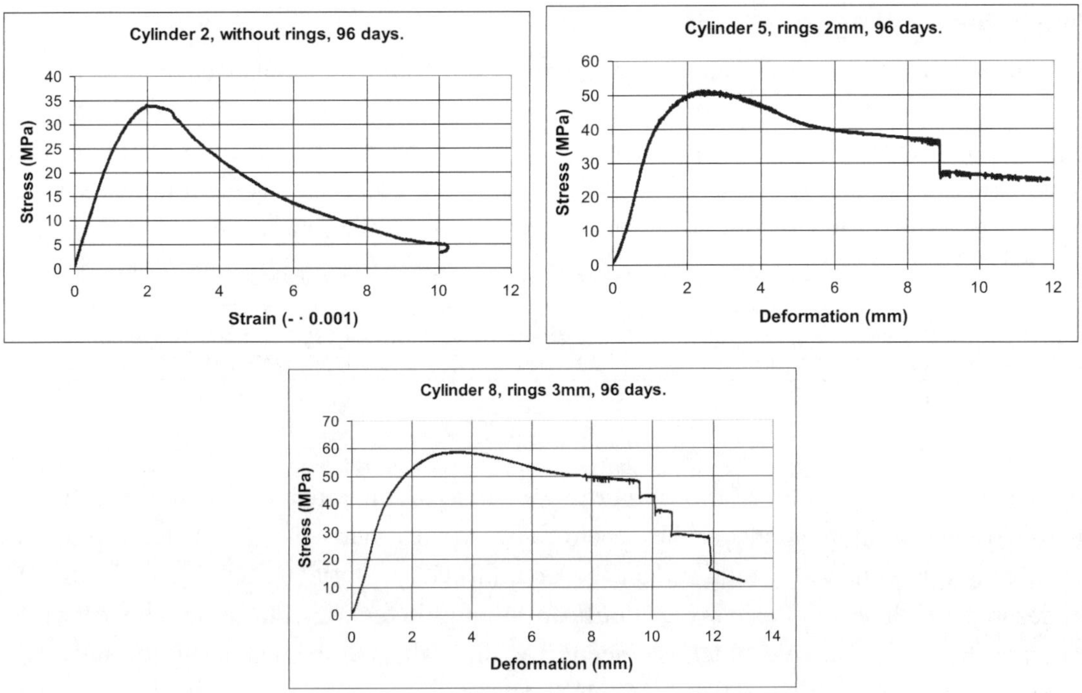

Fig. 2 Working diagrams created from contact measurement for specimen without rings (up left), for specimen with 2 mm rings (up right) and specimen with 3 mm rings (below).

4 Results

Fig. 2 shows working diagrams for specimens aged 96 days. The first is determined from extensometers and the horizontal axis is expressed as strain from extensometers. The second and third graphs, which are measurements on cylinders with reinforcing rings, are drawn using the total deformation of specimens 200 mm high.

Fig. 3 shows the steps in the processing of recorded DIC images. As an example, specimenss aged 96 days, cylinder No. 2 without reinforcing rings and cylinder No. 8 with reinforcing rings of 3 mm thickness are used. The colored field captures the total deformation (in horizontal and vertical direction) of the selected area. In the defined field, it is possible to enter any points for the definition of the distance measurement and also the proportional deformation. In the figure, they are indicated by red lines in the defined field.

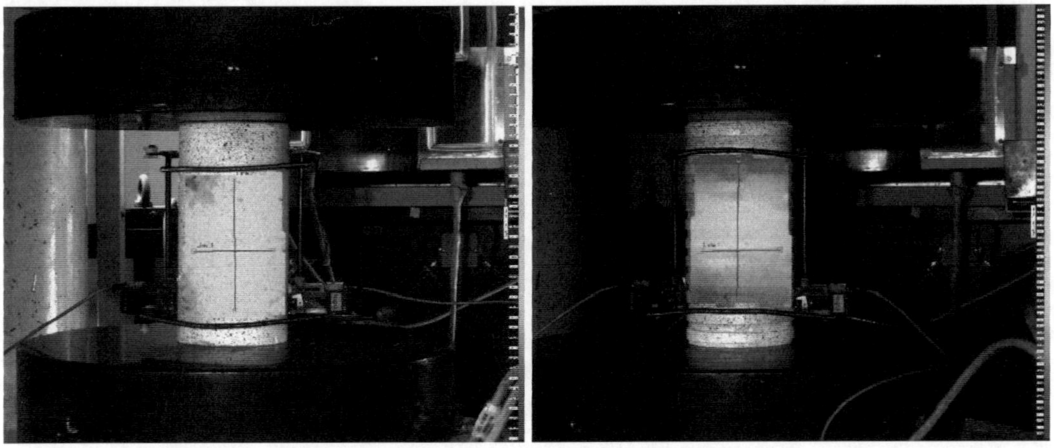

Fig. 3 Processing measurements using DIC. On the left, the total deformation of the body No. 2, on the right, the total deformation of the cylinder 8. Fitting extensometers.

Conclusions

The results of contact and non-contact measurements yielded very consistent results both in terms of the quality of the measurements and information on the use of a specific type of additional reinforcement. A very positive aspect can be found to be a significant increase in the compression strength of concrete already when using 2 mm thick rings. Furthermore, an almost 30% increase in strength is evident in 2 months from the first testing.

Table 2. Compression strength of tested concrete cylinders.

Age	Reference [MPa]	Cylinders with rings 2mm [MPa]	Cylinders with rings 3 mm [MPa]
28	25.1	50.8	57.9
96	34.0	58.3	67.6

As can be seen in Table 2, the compressive strength of concrete when using 3 mm circumferential reinforcement reaches even higher values than with 2 mm reinforcement. This strength is almost 2 x higher than the unreinforced material at the age of 3 months. Trends in material behavior can bring significant savings in the use of the amount of concrete, or cement binder in concrete which is evident from the values of the increase in strength over time when using less energy-intensive CEM II cement.

Acknowledgements

Thanks to the support from the SGS project No. SGS 22/089/OHK1/2T/11 Buildings materials and construction under view of experiment.

References

[1] Liao W.CH., Perceka W. and Wang M. "Experimental study of cyclic behavior of high-strength reinforced concrete columns with different transverse reinforcement detailing configurations," *Eng Struct*, **153**, pp. 290-301 (2017).

[2] ČSN EN 197-1 ed.2, Cement – Part 1: Composition, specifications and conformity criteria for common cements.

[3] ČSN EN 12390-3 (731302), Testing hardened concrete – Part 3: Compressive strength of test specimens, 2020.

Experimental Verification of the Material Properties of Concrete 50 Years Old Wastewater Treatment Plant

Aleš Palička[1,a], Pavel Tesárek[1,b] and Zdeněk Prošek[1,c]

[1]*Department of Mechanics, Faculty of Civil Engineering, CTU – Czech Technical University in Prague, Thákurova 7, 166 29 Prague, Czech Republic;*

[a]*ales.palicka@fsv.cvut.cz*, [b]*pavel.tesarek@fsv.cvut.cz*, [c]*zdenek.prosek@fsv.cvut.cz*

Abstract: The article focuses on the investigation of the environmental impact of contaminated by colloidal aggressive substances on a 50-year-old wastewater treatment plant, where the sampling of core boreholes of DN150mm × 370 mm in the tank was made. The properties investigated were the microstructure and compressive strength of the concrete. The results showed that the aggressive environment had no effect on the resulting compressive strength of the concrete.

Keywords: Degradation of concrete; Corrosion; Aggressive environment

1 Introduction

The aggressive environment in wastewater treatment plants is a constant problem due to the observance of hygiene standards in the elimination of the spread of bacteria, fungi, and viruses. So that the ecosystem is not damaged. The influence of wastewater has an impact on the construction part in the treatment stations, and subsequently it cannot be done without local remediation. Nowadays, the design of monolithic constructions mainly uses the watertight concrete system (WCS), "white tub system", for its waterproof properties with the extension of the curing time to reduce microcracks in concrete constructions [1]. Wastewater treatment plants cannot do without monolithic constructions with regard to their resistance to other materials. Their maintenance is appropriate due to the influence of attacking the aggressive environment with polluted water due to the chemical composition and degradation processes. The resulting value for pre-purified water is monitored and projected to the values of BOD (Biochemical Oxygen Demand), COD (Chemical Oxygen Demand), US (Undissolved Substances), N-NH4 (Ammonia Nitrogen), P_{total} (Total Phosphorus), N_{total} (Total Nitrogen) [2]. The main component of pollution are substances from human activity (humanoid substances), micropollutant substances and unwanted substances. These attributes are reduced with the help of mechanization, separated (mechanical cleaning) and further stabilized by a biological cleaning process [3]. Using the microorganism in which they live, they grow and reproduce. As a result of this biomass, CO_2, and hydrogen sulphide H_2S are released, which supports the degradation of concrete structures due to CO_2. The effect of poor maintenance of wastewater treatment plant managers with insufficient oxygenation is the conversion of H_2S to H_2SO_4, which reduces the alkalinity of hardened concrete supported by steel corrosion. The treatment process is supported by stabilization using nitrification to nitrates with the help of a phosphorus precipitation stabilizer [4]. This has the effect of promoting penetration into the pore structure of concrete, where crystallization and volume increase occur. With low strength, microcracks may occur on the surface of the concrete. In an environment in which concrete structures are stressed, the penetration of undesirable substances occurs and consequently the degradation of concrete structures, in extreme cases, chemical corrosion of steel occurs when there is insufficient coverage [5]. Another factor of corrosion due to carbonation supported by carbon dioxide. Chemical action and microstructural aggressive environment caused by organic acids reducing the alkalinity of concrete and soluble salts reducing the durability of concrete [6].

The research dealt with the influence of an aggressive environment on monolithic structures at a specific treatment plant, where reconstruction was taking place to increase the capacity of the wastewater treatment plant with the modernization of the technological part. The samples were taken from the place of constant influence of an aggressive environment, in which the cleaning cycle (aerobic and anaerobic) took place, according to the needs of the user. The selected site was chosen for its age and the effects of several factors without precipitation of phosphorus.

2 Material and samples

The wastewater treatment plant is in the village of Horní Bezděkov, where intensification and capacity increases were carried out. As part of the demolition work, the existing tank was used, and its volume was increased. (Fig. 1). The tank was used for 50 years, for the sludge separation process and for the areobic process (chemical cleaning), as needed by the plant manager. The individual boreholes were chosen in heights of tank, according to the frequencies of environmental contamination. In total, there were 7 bores. Of this, Core_4, broke off during the drill and was replaced at the same height of tank by Core_4x. Optical microscopy was carried out on the lowest outcrop Core_1, it was the most contamination site of the permanently flooded. The location of the individual core drillings can be seen in (Fig.1). Also, in the Fig. 1 you can see the maximum and minimum level of contaminated water during operation. The height is given from the sea level of the Baltic system.

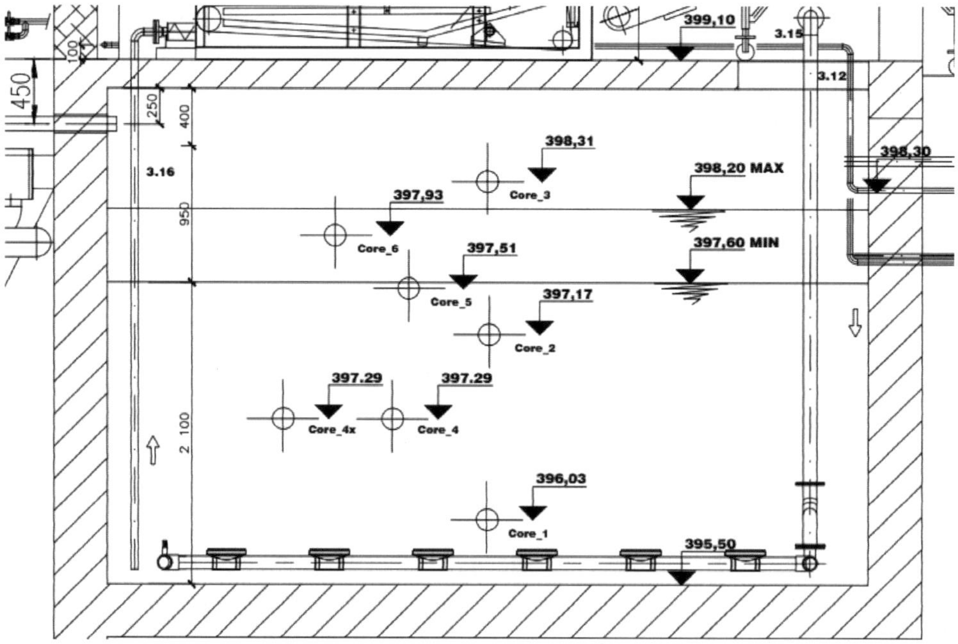

Fig. 1 Cut through the sampling point in the tank, scale 1:50.

Core drilling was carried out with the help of a DN150 mm diameter diamond core coring machine with cooling and pumping of drink water. The resulting individual dimensions of core were approximately 144.60 mm and 370 mm in height for a total of 7 pieces.

3 Experimental methods

Optical microscopy was used to describe to the microstructure of the concrete by ZEISS Axio Zoom V16 Stereo Zoom Microscope. The microscopic images were taken on core of borehole 1 (Core_1). The area examined was on a longitudinal section of the sample Core_1. The examined area had an area of 540,20 cm^2 at a magnification of 3.5 ×. For this reason, a 9 × 9 image matrix was used. The compressive strength was determined by uniaxial compressive test on the individual core of borehole. The specimens were moulded so that their height was fixed at the same value of 300 mm in order to comply with the condition of twice the diameter according to EN 12390-3 [7]. The testing was displacement controlled at a constant rate of 5 mm/min.

4 Results and discussions

In Fig. 2 we can see the microstructure of the Core_1 sample detected by optical microscopy. The sample (Core_1) shows the effect of the aggressive environment on the concrete structure. Part A in Fig. 2 shows the microstructure towards the soil showing a covering layer of cement paste of approximately 950micron thickness and this layer in Fig. 2 Part B cannot be seen due to environmental degradation contaminated by colloidal substances. The structure remains undisturbed in the longitudinal section direction.

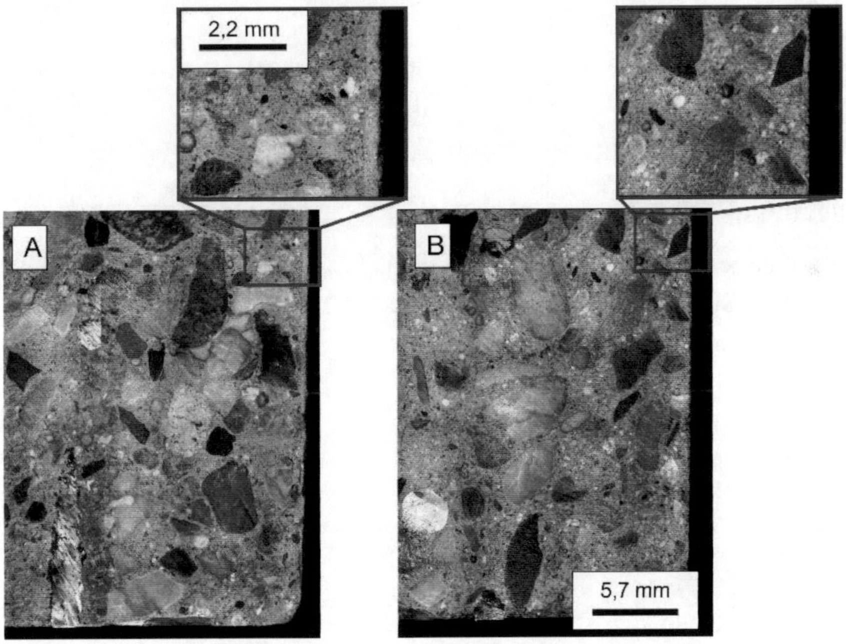

Fig. 2 Image for optical microscopy magnification 3.5x.

Table 1 shows the results of mechanical destructive testing of the individual samples, the range of compressive strength values was from 35.85 to 43.32 MPa. The Core_4x sample was removed from the statistics due to non-standard testing. The average strength of the concrete was at 37.49 ± 5.03 MPa. Considering the standard deviation, it can be specified that all the speciement had the same compressive strength values. The result shows that the aggressive environment has an effect on the cover layers and does not affect the final strength of the concrete.

Table 1. Compressive strengths of samples analysed.

Set	Diameter [mm]	Area [mm²]	Height [mm]	Weight [g]	Load [kN]	Compresive strenght [MPa]
Core_1	-	-	-	-	-	-
Core _2	144.73	16451.55	297.00	2335.80	670	40.73
Core _3	144.27	16347.15	297.00	2249.38	612	37.44
Core _4	144.66	16435.64	296.00	2364.46	712	43.32
Core _5	144.62	16426.55	297.50	2301.05	495	30.13
Core _6	144.64	16431.10	296.50	2277.59	589	35.85
Core _4x	144.19	16329.02	185.50	2273.00	427	26.15

Conclusion

Concrete samples (core boreholes) taken from the overburden were examined as part of the research. Over a period of 50 years of exposure to an aggressive environment The results showed the properties of concrete and their compressive strength at 37.49 ± 5.03 MPa, this is the strength in the entire profile with regard to the selection of a sample from one-way places of different exposure to an aggressive environment. Optical microscopy showed the occurrence of attack to the upper depth of the concrete.

Future research will address comparison of new samples supported by phosphorus precipitation and the effect of microorganism in monolithic constructions. Growth, propagation and settling in a porous structure.

Acknowledgements

This paper was financially supported by the Czech Technical University in Prague. a project under the number SGS22/089/OHK1/2T/11 and TACR the number SS03010302. The authors also thank company SMP Vodohospodářské stavby a.s.

References

[1] Radziejowska A. and Linczowski K., "Selection of "white tub" technology using milti-criteria analysis – a case study," *Czasopisma Polska Academia Nauk*, **68**, pp. 446-449 (2022).

[2] Hassen E. B. and Asmare A. M., "Modeling and monitoring of treated wastewater based on water quality assurance parameters," *International Scientific Organization, Chemistry International*, **5**(1), pp. 87-96 (2019).

[3] Nasr F.A, Doma H.S., Abdel-Halim H.S. and El-Shafai S.A., "Chemical industry wastewater treatment," *Water Pollution Research department*, **30**, pp. 1183-1206 (2004).

[4] Biological Wastewater Treatment – What Is It and How Does It Work? Water Treatment, Wastewater, Air and Energy, "Water Treatment Services", *Available from* https://watertreatmentservices.co.uk/biological-wastewater-treatment-how-does-it-work/ *Accessed:* 2023-04-02

[5] Chakraborty S., Mandal R., Chakraborty S., Guadagnini M. and Pilakoutas K., "Chemical attack and corrosion resistance of concrete prepared with electrolyzed water," *Journal of Materials Research and Technology*, **11**, pp. 1193-1205 (2021).

[6] EN 12390-3 (731302), Testing hardened concrete, *Compressive strength of test bodies – part 3,* (2020) ICS 91.100.30

Development and Verification of the Child Headform Numerical Impactor

Peter Palička[1,a], Róbert Huňady[1,b], Martin Hagara[1,c] and Michal Fabian[2,d]

[1]*Technical University of Košice, Faculty of Mechanical Engineering, Department of Applied Mechanics and Mechanical Engineering, Letná 1/9 Košice-Sever 042 00, Slovakia;*
[2]*Technical University of Košice, Faculty of Mechanical Engineering, Department of Automotive Production, Mäsiarska 74, Košice 040 01, Slovakia;*

[a]*peter.palicka@tuke.sk*, [b]*robert.hunady@tuke.sk*, [c]*martin.hagara@tuke.sk*, [d]*michal.fabian@tuke.sk*

Abstract: Mechanics is a science that studies motion and is therefore a fundamental building stone for many other scientific disciplines. These other disciplines include human biomechanics and therefore the response of the human body to some mechanical stress. This article is devoted to the creation and verification of the child headform impactor, which is used in the automotive industry in the development of new vehicles. The child headform impactor is a test object representing a child's head and is used to verify the safety of a vehicle in the event of a collision with a child. The so-called Head Injury Criterion (HIC) is evaluated on the impactor and the deformation zones on the bonnet of the vehicle are then modified according to this criterion in order to avoid serious or critical head injuries. The development and verification of the numerical model of the child headform impactor was based on experimental measurements. Physical impact tests were performed with the headform impactor fired at a bent steel plate, evaluating and comparing the HIC with the results from the simulations. The certification of the impactor was based on the rules of the certification test, i.e. free fall of the impactor on a rigid steel plate and the acceleration of the impactor was measured.

Keywords: Impact test; Child headform impactor; FEM; Pedestrian protection

1 Introduction

The aim of car manufacturers is to minimise the number of traffic accidents by using active and passive safety systems. Active safety systems try to prevent crisis situations and avoid accidents. These systems include e.g. ABS (Anti-lock Braking System), ESP (Electronic Stability Programme) and many others. Despite the sophistication of active safety systems, traffic accidents cannot be completely prevented. In cases where collisions cannot be prevented, passive safety systems come into action. The aim of these systems is to minimize the risk of serious injury to a person health. In a collision with a pedestrian, the pedestrian's head hits the bonnet or windshield of the car. The head impact is the most critical in terms of pedestrian safety and it is simulated with headform impactor. The acceleration time history curve of the headform impactor after contact with the bonnet is evaluated and the HIC (Head Injury Criterion) is then calculated from this curve. The aim of this work is to develop and verify a numerical model of the child headform impactor that would match as closely as possible the results obtained from experimental tests [1,2].

2 Experimental and numerical investigation

There are studies that deal with a similar problem of creating and validating a numerical impactor of a child or adult head. Researchers are also focusing on the material properties of rubber and many other areas related to the pedestrian protection [3-5]. A headform impactor (see Fig. 1) is an assembly of multiple components designed to meet given biomechanical properties similar to those of the human head. It consists of an aluminium hemisphere and a synthetic cover. Its outer diameter shall be 165 mm ±1 mm and its weight shall be 3.5 kg ± 0.07 kg. The cut-out in the hemisphere shall contain one tri-axial accelerometer or three uniaxial accelerometers. Their first natural frequency shall be greater than 5000 Hz. The position of the accelerometers from the centre of the hemisphere shall be defined for a direction perpendicular to the direction of measurement ±1 mm and in the direction of measurement ±10 mm. The Channel Frequency Class (CFC) response value shall be 1000 Hz and the Channel Amplitude Class (CAC) response value shall be 500 g for acceleration [6].

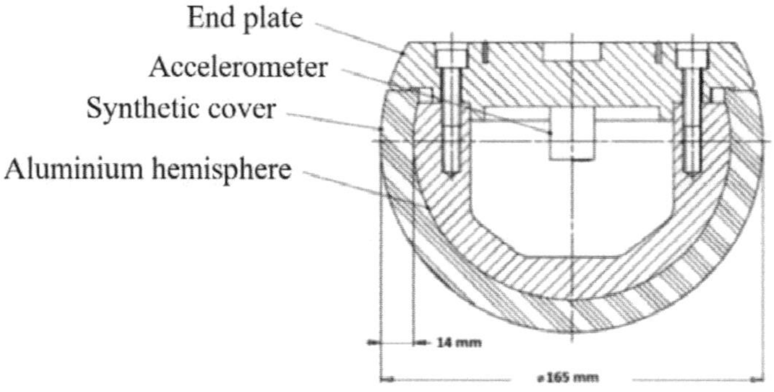

Fig. 1 Main parts of the child's headform impactor [6].

The aluminium hemisphere and the end plate were meshed with bigger quad elements. The synthetic cover was meshed with the smaller quad elements as the synthetic cover is in direct contact with the sheet metal. The new numerical model of child headform impactor is shown in Fig. 2 and the cut through the impactor where the sensors are placed can be seen in Fig. 3. Viscoelastic Ogden Rubber for Solid Elements (Ogden-Based Viscous Response) was chosen as material model for the synthetic cover.

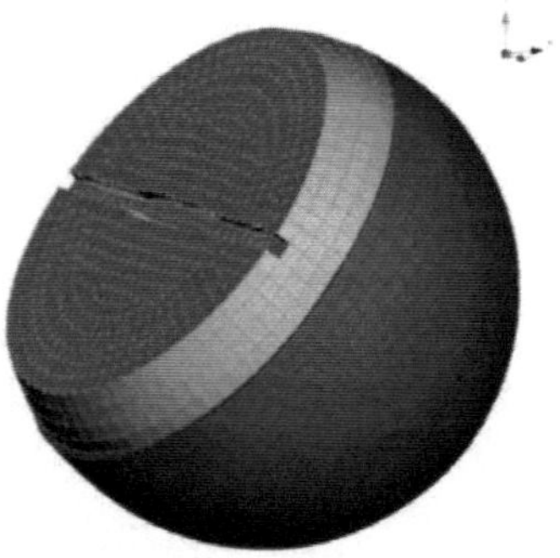

Fig. 2 Numerical model of the child headform impactor.

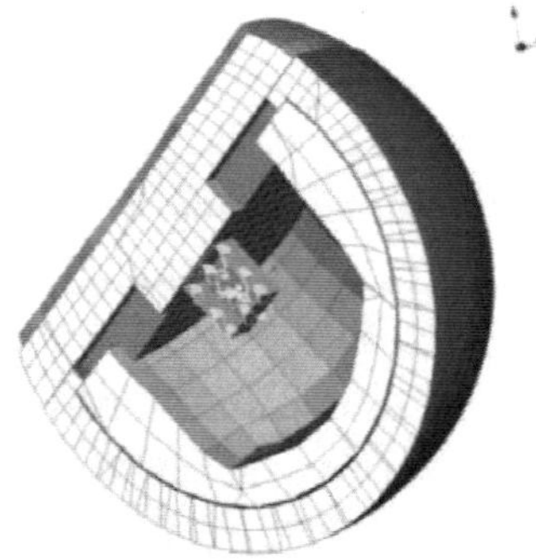

Fig. 3 A cut through the impactor model.

Table 1 and Table 2 show the values of the masses, the coordinates of the centre of gravity (CG) as well as the moments of inertia around the individual axes of the current and new computational model.

Table 1. Comparison of the masses and coordinates of the center of gravity of the current and new impactor.

	Mass [kg]	CGx [mm]	CGy [mm]	CGz [mm]
Current impactor	3.452	-25.299	0.00017	97.486
New impactor	3.504	-26.473	-0.41295	98.429

Table 2. Comparison of the moments of inertia of the current and new impactor.

	I_{xx} [kg.mm^2]	I_{yy} [kg.mm^2]	I_{zz} [kg.mm^2]	I_{xy} [kg.mm^2]	I_{yz} [kg.mm^2]	I_{xz} [kg.mm^2]
Current impactor	9360.63	8688.04	9664.03	0.00597	-0.20834	784.90
New impactor	9585.57	8869.70	9859.68	-0.01155	-3.76055	777.89

The certification test shall be carried out by free fall of the impactor (see Fig. 4). The impactor shall be allowed to fall freely from a height of 376 mm ±1 mm at an angle of 50° ±2° onto a rigid steel plate at least 50 mm thick and 300 mm x 300 mm in dimensions. This test shall be carried out three times for the impactor, each time rotating the impactor 120° about the axis of symmetry. The test shall be carried out at a temperature of 20 °C ± 2 °C, the temperature deviation being valid for a relative humidity of 40 % ± 30 %. The resulting measured values from the certification test shall be in the range 245 ÷ 300 g, which, when converted, gives acceleration values in the range 2 403 ÷ 2 943 m/s^2 [6].

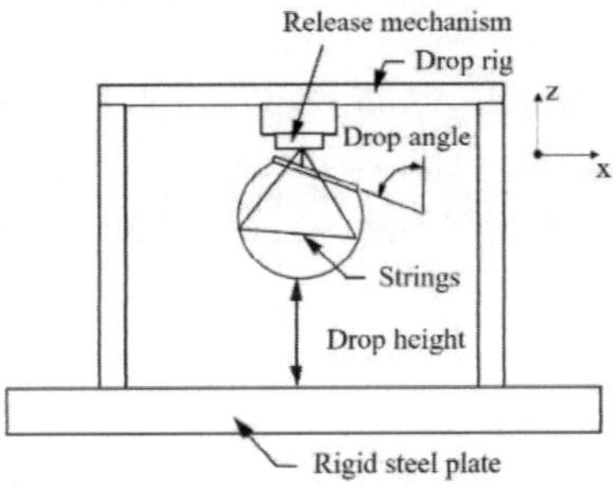

Fig. 4 Child headform impactor certification test set-up [6].

The free sheet impact test (see Fig. 5) is carried out by throwing the impactor into a sheet of 1120 x 1120 mm. The plate is 0.7 mm thick and has a free area of 1000 x 1000 mm. The sheet metal is folded by 20 mm over the free width of 1000 mm and its angle to the horizontal support is 15°, which is intended to approximately simulate the shape and position of the bonnet of a car. The sheet is loosely clamped on a rigid frame with rounded edges. The sheet is initially flat, but when mounted on the rigid frame, it is clamped under the rails on the right and left edges, which creates an elastic deformation of the sheet and achieves the desired 20 mm bending shape. The sheet metal clearance under the rails is 0.3 mm. Its position in the longitudinal direction is secured by pins. The impactor is thrown against the plate at an angle of 50° at a speed of 11.3 m/s, which represents a speed of 40.68 km/h. The contact point of the impactor with the sheet metal is in the centre of the sheet metal. During the test, the time history of the accelerations is recorded and the HIC_{15} (Head Injury Criterion) value is calculated, which indicates the severity of the head injury at impact [6].

Fig. 5 Simulation of head impact test.

The HIC_{15} allows the severity of head injuries in a crash to be quantified. The relationship for calculating the HIC_{15} value, as well as the limit values, are based on experimentally obtained data

$$HIC = \left[\frac{1}{t_2 - t_1} \int_{t_1}^{t_2} a \, dt \right]^{2.5} (t_2 - t_1), \tag{1}$$

where

$$a = \sqrt{a_x^2 + a_y^2 + a_z^2}, \tag{2}$$

where a_x^2, a_y^2, a_z^2 represent the accelerations in each axis and a is the resulting acceleration in multiples of g $(1 \text{ g} = 9.81 \text{ m/s}^2)$. The values t_1 and t_2 are the two time instants during the impact that determine the start and end interval of the recording, which is a maximum of 15 ms.

After creating a computational model of the new impactor and after defining all boundary conditions and creating a simulation, the next step was to compare the results from the simulation with the results from the real tests. Acceleration curve and HIC_{15} values from tests and simulations of the current and new impactors are shown in Fig. 6.

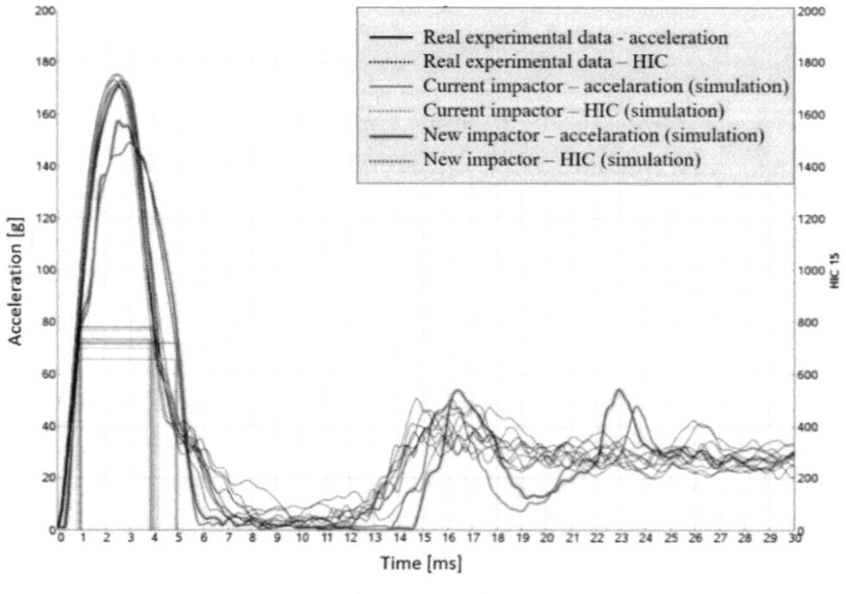

Fig. 6 Comparison of time histories from tests and simulations of the current and new model of the impactor.

3 Adjusting a new model based on test results

In the basic setup, the free sheet had a thickness of 0.7 mm, a density of 7850 kg/m^3, the clearance under the rail was 0.3 mm, the coefficient of shear friction between the steel parts was 0.1 and between the sheet and the impactor was 0.65. To adjust the new numerical model of the impactor to the results of the experimental tests, several variants were created, where different test parameters were changed such as the coefficient of friction between the steel parts (from 0.1 to 0.3), the mesh on the contact surfaces was refined, and the material models of the individual parts of the impactor were changed. The effect of each change was monitored and evaluated by simulating the impact of the impactor on the sheet metal and also compared with the experimental results. Several such resulting curves from the simulations can be seen in Fig. 7.

During the impactor adjustment process, two paths were created for subsequent impactor adjustment. In the simulations, it was shown that there is sliding of the synthetic cover between the cover and the aluminium hemisphere of the impactor as well as between the synthetic cover and the sheet metal. As a result of eliminating this error, a pre-tension between the cover and the aluminium hemisphere was defined to simulate the stretching of the cover on the hemisphere. Thus, the impactor labelled IMP_01 was an impactor without pre-tension and IMP_02 was an impactor with pre-tension.

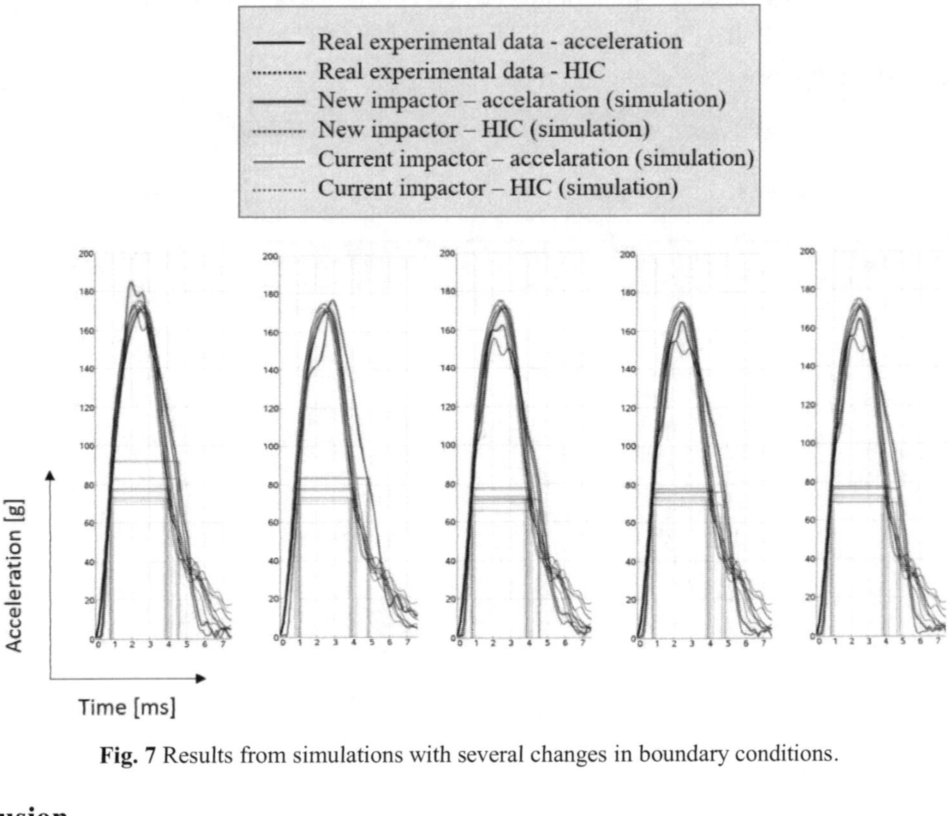

Fig. 7 Results from simulations with several changes in boundary conditions.

Conclusion

Two numerical models of the child headform impactor were created and simulations of the impact tests were performed on a slightly bent sheet plate to simulate the bonnet of a car. The results of the numerical analysis were compared with the results from experimental measurements. Fig. 8 shows the agreement in the obtained results and therefore it is possible to conclude that both impactors are describing the behaviour of the real impactor in experimental tests with high accuracy.

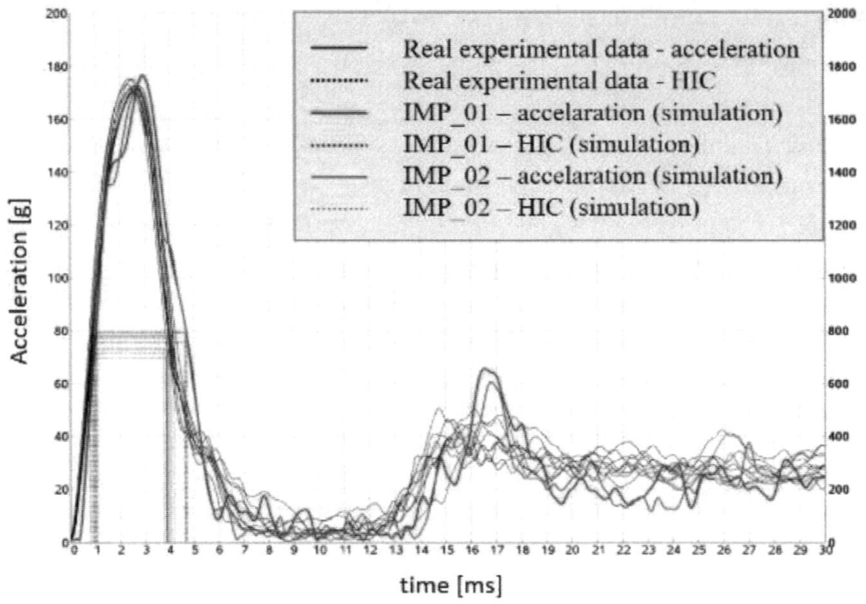

Fig. 8 Comparison of experimental and numerical results.

Fig. 9 shows the results of the acceleration of the impactors from the certification test. The impactor IMP_02 has exceeded the upper allowed acceleration limit and therefore failed the certification test. IMP_01 has passed the certification test and can be used in pedestrian protection simulations.

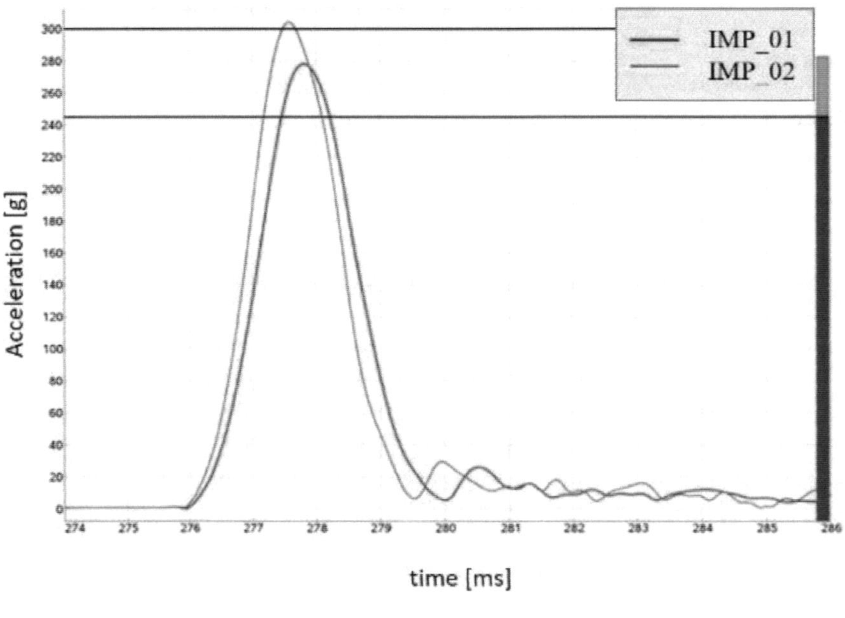

Fig. 1 Results from numerical simulation of certification test of child headform impactor.

Acknowledgements

This work was supported by the Ministry of Education, Science, Research and Sport of the Slovak Republic under the projects VEGA 1/0516/22 and VEGA 1/0500/20.

References

[1] Untaroiu, C., Shin, J., Crandall, J. and Crino S. "Development and Validation of a Headform Impactor Finite Element Model with Application to Vehicle Hood Design for Pedestrian

Protection," 9th International LS-DYNA Users Conference, Detroit, 2006, *Available from* https://www.researchgate.net/publication/252741790_Development_and_Validation_of_a_Headform_Impactor_Finite_Element_Model_with_Application_to_Vehicle_Hood_Design_for_Pedestrian_Protection *Accessed:* 2023-03-31.

[2] Sulzer, J., Kamalakkannan, S.B., Wiechel, J., Guenther, D., Tanner, C.B. and Morr, D.R., "Simplified MADYMO Model of the IHRA Head-form Impactor," *Available from* https://www.researchgate.net/publication/242356520_Simplified_MADYMO_Model_of_the_IHRA_Head-form_Impactor *Accessed:* 2023-03-31.

[3] Teng, T. and Ngo, V. "Analyzing pedestrian head injury to design pedestrian-friendly hoods," *Int J Auto Tech Kor*, **12**, pp. 213-224 (2011).

[4] Choi, J.H., Park, B.C. and Kim J.S. "Development and Validation of FE Adult Headform Impactor for Pedestrian Protection," *Transactions of the Korean Society of Automotive Engineers*, 20(2), pp. 64-69 (2012).

[5] Lee, C. and Jeong, H.Y. "Development of Headform Impactor Finite Element Model Considering the Hyperelastic and Viscoelastic Responses of Rubber," *Int J Auto Tech Kor*, **19**, pp. 523-534 (2018).

[6] United Nations Economic Commission for Europe, Regulation No. 127, *Available from* https://unece.org/fileadmin/DAM/trans/main/wp29/wp29regs/2018/R127r2e.pdf *Accessed:* 2023-03-31.

Failure of Steam Piping Structure

Miroslav Pástor[1,a], Pavol Lengvarský[1,b], Martin Hagara[1,c] and Róbert Huňady[1,d]

[1]Technical University of Košice, Faculty of Mechanical Engineering, Department of Applied Mechanics and Mechanical Engineering, Letná 1/9 Košice-Sever 042 00, Slovakia;

[a]miroslav.pastor@tukle.sk, [b]pavol.lengvarsky@tuke.sk, [c]martin.hagara@tuke.sk, [d]robert.hunady@tuke.sk

Abstract: The aim of paper is an analysis of the causes of an accident that occurred in the piping system as a result of the failure of the control system and subsequent uncontrolled operation of the steam boiler for several hours. Cracks were found in many places of the pipe system of the boiler after it was shut down, and in one place the pipe was completely destroyed. The temperature change of a piping system was performed by the finite element method. Quantification of residual stresses by the hole drilling method were performed to assess possible causes of the failure of the piping system. From the obtained results, it is clear that the temperature change in the piping system cannot be considered the main cause of the pipe system failure. High levels of residual stress was identified at the site of the breach were found to be the most important cause of the failure.

Keywords: Welded structure; Failure; Stress analysis

1 Introduction

The steam boiler, waste collection bunker and the flue gas cleaning equipment are the main technological units of the waste incinerator for energy recovery of municipal waste. The heat produced in the form of steam is used for technological needs of the incineration plant as well as for the central heat supply of residential buildings and also for generation of electric power. In order for the complex technological units to function properly, it is necessary to constantly monitor and evaluate the operational parameters. The most common method used to set optimum operating modes of steam boilers is to control changes in temperature and pressure, where pressure of the medium can change independently of its temperature. Thus, acceptable changes in the temperature of the medium used during startup and shutdown processes are to be determined [1]. Modern automated control systems of steam boilers must also enable to solve the problem of combustion of combustible (explosive) materials contained in municipal waste during operation. For this reason, the design of steam boilers already places great emphasis on the analysis of temperature fields. For steel welded structures operated at high temperatures or in a humid environment, it is necessary to pay close attention to issues related to corrosion, creep. Prolonged exposure to elevated temperatures of ferritic steel boiler pipes reduces their service life and is often a major cause of damage [2]. Most cracks occur in the bent parts of the tubes which confirms the possible presence of residual stresses caused by bending. Insufficient heat treatment (e.g. annealing) does not release stresses [3,4]. As the knowledge gained from practice shows, failure of load-bearing structures may occur not only due to operating load, material fatigue, corrosion, but also due to reconstructions. Duda et al. [5] describes the procedure of the boiler failure analysis that occurred during the reconstruction works. The numerical and experimental modelling methods are used to document the conclusions. Although modern computer simulations are used in the design of complex devices, it is important to point out that defining boundary conditions corresponding to operating conditions is very difficult in such cases. The quantification of some effects in these situations by other than experimental methods is impossible. In identifying failure in welded steel structures, comparative analysis of the material taken from the location of the failure with the new material is often used to determine the causes. In this case, however, the equipment operator had no new material available for comparison. Numerical modelling is very difficult to use for the assessment of the operation safety of a steam boiler after its failure, as it is a steel welded structure. In these cases, it is almost impossible to define the boundary conditions so that they correspond to the actual conditions, i.e. to take into account possible assembly and manufacturing inaccuracies, the operating load history as well as thermal analysis. It should be pointed out that the analysed steam boiler (Fig. 1) is a part of the incinerator, where unforeseen thermal shocks such as burning of explosive material and the like often occur. For this reason, the authors proposed a methodology for assessing the further safe operation of the steam boiler on the basis of available data, i.e. supplied drawing documentation, description of the accident, material taken from the place of failure. First of all, measurement of the residual stresses in the materials collected from the damaged area were carried out. The numerical modelling was performed using the submitted design documentation, visual inspection and information on the process of the failure.

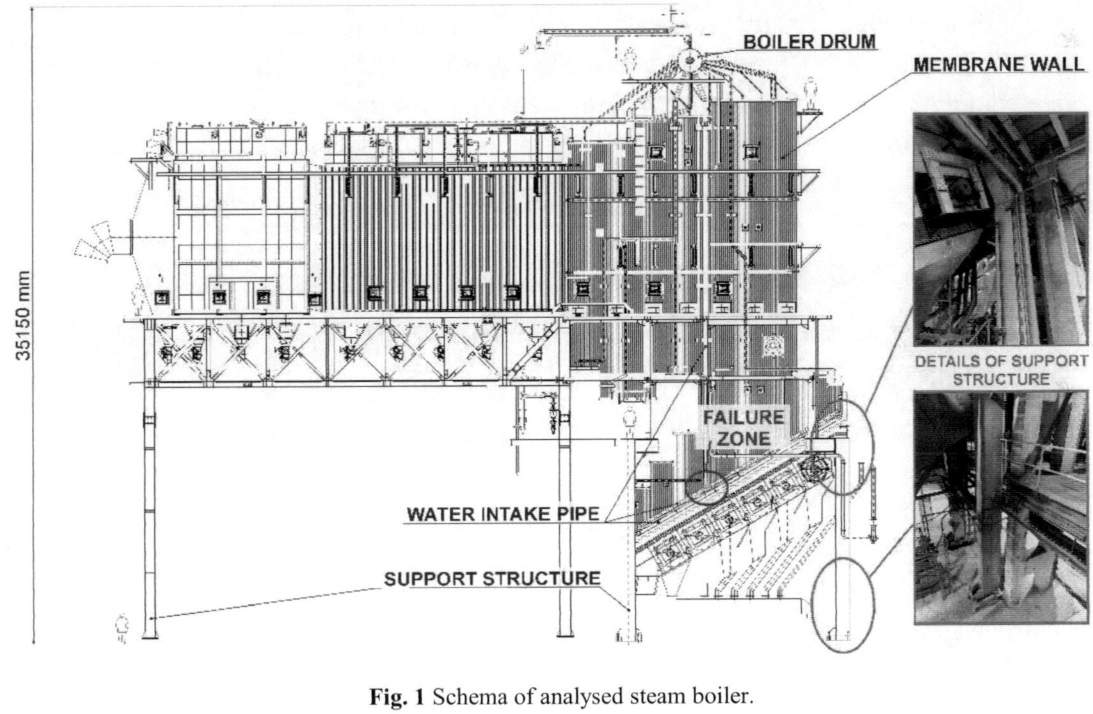

Fig. 1 Schema of analysed steam boiler.

2 Description of the steam boiler accident

Piping systems of steam boilers are loaded by internal pressure and also by thermal stresses. These are welded steel structures for which, when assessing their durability and safe operation, it is necessary to consider the increased levels of tensile stresses in the vicinity of welds caused by welding technology. In many cases, a failure occurs only around the welds. The steam boiler under consideration was in operation for less than four years. During the previous operation of the steam boiler there was not observed any significant improper functioning, which could predict a dangerous situation or failure. Despite these facts the failure of the steam boiler occurred within a few hours. Firstly, there occurred a failure in the control unit followed by the loss of information on the key parameters of the combustion process as well as boiler control itself, namely, information on the water level in the drum, the steam flow, pressures and the temperature in the combustion chamber followed, and then failure of the cooling towers control and a water intake control valve. As the control unit was not functional, the amount of water in the piping system decreased because the supply valve was blocked. This fact was confirmed by the drop of water level in the boiler drum below the critical level within 90 seconds of the control unit failure. After the control unit failure, it was found that 16 input-output cards of the control unit were inactive. Based on the diagnostics of the control system it was found that the control unit failure was caused by the damaged converter. After diagnostics and necessary repairs, the system was ready to restart in less than two hours. After the system was started, the water intake valve suddenly fully opened due to the failure of the system's security features. It should be pointed out that if the water level in the boiler drum falls below a critical value (this state occurred immediately after the control unit failed), the water piping system must not be filled with water. Within two minutes the water valve was manually shut and the shutdown of the steam boiler started. The above-mentioned procedure was in accordance with the operating manual. Despite fast intervention of the service personnel within approx. 10 minutes after the system start-up, a leak of steam on both sides of the boiler was observed. Works on the emergency shutdown of the steam boiler continued until the boiler pressure was reduced to atmospheric pressure. This process took about 7 hours. After the boiler had cooled down, the side walls of the boiler were uncovered. Fig. 2a shows a part of the membrane wall and water intake pipe. A lot of cracks were identified during the inspection of the piping system, especially on the welds of the water intake pipe (Fig. 2b).

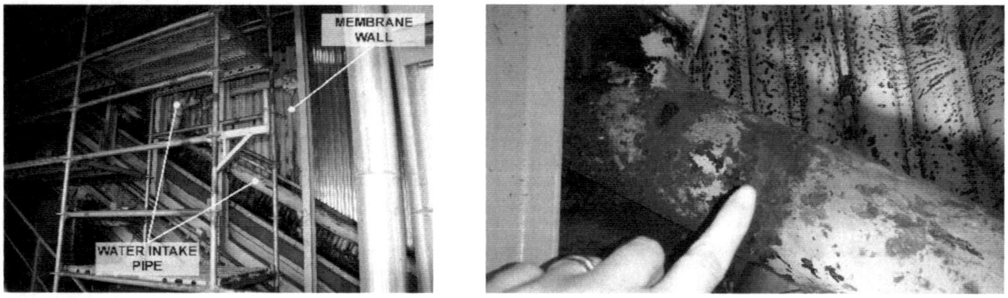

(a) (b)

Fig. 2 (a) Membrane wall and water intake pipe, (b) cracks of welds on the water intake pipe.

On the right side of the boiler, the connecting part of the pipe leading from the water intake pipe to the pipe of the membrane wall was completely torn off (Fig. 3a). Fig. 3b gives a detailed view of the pipe torn off from the connection part.

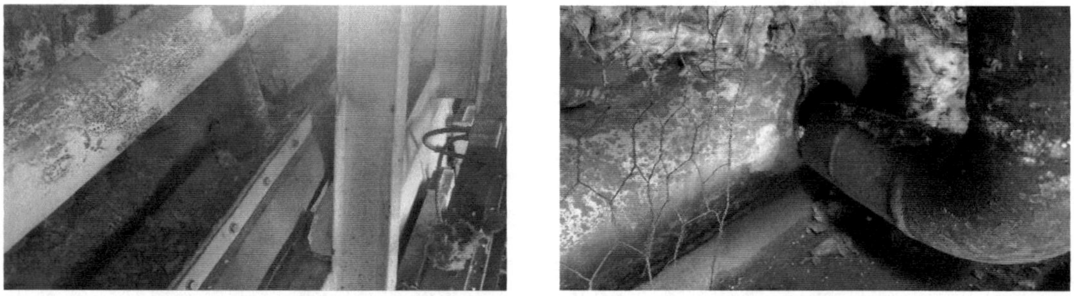

(a) (b)

Fig. 3 Weld failure of the water intake pipe: (a) general view, (b) detailed view.

The initial conclusions on what possibly caused the boiler piping system failure were drawn based on the above-mentioned facts. The most probable cause was failure of the control unit, which caused blockage of the water intake valve resulting in the lack of water in the piping system. The boiler subsequently operated in an uncontrolled mode (further waste burning). During that time individual parts of the piping system were exposed to uneven thermal stress leading to the initiation of cracks and subsequent failure of some welded joints. Due to the fact that the damaged parts of the piping system were repaired and the operation was restarted within several tens of hours, it was not possible to carry experiments on the damaged pipeline immediately. For this reason, the authors used numerical modelling method to confirm or refute the above hypothesis. The authors considered it essential to know actual mechanical properties of the used material (WNr. 1.0570), but also the residual stresses in the parts taken from the damaged piping system in order to assess their possible impact on the occurrence of the accident.

3 Numerical analysis of failure part of piping system

Numerical finite element modelling is used for stress and strain analysis in the design of welded steel structures. When analysing the steam boiler failure by numerical modelling, it is almost impossible to define boundary conditions so that they correspond to actual conditions, i.e. take into account possible assembly and manufacturing inaccuracies, the operating load history as well as thermal analysis. It should be pointed out that the analysed steam boiler is part of the incinerator, where unforeseen thermal shocks occur such as burning of explosive material. Numerical modelling is reasonable especially at the design stage of machines and equipment. Therefore, the authors performed numerical modelling on a model case where the boundary conditions were defined on the basis of the available information, i.e. drawing documentation, visual inspection and description of the accident. The impact of the temperature change was defined as the unit load, as the information on the temperature in individual parts of the steel structure is not registered by the operator. The change in the temperature was calculated in view of the change in the length of the vertical part of the water pipe, which caused the load of the analysed node III (Fig. 4). In numerical modelling, the material properties of the model were defined according to the drawing documentation. The numerical modelling was used to model only part of the water intake system where the weld failure occurred. Based on visual inspection it was found that the interconnecting pipe was welded at the locations as shown in Fig. 4. The numerical analysis was used for the verification of the effect of temperature on stress levels in the concentrator area. For this reason, only the linear static analysis was performed.

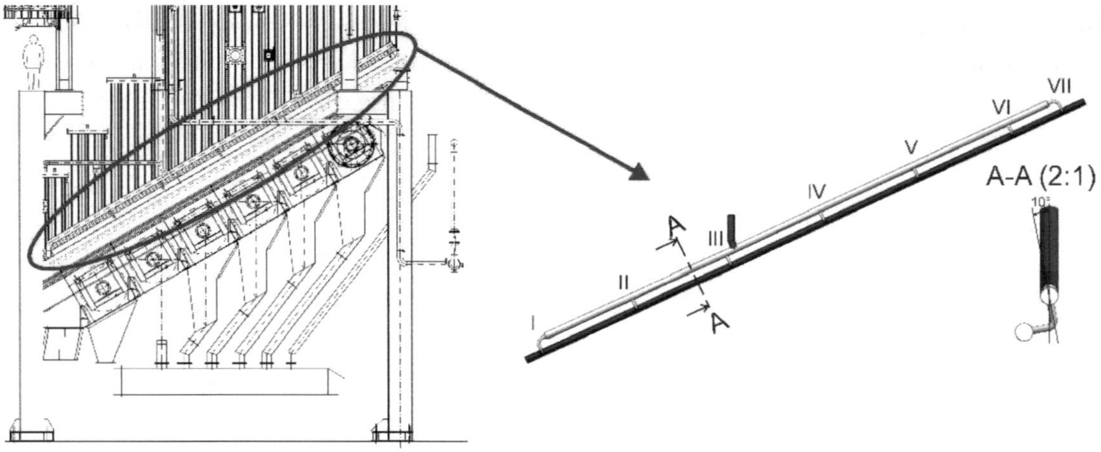

Fig. 4 Model of analysed part of water intake system.

When defining the boundary conditions, it was taken into account that one part of the piping system was fixed (the blue tube welded to the membranes, Fig. 4), and the loading was caused by displacement of the pipe by 1 mm in the vertical direction (Fig. 4, red tube). Ansys software was used for the calculation. The finite element model was made from 987 968 nodes and from 494 709 volume finite elements with quadratic approximation. Fig. 5 shows a field of reduced stresses of the analysed part of the piping system. Fig. 6 gives a detail of the field of the reduced stresses at the site of the failure.

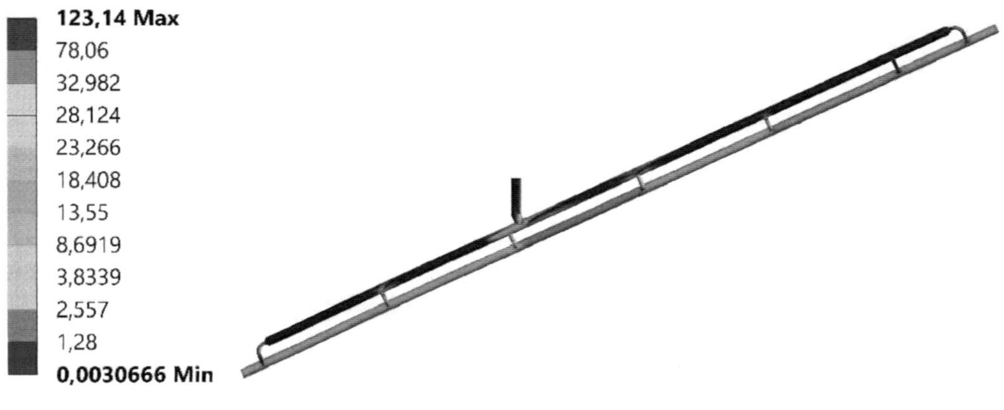

Fig. 5 Field of the von Mises stresses of the analysed part of the piping system (MPa).

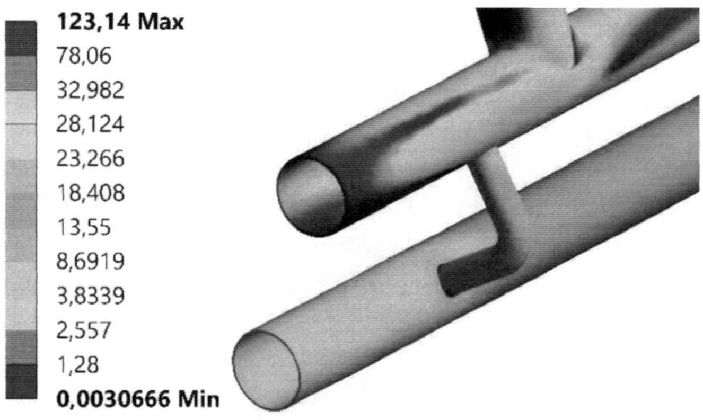

Fig. 6 Detail of the field of the reduced stresses at the failure point III (MPa).

As can be seen from the results of numerical modelling for a simplified model case, the most significant increase in stress levels occurred in the interconnecting pipeline at point III (Fig. 4). The maximum value of the reduced stress corresponded to the failure point in the part of the interconnecting pipe in that location (Fig. 4). Increased levels were also identified in other locations by the FEM, where cracks in the welded joints also occurred. It should be noted that despite the good agreement of the results of numerical modelling with the locations of weld failure, the achieved numerical values of stresses cannot be considered accurate, because it was a simplified model case. It is a qualitative rather than a quantitative analysis.

4 Residual stress quantification

When assessing the lifetime and reliability of structures, it is important to know the levels of the residual stresses often associated with their failure or damage. The state of residual stress in supporting structure elements is the key factor of its behaviour in service. Therefore, knowledge of residual stress may predict signs of component failures and thus prevent and avoid accident. The residual stresses quantification by the hole drilling method was performed on a sample taken from the damaged piping system (Fig. 7). The drilling was performed by SINT MTS 3000 drilling machine (Fig. 8). The drilling process was performed in twenty steps for a blind hole with the total depth 2 mm in accordance with ASTM E-837-13a Standard [6]. The hole diameter was 1.6 mm. After each step, the magnitude of the released strains was recorded by the SPIDER measuring device with the strain gauge RY61-1.5/120S. Fig. 9 shows the drilling process on the inside of the pipe. The residual stresses and the directions of principal normal stresses were determined using Eval 7 program according to ASTM E 837-13a Standard. The view on all locations of strain gauges is given in Fig. 9.

Fig. 7 Location of the sampling from the damaged piping system.

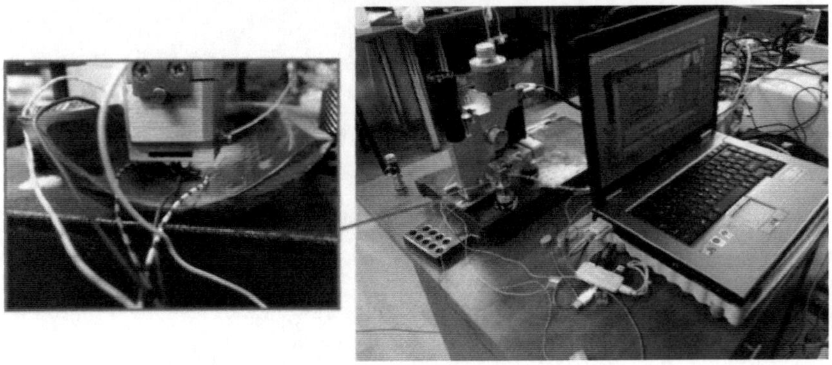

Fig. 8 Drilling process on the inner side of the pipe.

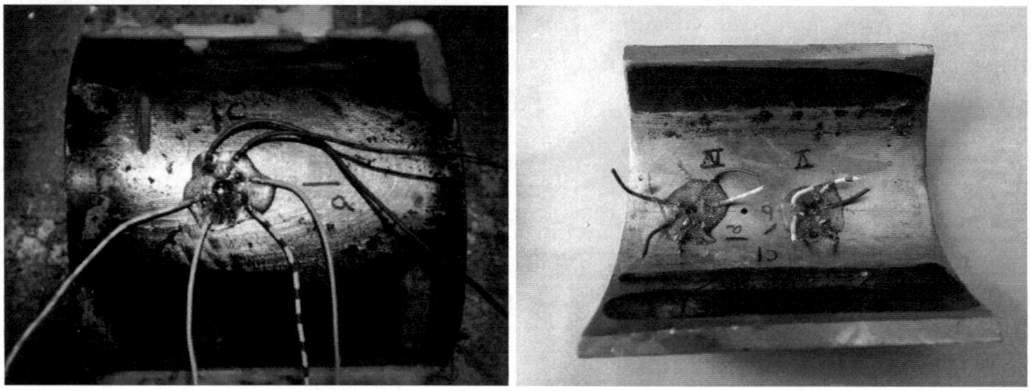

(a) (b)

Fig. 9 Locations of strain gauges: (a) outer side, (b) inner side.

Table 1 gives values of the residual stresses and their directions for each measurement point determined according to ASTM E 837-13a Standard. The angle φ represents the deviation of the direction of the maximum principal stresses from the axis parallel to the pipe axis.

Table 1. Magnitudes of residual stresses and their directions according to ASTM E 837-13a Standard.

Measured point	Tube side	σ_{max} (MPa)	σ_{min} (MPa)	φ (°)	σ_{red} (MPa)
I	external	44,19	-48,47	-82,77	80
IV	internal	347,26	169,74	42,84	300
V	internal	246,14	126,01	57,61	213

High tensile stress levels were found by measuring residual stresses in the selected locations. Based on the above we can state that the values of the reduced stresses according to von Mises reached around 300 MPa, thus, achieving the yield strength of the used material. Despite the measured values of the residual stresses being relatively high, under normal operating conditions even such stress levels would not cause a steam boiler failure considering the fact that the steam boiler has been in operation for about 4 years. As a result of a sudden change in the temperature in the piping system due to the failure of the control unit, the resulting superposed stresses exceeded the permissible values, which led to the failure of the welded joint. All the above-mentioned shows that the high levels of the residual stresses in combination with the emergency condition had the highest impact on the failure.

Conclusion

In the event of damage in welded steel structures, a comparative analysis of the material taken from the failure location with the new material is often used to determine the causes. In this case, however, no new material was available to the plant operator. For this reason, the authors proposed a methodology for assessing the further safe operation of the steam boiler on the basis of available data, i.e. supplied drawing documentation, description of the accident, material taken from the place of failure.

The numerical modelling on a model case was performed, where:

- The boundary conditions were defined on the basis of available information and the material properties of the model were defined according to the drawing documentation.

- The effect of temperature change was defined as the unit load, as the information on the temperature in individual parts of the steel structure is not registered by the operator.
- The good agreement of the results of numerical modelling with the locations of weld failure. However, the achieved numerical values of stresses cannot be considered accurate, because it was a simplified model case. It is a qualitative rather than a quantitative analysis.

Although the determination of residual stresses is currently well known, the authors consider their determination to be a quick tool to assess the current state of the structure after an accident. The main advantage is that, in addition to production technology, assembly inaccuracies and the like, they also take into account stresses from previous operation. Based on analysis of the residual stresses can be concluded:

- Despite the fact that the measured values of residual stresses are relatively high (300 MPa), under normal operating conditions even such a stress level would not cause a steam boiler accident, which is documented by the fact that the steam boiler has been in operation for about 4 years.
- As a result of a sudden change in the temperature in the piping system due to the failure of the control unit, the resulting superimposed stresses exceeded the allowed values, which led to the failure of the welded joint.
- However, high levels of residual stresses are determined, it can be stated that further operation of the device is not safe, or that the assessment will require further in-depth analyses, the implementation of which is often more time-consuming and costly.

The paper showed the steam boiler failure based on the measurement of the residual stresses in locations of the damaged structure, in this case of the piping system. Detection of possible causes by analytical or numerical modelling in such cases is almost impossible as it depends on many unknown factors entering the calculation. From the above conclusions as well as taking into account the previous trouble-free operation of the steam boiler, it can be stated that after the repairs of the identified cracks, the steam boiler can be operated safely.

Acknowledgements

This research was funded by Scientific Grant Agency of the Ministry of Education and Science of the Slovak Republic and the Slovak Academy of Sciences, grant numbers VEGA 1/0500/20 and VEGA 1/0516/22.

References

[1] Taler, J. and Dzierwa, P. "A new method for optimum heating of steam boiler pressure components," *International Journal of Energy Research*, **35** (10), pp. 897-908 (2016).

[2] Prabu, S.S., Choudhary, A., Mittal, N., Gupta, S., Ramkumar, D. and Natarajan, A. "Failure evaluation of SA 210C riffle water wall tubes in 70 MW CFBC boiler," *Engineering Failure Analysis*, **95**, pp. 239-247 (2019).

[3] Duarte, C.A., Espejo, E. and Martinez, J.C. "Failure analysis of the wall tubes of a water-tube boiler," *Engineering Failure Analysis*, **79**, pp. 704-713 (2017).

[4] Ahmad, J., Purbolaksono, J. and Beng, L.C. "Thermal fatigue and corrosion fatigue in heat recovery area wall side tubes," *Engineering Failure Analysis*, **17**, pp. 334-343 (2010).

[5] Duda, P., Felkowski, L. and Dobrzański, J. "An analysis of an incident during the renovation work of a power boiler superheater," *Engineering Failure Analysis*, **57**, pp. 248-253 (2015).

[6] ASTM E387-13a Standard test method for determining residual stresses by the hole-drilling strain-gage method, 2013. Annual Book of ASTM Standards American Society for Testing and Materials West Conshohocken.

Increasing the Carrying Capacity of Anchor Bolts by Design Modification of the Nut

Miroslav Pástor[1,a], Pavol Lengvarský[1,b], Martin Hagara[1,c], Alžbeta Sapietová[2,d] and Štefan Gašpar[3,e]

[1]*Department of Applied Mechanics and Mechanical Engineering, Faculty of Mechanical Engineering, Technical University of Košice, Letná 1/9, 042 00 Košice-Sever, Slovakia;*
[2]*Department of Applied Mechanics, Faculty of Mechanical Engineering, University of Žilina, Univerzitná 8215/1, 010 26 Žilina, Slovakia;*
[3]*Department of Design and Monitoring of Technical Systems, Faculty of Manufacturing Technologies, Technical University of Košice, Štúrova 31, 080 01 Prešov, Slovakia;*

[a]*miroslav.pastor@tuke.sk;* [b]*pavol.lengvarsky@tuke.sk;* [c]*martin.hagara@tuke.sk,* [d]*alzbeta.sapietova@fstroj.uniza.sk,* [e]*stefan.gaspar@tuke.sk*

Abstract: Prestressed bolted connections are commonly used for anchoring machinery and equipment, where prestress is induced by mechanically tightening the nut to the required torque. To ensure safe and reliable operation, the prescribed tightening procedure must be followed, especially in the case of anchor bolts embedded in a concrete foundation. The paper deals with the problem of anchoring a new casting stand using the original anchor bolts subjected to cyclic loading during rotation of the casting stand base. The aim was to propose a methodology to verify the reliable and safe operation of the original anchoring system, taking into account the current condition of the bolts. In order to assess the actual condition of the anchorage, as well as to determine the necessary prestressing, the authors proposed the use of a modified nut with a lightened first thread, whose shape and dimensions were determined on the basis of the results of numerical modelling using the finite element method. The suitability of the proposed methodology was experimentally verified by carrying out measurements under different loading modes, which took into account the actual operating conditions. The interpretation of the obtained results, including a comprehensive view on the safety of the anchorage, provides relevant evidence of the functionality and effectiveness of the proposed anchorage solution for the new casting rack.

Keywords: Prestressed bolts; Operational measurement; Stress analysis

1 Introduction

Numerous standards deal with the design and connection of bolts (anchor bolts) with concrete [1,2]. In recent years, anchor bolt failures have been studied extensively in many ways, including experimental testing, numerical simulations, and verifications. Delhomme et al. [3] studied the experimental behaviour of anchor bolts under static tensile loading and also analysed the relaxation of these bolts. Li and Liu [4] studied the shear behaviour of the fully grounded bolts through experimental testing, during which the direct shear tests were carried out on three groups of specimens with different bolting angles or grout strengths, measuring shear loads, shear displacements, and strains in bolts. They observed that the bolts failed near the concrete foundation. Gong et al. [5] studied the failure analysis of anchor bolts used to fix a seawater booster pump in a nuclear power plant. They concluded that the failure was due to screw corrosion. The failure modes of anchor bolts under cyclic loading were investigated by Feng et al. [6]. They analysed the mechanical responses of bolts with different diameters and applied different cyclic loads. The behaviour of the bolted joints is also affected by the preload in the bolt. The preloaded anchor bolts were investigated and described in paper [7]. To investigate the bolts' inner force quantities and displacement,

Huang et al. [8] analysed the influence of the anchorage length using the load transfer method. The obtained results confirmed that the increasing anchorage length caused significant unevenness of the interfacial shear stress distribution, and a greater bearing capacity of the anchoring bolt. Yang [9] investigated the mechanism of force evolution in bolts and found that bolts with different anchorage lengths should be used to support roadways with different deformation characteristics. His research concluded that the bolts should be kept in the deformation phase of the anchoring force as much as possible when the surrounding rock deforms. Liu et al. [10] proposed a stress reduction calculation method to calculate the actual stress of the bolt from the measured value registered by the pulse pre-pumped Brillouin optical time domain analysis technology of different types of bolts embedded in concrete with resin as an anchoring agent. However, dynamic loading also affects the lifespan of the bolt joint [11,12]. Trebuňa et al. [13] considered the possibilities of reducing the vibration amplitudes of the converter base through design changes and changes in the pre-stress of the screw connections. However, the high stress in anchoring bolts, among other things, causes serious deformation and damage of equipment, as well as steel constructions [14-16]. Lin and Wu [17] investigated the maximum axial force in bolts, employing conservative failure probability assumptions made by the finite element software. Using numerical simulations performed in Matlab, Zheng et al. [18] analysed the influence of different anchoring methods, changes in surrounding rock properties, bolt properties, and pre-tightened forces on the pre-stress distributions in the bolt body and surrounding rock. Miao [19] reviewed methodologies concerning bolt tightening force measurements and monitoring for loose bolts.

The paper deals with anchor bolts embedded in concrete, which are part of the existing technological chain. For this reason, it is not possible to perform laboratory measurements. This problem was solved by using a modified slotted nut to relieve the first thread, which is the most stressed and is the main cause of failure of the bolted joint. The modification of the slotted nut is primarily associated with an increase in the dimensions of the nut, i.e. its height and external dimensions have been increased.

2 Introduction to the anchoring of the casting pedestal

The continuous casting of slabs is currently the most widely used method, in which the solidification of steel leads to the production of long steel products. Its fundamental principle is that the liquid steel is first infused from a ladle into a tundish and then distributed through many sprues in a vertical direction into water-cooled copper moulds. The ladles, transported by an overhead bridge crane and a casting pedestal, are used to bring the liquid steel from the converter to the tundish. The casting pedestal is used for constraining the casting ladles and their rotation from the reserve position to the casting position, and vice versa. It also serves to drain the molten steel into the emergency containers in case of a failure of the slide cap, and for lowering and lifting the casting ladles during various technological operations. It is a two-position pedestal realised as a lifting and rotating pedestal with a load cell. The supporting structure of the casting pedestal is formed mainly by steel weldments. Their dimensions and shapes are designed in such a way that after the casting ladle filled with liquid steel is placed in the reserve position, the casting pedestal ensures its transfer around the axis of the casting pedestal to the casting position. After casting the steel from this ladle into the tundish to about 25% of the liquid steel content, the second full pan can be placed on the other/opposite traverse side (in the reserve position). If there is only molten slag left in the casting ladle (approximately 7000 kg), the pedestal is rotated so that the full ladle is moved from the reserve position to the casting position. Selected technical characteristics of the assessed casting pedestal (Fig. 1) are:

- Maximum weight of the casting ladle with metal (steel) 260 000 kg;
- Weight of empty ladle 70 000 kg;
- Ladle lift height 800 mm;
- Pedestal angular velocity 1 rpm;
- Maximum angle of rotation 230°;
- Roller track mean diameter 6 400 mm.

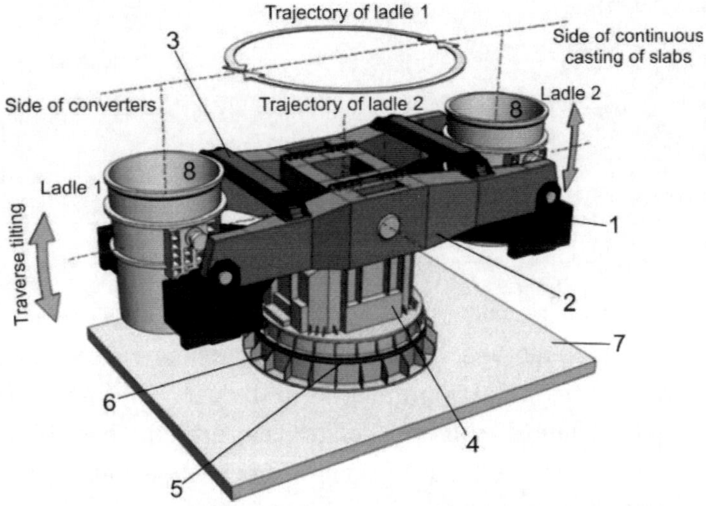

Fig. 1 Basic parts of the casting pedestal: 1—hinging supporting baskets; 2—traverse beams; 3—connecting rods; 4—bearing support pedestals; 5—roller bed; 6—supporting structure; 7—concrete foundation; 8—casting ladles.

As part of the overhaul of the casting pedestal, some supporting members were replaced, but replacement of the anchor bolts embedded in the concrete foundation was not possible. When the original structure was disassembled, damage to the first threads on the concrete foundation side of some of the anchor bolts was found. As already mentioned, their replacement was not possible, or very difficult. Therefore, it was necessary to design another solution to achieve safe operation at the required bolted joint preload. By numerical modelling, a design modification of the nut was proposed to lighten the first threads. Subsequently, experimental measurements were carried out, taking into account real operational conditions. Based on the results obtained, conclusions were drawn on the further operation of the casting pedestal. The value of the considered preload of 180 kN in the bolted joints was taken from the original static calculation. Fig. 2 shows the layout of three types of anchor bolts (34 anchor bolts in total) with basic dimensions, which are used to fix the casting pedestal to the concrete foundation.

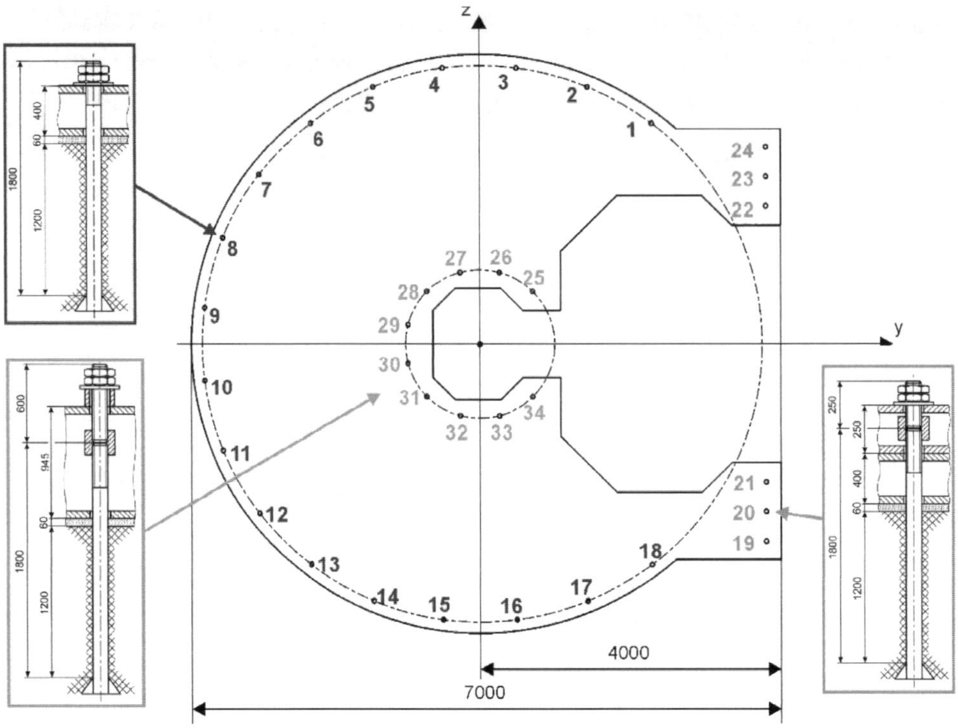

Fig. 2 Basic dimensions of anchor bolts and their layout.

It was necessary to design the methodology of re-anchoring the casting pedestal so as not to damage the bolted connections (releasing the bolts in the concrete foundation), the repair of which would be not only technically and financially very difficult, but also time-consuming. In addition to the machining method, the lifespan of the anchor bolts under investigation is influenced, in particular, by the impact of notched effects, which is most significant and unfavourable during the fatigue loading of machine parts and structures. In the case of the bolt and nut, according to Fig. 3, the decisive design notches are in the following places:

- Location 1—passage of the cylindrical part of the bolt to the head—this location is absent from the anchor bolts;
- Location 2—thread runout;
- Location 3—thread;
- Location 4—bolt thread at the first nut's support thread (the most common cause of failure).

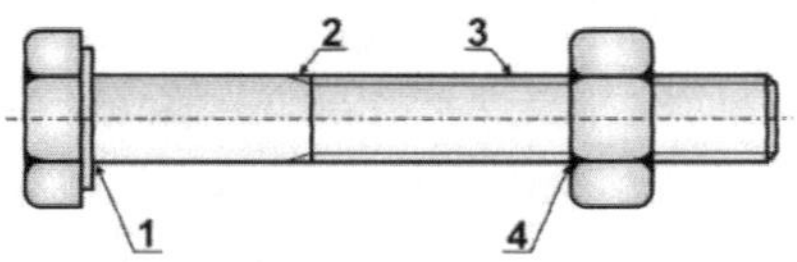

Fig. 3 Notch locations on the bolt.

The fatigue notch factor is most significant in the bolt thread at Location 4. This fact is related to the stress distribution in the bolt thread, which changes significantly in the part of the thread bolted into the nut. The change in stress distribution is caused by the uneven force distribution in the bolt on the individual threads (Fig. 4) due to the different deformations of the bolt and nut.

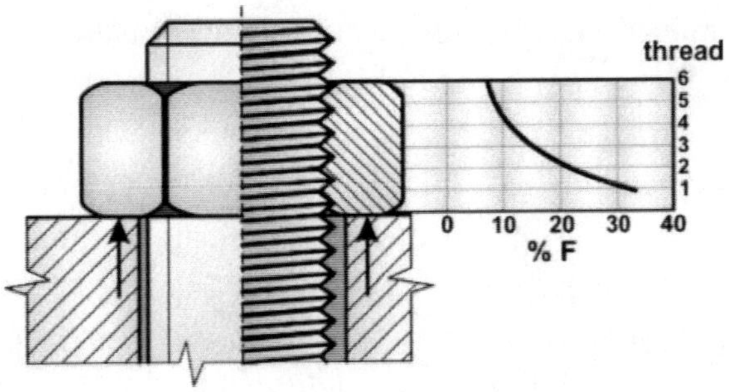

Fig. 4 The uneven force F distribution in the threads.

The first thread of the nut and the corresponding cross-section of the bolt are the most loaded. As a result, the first load-bearing thread exhibits the most significant notch effect, which is caused by the load from the force flow in the shank and the local bending of this most loaded thread. The fatigue notch factor for the threaded part of the bolt depends on the mechanical properties of the bolt, the technologies of its production, the shape of the thread profile, its surface, and the type and height of the nut. From the tests of bolted joints in which the fracture occurs at Location 4 (Fig. 3), it is possible to derive a four- to eight-fold notch effect.

In this paper, a methodology for achieving the required bolted joint preload is proposed and experimentally verified, taking into account the damage to the threads on the bolt. The proposed design modification of the nut achieved the relieving of the first threads on the bolt, thus ensuring the safe and reliable operation of the casting pedestal anchoring.

3 Materials and methods

The aim was to design and experimentally verify a methodology for applying the required preload in the anchor bolts of a casting pedestal that had been in operation for less than 30 years. The authors assumed real operational conditions where it was impossible to perform the test tightening of nuts exceeding the allowable values of axial forces in the bolts, as they could not be released from the concrete foundation. In such a case, the new casting pedestal, and thus the entire production, would be undesirably shut down. When proposing the methodology, the authors utilised proven knowledge obtained from the failure of bolted joints, where in similar cases, the damage most often occurred at the place of the first load-bearing thread (see Fig. 3). Nuts with notches were designed and subsequently fabricated using numerical modelling based on the finite element method to achieve the required preload. The notch was used to provide a relieving of the first threads (Fig. 5b). A methodology for measuring the axial forces in the anchor bolts under the operational load was proposed and experimentally verified using dynamometers to verify the functionality of the bolted joint (set preload) and assess the safe operation of the casting pedestal. Before tightening the nuts, a strength calculation of the preloaded anchor bolts was performed to assess their safe operation, taking into account the loading history to date.

4 Proposal of nuts modification using finite element method (FEM)

To reduce the significant notch effect of the normalised thread as much as possible, it was necessary to ensure the best possible quality of its surface at the root of the profile, and to reduce the relative load on the first and second threads. This can be done through structural modifications based on the idea of changing the deformation stiffness of the nut around the first threads. By reducing the cross-sectional area of the nut (contact of the nut with the washer), the conditions for the bending expansion of the nut were created. In addition, when the normalised nut is tightened to higher torque, plastic deformations and the compression of the first co-locking threads on the nut and bolt occur. Depending on the magnitude of these undesirable phenomena, damage to the bolts occur, which in the case of anchor bolts are extremely technically demanding to replace. Therefore, the authors' attention was on achieving an optimum stress distribution of the co-collecting threads of the nut and bolt. Fig. 5a shows the model of the normalised nut M56, and Fig. 5b shows the nut with the notch formed. As the first threads were relieved, to create a notch with the required dimensions, it was necessary to change the outer dimensions and the total number of threads in the nut, as shown in Fig. 5b. The proposal of the outer dimensions was based on the dimensions of the torque wrench head (Enerpac HXD 120-CC1290).

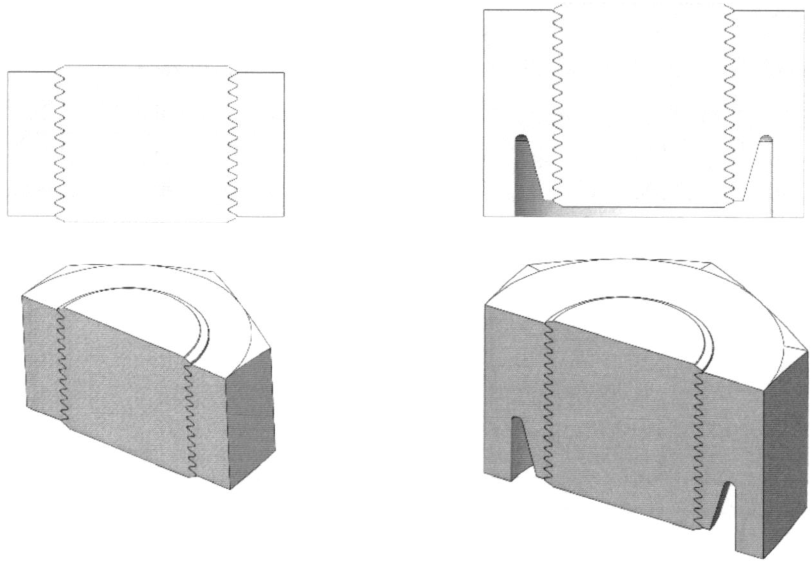

(a) (b)

Fig. 5 Sections through the analysed matrices: (a) normalised nut; (b) modified nut with the designed notch.

Fig. 6a,b show the bolted joint model with the finite element mesh formed and the boundary conditions defined. By using symmetry, only 1/12 of the overall model was modelled. The nut was simply supported and the bolt was loaded by the force equal to 15 000 N. The mesh of normalised nut M56 consisted of approximately 155 000 volume finite elements with quadratic approximations, and approximately 230 000 nodes. The mesh of the nut with the notch consisted of approximately 456 000 volume finite elements with quadratic approximations, and approximately 668 000 nodes.

The FEM was a qualitative comparison of stress levels for different nut shapes, so for this reason only a static analysis with linear elastic material behaviour was performed. The material properties of the bolt and nut were the same, i.e., the yield strength was 355 MPa and Young's modulus was $E = 2 \cdot 10^5$ MPa. Fig. 7a,b show details of the equivalent stress fields around the first relieved threads. From comparing the values of the maximum equivalent stresses around the thread root (from 379 MPa to 226 MPa), it is clear that the above modification resulted in a reduction of equivalent stress by approximately 40%. When designing the dimensions and shape of the notch, the technological procedure of its production was taken into account.

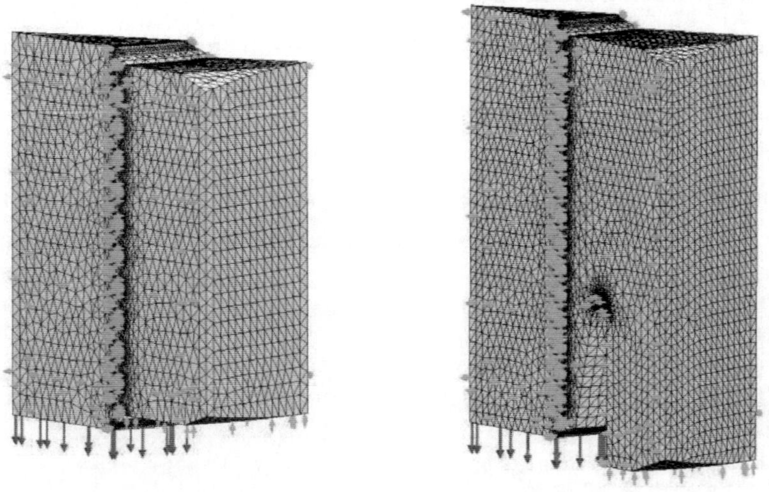

a) b)

Fig. 6 1/12 of the overall model: (a) normalised nut; (b) designed nut.

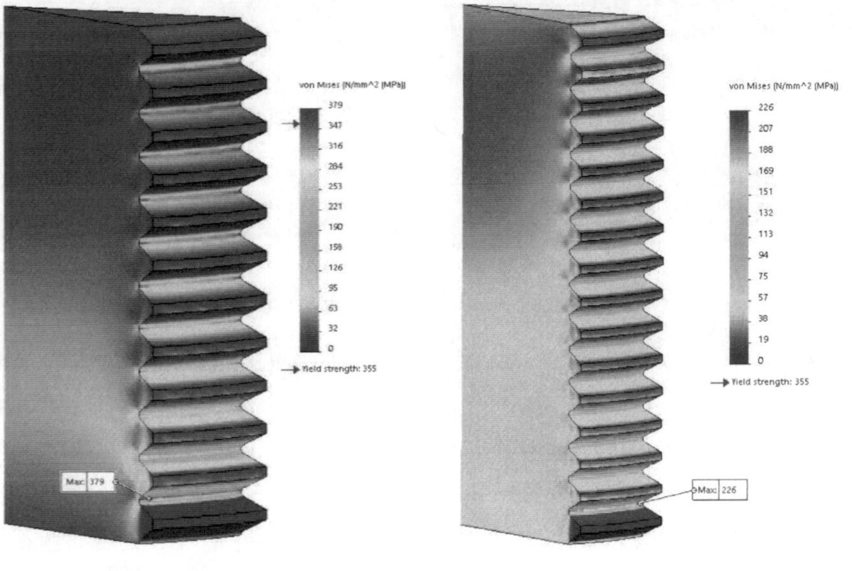

(a) (b)

Fig. 7 Fields of equivalent stress around the root of the first thread on the anchor bolt when using: (a) normalised nut; (b) designed nut.

Fig. 8 shows a top and bottom view of the modified nut proposed by the authors.

Fig. 8 Top and bottom view of the modified nut.

5 Measurement at operational loading

As already mentioned in the introduction, the first step was to set the required value of the axial force in the prestressed bolted joints. For the adjustment, prototype dynamometers designed by the authors were used, see Fig. 9, Fig. 10a. The dynamometer was designed in such a way that the responding magnitude of the axial force in the bolt induced by tightening the nut could be determined based on its deformation measured by strain gauge sensors. The same dynamometers were used in the second stage for the operational measurement of axial forces during the manipulation of the casting stand. The most stressed bolts No. 8, No. 11, No. 19 and No. 24 (see Fig. 2) were chosen for the operational measurements. The authors had prepared four dynamometers, but at the beginning of the operational measurements one of them was damaged by the operator. Therefore, the operational measurements continued on three anchor bolts.

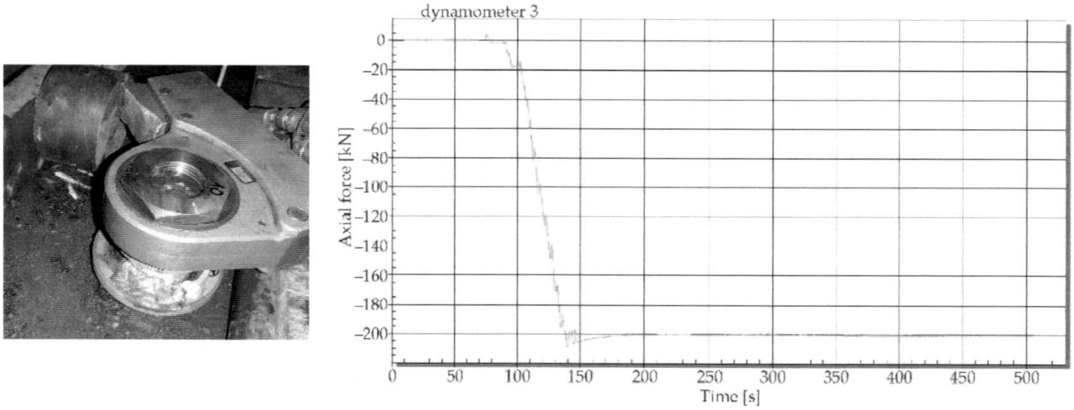

(a) (b)

Fig. 9 Setting the required preload value (200 kN) in the anchor bolt No. 10.

After a detailed analysis of the condition of the anchor bolts, as well as of the measures taken to relieve the first and the other three bolt threads at the nut location, and also taking into account the previous history of the loading of the anchor bolts, the authors determined the preload forces in the individual anchor bolts as follows:

- Anchor Bolts No. 1 to No. 24 should be preloaded to an axial force corresponding to 200 kN, which corresponds to a given state of bolts, nuts, and washers when defining a pressure value of 120 bar;
- Anchor Bolts No. 25 to No. 34 should have a preload with regard to the passage of the thread into the shank as well as the flexibility of the supporting structure, and a lower range of operational axial forces of 180 kN, i.e., approximately 110 bar.

Dynamometers 1 to 4 were used to measure the time changes of the axial forces in the anchor bolts on the casting pedestal. When selecting the anchor bolts under consideration, effort was made to remove the most, or nearly the most loaded bolts. Based on the above requirements, anchor Bolts No. 8, No. 11, No. 19 and No. 24 were selected, which are located on the outer contour of the supporting structure, therefore the preload value was selected at 200 kN. After putting the casting pedestal into testing operation and balancing the strain gauge apparatus connected to the completed measuring chain, the strain gauge measurement of changes in the axial forces in the anchor bolts began. Unfortunately, the first measurements showed that the cabling on Dynamometer 4 was damaged by the operating personnel. Due to the time duration of the planned outage of the casting pedestal, it was not possible to perform its repair. Therefore, the operational measurement was continued by using only three dynamometers.

The first operational measurements, which were performed only when rotating the casting pedestal without the ladle in the first and second basket, showed that the changes in the axial forces in the bolts did not exceed 6 kN in these measurements. Fig. 10 shows the time courses of changes in the axial forces in Bolts No. 8 (dyn 3), No. 19 (dyn 1), and No. 24 (dyn 2) when rotating the casting pedestal without a ladle in Basket 1 and Basket 2, while the beginning of rotation at approximately 2000 s is from the position of Basket 1 from the reserve to the casting position. After a short interruption of the rotation, the rotation continues and the traverse is lifted. It can be seen from the course that the axial force increments do not return completely to zero after the return of Basket 1 to the reserve position (time 5000 s), which may be caused partly by inaccuracies in the return of the traverse to the same position, and also by the changed geometry due to the position of the rollers on the roller track.

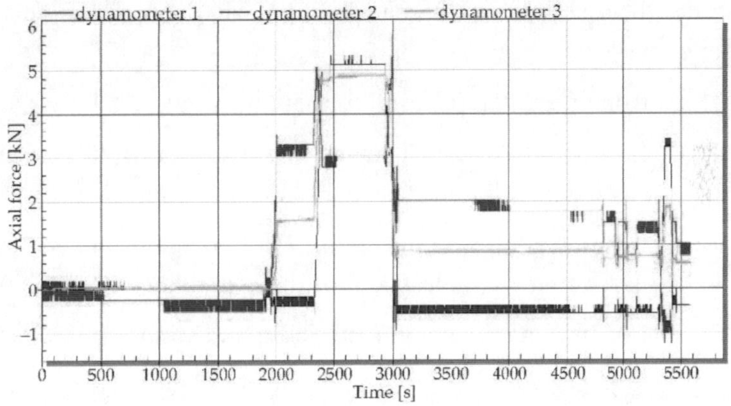

(a) (b)

Fig. 10 Recording of time changes of axial forces when rotating the casting pedestal with empty baskets.

Although these are forces that cause stress in bolts below 3 MPa, i.e., absolutely negligible components in terms of the resulting stress, they testify not only to the condition of the roller track but also to the stiffness of the supporting structure of the casting pedestal.

The following section shows the selected time changes of the axial forces in Bolts No. 8, No. 19, and No. 24. The operator used ladles weighing 70 300 kg (empty) and 241 600 kg (full). Fig. 11 shows the time changes of the axial forces when loading the empty ladle into Basket 1 in the reserve position at about 50 s after the start of the measurement. It can be seen that a positive increase in axial forces occurred in Bolts No. 19 and No. 24, which is caused by the fact that, after loading the empty ladle, the preload in Bolts No. 19 and No. 24 is reduced, and this is reflected by an increase in the deformation part of the dynamometer, resulting in a positive reading on the strain gauge apparatus. Simply put, if the preload in the bolts were not set, it would not be possible to register a decrease in the axial force. At a time of about 100 s, the rotation of the casting pedestal with an empty ladle in Basket 1 started from the reserve position to the casting position, which caused changes in the polarity of the force increments in the bolts when Basket 1 was moved to the casting position. It is evident from the courses that in this operation the axial force range in Anchor Bolt No. 8 is 25 kN, in Anchor Bolt No. 19 it is 13 kN, and in Anchor Bolt No. 24 it is 19 kN. At the end of the time period (Fig. 11), Basket 1 with an empty ladle was in the casting position.

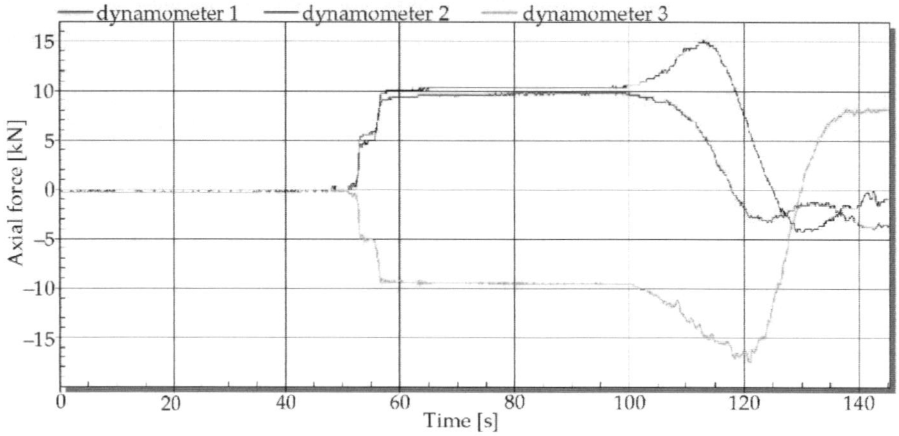

Fig. 11 The time courses of axial forces in Bolts No. 8, No. 19, and No. 24 when loading the empty ladle into Basket 1 and rotating the pedestal to the casting position.

In order to assess the safe operation of the casting pedestal, the following operational modes were examined:

I. Rotation of the pedestal without a ladle in Baskets 1 and 2;

II. Loading the empty ladle with subsequent rotation, and lifting of the ladle in the casting position in Basket 1 and subsequently in Basket 2;

III. Loading a full ladle weighing 241 600 kg into Basket 1 and subsequently into Basket 2, rotating the pedestal, and lifting and launching the basket with the ladle;

IV. Loading the empty and full ladle and handling the pedestal, including the replaced ladles in baskets.

After analysing the time records of changes in the axial forces in the bolts in the above-mentioned operational modes of the pedestal, it was found that the maximum range of changes in the axial forces ΔN_i in the measured bolts is the following: $\Delta N_8 = 61$ kN, $\Delta N_{19} = 60$ kN, $\Delta N_{24} = 61$ kN.

Conclusion

At the outset it should be noted, that as mentioned above, the operational measurements of the time changes of the axial forces were carried out with an empty ladle weight of 70 300 kg and a full ladle weight of 241 600 kg. In the technical characteristics of the casting pedestal defined by the operator, the values of an empty ladle at 70 000 kg and a full ladle at 260 000 kg were defined, which represents a difference of 0.43% in the case of an empty ladle, but in the case of a full ladle its maximum possible weight is 7.1% higher, thus the stress and strain fields in the elastic behaviour of the structure will be higher according to the stated difference, including axial forces. Therefore, in the subsequent analysis, it was necessary to consider their increase by 7.1% at the measured maximum values of an axial force range of 61 kN, i.e., with increased range $\Delta N_{\max} = 65$ kN.

From the analysis of the critical points on the anchor bolts, it can be seen that the replacement of the original nuts with new nuts played a very important role, as the first thread of the bolt in the nut is more than twice as sensitive to the occurrence of a failure as the thread termination point. The anchor bolt's failure limit is determined by a location which, despite the frequent occurrence of the failure, is a much safer cross-section.

The experimental determination of axial force ranges has shown that the calculation of the forces in the anchor bolts ensures a high value of anchor safety. On the other hand, the force measurement methodology and the interpretation of the results achieved, including a comprehensive view of anchoring safety, provides relevant evidence of the functionality and effectiveness of the proposed solution.

In order to ensure safe operation of the anchoring during the lifespan of new parts for the casting pedestal, and taking into account the accumulation of damage to the anchoring bolts to date, it is necessary to apply the current recommendation of the authors that, when placing a full ladle in the casting pedestal and when rotating it, there should always be an empty ladle or a ladle with molten slag in the other basket. In accordance with the fatigue curve for the material values of the bolt and in the implementation of 35 melts per day, the lifespan of the bolts of the casting pedestal will be approximately 10 years. Otherwise, if a full ladle is always loaded into the casting pedestal and there is not at least an empty ladle on the other side, the lifespan of the anchoring will be reduced to about one third.

Acknowledgements

This paper was supported by the Ministry of Education of Slovakia Foundation under grant projects VEGA No. 1/0516/22, VEGA No. 1/0141/20 and VEGA No. 1/0509/23.

References

[1] EN 1993-1-8:2005—Eurocode 3: Design of Steel Structures Part 1-8: Design of Joints. *Available from* https://standards.iteh.ai/catalog/standards/cen/312c51d1-0813-4e29-b1dd-5db5f15e2497/en-1993-1-8-2005 *Accessed:* 2021-06-21.

[2] Design Guide 1: Base Plate and Anchor Rod Design, 2nd, ed.; American Institute of Steel Construction. *Available from* https://www.aisc.org/Design-Guide-1-Base-Plate-and-Anchor-Rod-Design-Second-Edition *Accessed:* 2021-06-21.

[3] Delhomme, F., Debicki, G. and Chaib, Z. "Experimental Behaviour of Anchor Bolts under Pullout and Relaxation Tests," *Constr Build Mater*, **24**, pp. 266-274 (2010).

[4] Li, Y. and Liu, C. "Experimental Study on the Shear Behavior of Fully Grouted Bolts," *Constr Build Mater*, **223**, pp. 1123-1134 (2019).

[5] Gong, Y., Ding, Q. and Yang, Z.G. "Failure Analysis on Premature Fracture of Anchor Bolts in Seawater Booster Pump of Nuclear Power Plant," *Eng Fail Anal*, **97**, pp. 10-19 (2019).

[6] Feng, X., Zhang, N., Yang, S. and He, F. "Mechanical Response of Fully Bonded Bolts under Cyclic Load," *Int J Rock Mech Min Sci*, **109**, pp. 138-154 (2018).

[7] Wang, D.S. and Liu, A.Q. "Study on the Performance of Full-Length Pre-Stressed Anchor Bolts," *Adv Mater Res*, **639-640**, pp. 765-769 (2013).

[8] Yang, S.S. and Cao, J.P. "Evolution mechanism of anchoring stress and its correlation with anchoring length," *J Min Saf Eng*, **27**, pp. 1-7 (2010).

[9] Liu, Q., Chai, J., Chen, S., Zhang, D., Yuan, Q. and Wang, S. "Monitoring and Correction of the Stress in an Anchor Bolt Based on Pulse Pre-Pumped Brillouin Optical Time Domain Analysis," *Energy Sci Eng*, **8**, pp. 2011-2023 (2020).

[10] Huang, M.H., Zhao, M.H. and Chen, C.F. "Influence of anchorage length on stress in bolt and its critical value calculation," *Rock Soil Mech*, **39**, pp. 4033-4041 (2018).

[11] Sapieta, M., Sapietova, A. and Gajdos, L. "Determine the Fatigue Life of Flange of Bearings Test Station," *Procedia Eng*, **177**, pp. 548-553 (2017).

[12] Wen, Z., Shi, Y., Cui, Z.D. and Wang, R.C. "Study of Stress Features of Fully Grouted Prestressed Anchors," *Rock Soil Mech*, **31**, pp. 177-181 (2010).

[13] Trebuňa, F., Šimčák, F. and Bocko, J. "Decreasing of Vibration Amplitudes of the Converter Pedestal by Design Changes and Changes in Prestress of the Bolted Joints," *Eng Fail Anal*, **16**(1), pp. 262-272 (2009).

[14] Choi, J. and Kim, B. "Failure Analysis of Anchor Bolt of Rail Fastening System for Direct Fixation Track," *Eng Fail Anal*, **112**, 104513 (2020).

[15] Tizani, W., Rahman, N.A. and Pitrakkos, T. "Fatigue Life of an Anchored Blind-Bolt Loaded in Tension," *J Constr Steel Res*, **93**, pp. 1-8 (2014).

[16] Shen, X., Lu, L. and Zeng, D. "Fatigue Failure Analysis of High Strength Bolts Used for High-Speed Railway Vehicle Braking Discs," *Eng Fail Anal*, **115**, 104661 (2020).

[17] Lin, S.R. and Wu, W.F. "Estimation of Maximum Axial Force of Anchor Bolts in Consideration of Random Bolt Failures," *Int J Press Vessel Pip*, **131**, pp. 52-59 (2015).

[18] Zheng, X.G., Zhang, N. and Xue, F. "Study on stress distribution law in anchoring section of prestressed bolt," *J Min Saf Eng*, **29**, pp. 365-370 (2012).

[19] Miao, R., Shen, R., Zhang, S. and Xue, S. "A Review of Bolt Tightening Force Measurement and Loosening Detection," *Sensors*, **20**, 3165 (2020).

Effect of Si-Based Waste Material on the Final Mechanical Properties of Cement Pastes

Zdeněk Prošek[1,a], Pavel Tesárek[1,b] and Aleš Palička[1,c]

[1]*Department of Mechanics, Faculty of Civil Engineering, CTU – Czech Technical University in Prague, Thákurova 7, 166 29 Prague, Czech Republic;*

[a]*zdenek.prosek@fsv.cvut.cz*, [b]*pavel.tesarek@fsv.cvut.cz*, [c]*palicale@student.cvut.cz*

Abstract: The article focusses on the effects of waste Si-based material on the mechanical properties of the resulting cement pastes. In this study we use Portland cement CEM I 42.5R and Si-based waste materials from the packaging industry from Recifa, a.s. The difference between individual samples was in the concentration of Si-based materials. We investigate the influence of five different amounts of SiO_2-based materials, namely, 10 wt. %, 20 wt. %, 30 wt. %, 40 wt. % and 50 wt. %. The determined mechanical properties were compressive and flexural strengths. Mechanical properties were investigated in 7 and 28 day samples with dimensions equal to 40 × 40 × 160 mm. The results showed a negative effect of Si-based materials on mechanical properties.

Keywords: Waste Si-based materials; Cement pastes; Mechanical properties

1 Introduction

Sustainability of concrete depends on its composition and origin of its components, being in general a binder, and fillers. Depletion of these natural resources hinders the sustainability of concrete structures. One of the possible strategies to mitigate the environmental impacts of concrete production is replacing quarried fillers with construction and demolition waste (CDW). The efficient utilization of CDW has become a major topic of research in civil engineering during the last decade [1]. The incorporation of larger CDW pieces (mostly aggregates, masonry, asphalt pavement, glass) has become a common practice in the construction industry, but the utilization of subsieve fractions (<0.250 mm) of silicon (Si) based materials has not been established despite the common use of microfillers, such as microsilica [2]. Si-based waste is one of the major environmental problems worldwide, because glass can be taken as non-biodegradable, when landfilled can trigger severe environmental pollution, and large deposits are not economically feasible. From the point of view of origin, it is further divided into packaging waste industry, construction industry, automotive industry, energy industry, etc. Each source brings with it different sources of micro-level additives and impurities that are likely to react in cement composites.

The use of waste Si-based materials as alternative fillers to concrete have attract interest of different researchers. However, when added in relevant amounts, the addition of these secondary materials is associated with deterioration of mechanical strength due to the formation of ITZ [3]. It turned out that more than 20 wt.% replacements of raw materials with the Si-based secondary fillers are not suitable for cementitious composites [4]. Smaller amounts are sufficient to fill pores [5] while disrupting the microstructure due to the formation of ITZs at an acceptable level. Waste Si-based materials are not optimal – their specific surface is relatively large, multiplying the negative effects. But the effects are always hard to estimate or predict, since micro-level additives exhibit complex interactions with the grains of binder, that are often larger and become coated by the filler particles. The ITZs usually formed due to transport of chemical elements [6, 7].

The present paper describes the basic effects of Si-based waste materials on a pure cement paste formed from Portland clinker cement.

2 Materials and samples

The test samples were composed of Portland cement CEM I 42.5R (Heidelberg Cement Czechia - Radotín) and different amounts of waste silica (SiO_2), which comes from waste materials from the packaging industry from Recifa, a.s. CEM I 42.5R had grain sizes up to 180 microns and contained large amounts of allite (C_3S) 74.6% and smaller amounts of other clinker minerals (C_2S 7.2%, C_3A 8.1%, C_4AF 8.5%). In contrast, the waste silica had a grain size of up to 80 microns, and therefore, the effect on the workability of the fresh mixture was investigated. Five concentrations of SiO_2 were tested: 10 wt. %, 20 wt. %, 30 wt. %, 40 wt. % and 50 wt. %. The samples were compared with reference samples composed of Portland cement only.

The amount of water was the same in all mixtures. The experiment investigated the possibility of substituting cement. Therefore, the water to binder ratio (w/b) was the same in all fresh mixes at 0.35. Compared to that, the water to cement (w/c) ratio varies from 0.37 to 0.7. The workability of the fresh mixtures was determined by flow table test during the sample production. The composition of individual mixes and the workability of fresh mixtures can be seen in Tab 1. Each set consisted of 6 samples with dimensions equal to 40 × 40 × 160 mm. The next day after production, the test samples were demoulded and stored in a water bath at temperature of 22±5 °C.

Table 1. Composition of the testing samples.

Set	CEM I 42.5R [kg]	SiO_2 [kg]	Water [kg]	Water/cement ratio [-]	Flow test [mm]
REF	3.0	0.0	1.05	0.35	205±5
S10	2.7	0.3	1.05	0.39	210±5
S20	2.4	0.6	1.05	0.44	205±10
S30	2.1	0.9	1.05	0.50	215±1
S40	1.8	1.2	1.05	0.58	205±1
S50	1.5	1.5	1.05	0.7	195±5

3 Experimental methods

Flexural and the compressive strengths were determined using the Web Tiv Ravestein devices on 7- and 28-days old samples. The resulting value was determined as the average of values excluding the lowest and highest values, and the standard deviation was determined.

The flexural strength was determined by a three-point bending test. The test is performed on a beam with dimensions of 40 × 40 × 160 mm supported on two supports 100 mm apart, which is loaded by force in the centre of the beam. Flexural strength was determined from the relationship between the maximum force achieved and the dimensions of the beam at the fracture point. Three samples of each mixture were tested. The testing was displacement controlled at a constant rate of 1 mm/min.

The compressive strength was determined by uniaxial compressive test on the half samples with dimensions of 40 × 40 × about 80 mm (broken parts of samples after bending test). The loading area was 40 × 40 mm. Six samples of each mix were tested. The testing was displacement controlled at a constant rate of 5 mm/min.

4 Results and discussion

The results show the effect of waste Si-based materials on the flexural strength (Fig. 1) and the compressive strength (Fig. 2) of cement pastes. The results show the negative effect of SiO_2 waste on the seven-day flexural strength and compressive strength of cement pastes. In the case of 7-day strengths 50 wt. % SiO_2, there was a decrease of 33 % in flexural strength and 58 % in compressive strength compared to the reference sample.

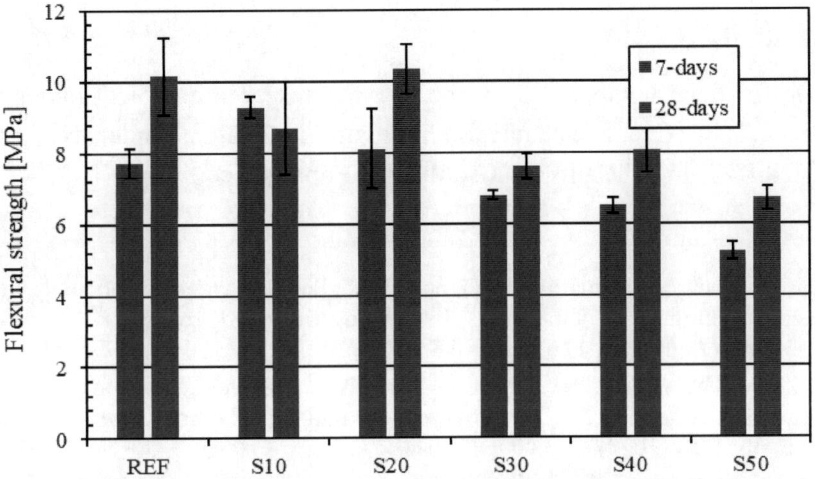

Fig. 1 Comparison of flexural strength (with standard deviations).

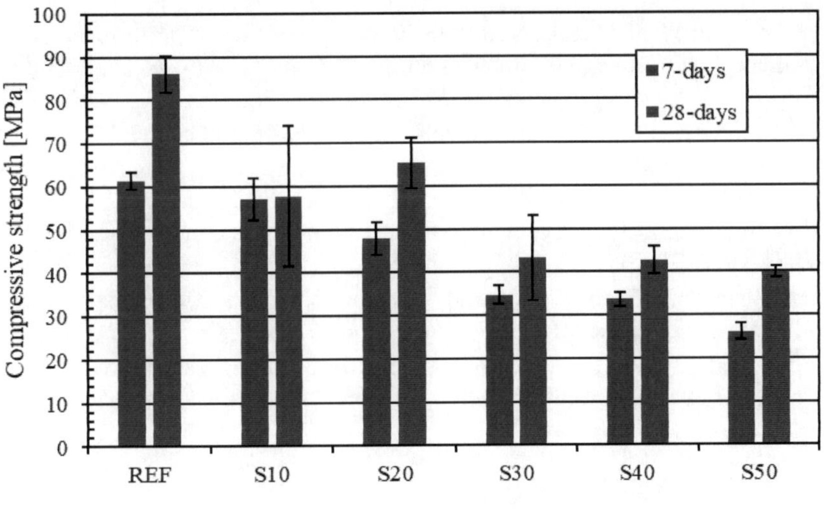

Fig. 2 Comparison of compressive strength (with standard deviations).

In the case of 28-day strengths, the difference in values is the same as in the case of 7-day strengths (flexural strength decreased by 34 %, compressive strength decreased by 54 %), which shows that the waste Si-based materials fulfills the function of a filler and does not participate in hydration [8].

Conclusions

This work was focused on the effects of waste Si-based materials on mechanical properties of resulting cement pastes. The tested samples were composed of Portland cement CEM I

42.5R and waste SiO_2 materials from the packaging industry from Recifa, a.s. The difference between individual samples was in the concentration of SiO_2. Based on the results, it can be concluded that the introduction of inert SiO_2 results in a less compact structure of cement pastes, leading to deterioration of the flexural strength and the compressive strength. A linear dependence of the decrease in strength with increasing amount of SiO_2 can be seen.

In the future, the research will focus on confirming effects of waste Si-based materials from the packaging industry on mechanical properties of cement pastes by using scanning electron microscope with X-ray microanalysis by confirms the inert effect of waste SiO_2 from the packaging industry on the phase composition of the composite.

Acknowledgements

This paper was financially supported by the GA CR research a project under the number GA23-05500S and by Czech Technical University in Prague under No. SGS project SGS22/089/OHK1/2T/11. The authors also thank company Recifa a.s.

References

[1] Jin, R., Li, B., Zhou, T., Wanatowski, D. and Piroozfar, P. "An empirical study of perceptions towards construction and demolition waste recycling and reuse in China," *Resources, Conservation and Recycling*, **126**, pp. 86-98 (2017).

[2] Rais, M. S. and Khan, R. A. "Strength and durability characteristics of binary blended recycled coarse aggregate concrete containing microsilica and metakaolin," *Innovative Infrastructure Solutions*, **5**, pp. 1-13 (2020).

[3] Máčalová, K., Václavík, V., Dvorský, T., Figmig, R., Charvát, J. and Lupták, M. "The use of glass from photovoltaic panels at the end of their life cycle in cement composites," *Materials* **14**(21), pp. 1-19 (2021).

[4] Penacho, P., de Brito, J. and Veiga, M.R. "Physico-mechanical and performance characterization of mortars incorporating fine glass waste aggregate," *Cement and Concrete Composites*, **50**, pp. 47-59 (2014).

[5] Du, H. and Tan, K.H. "Properties of high volume glass powder concrete," *Cement and Concrete Composites*, **75**, pp. 22-29 (2017).

[6] Nežerka, V., Hrbek, V., Prošek, Z., Somr, M., Tesárek, P. and Fládr, J. "Micromechanical characterization and modeling of cement pastes containing waste marble powder," *J Clean Prod*, **195**, pp. 1081-1090 (2018).

[7] Ren, X. and Zhang, L. "Experimental study of interfacial transition zones between geopolymer binder and recycled aggregate," *Constr Build Mater*, **167**, pp. 749-756 (2018).

[8] Prošek, Z., Trejbal, J., Nežerka, V., Goliáš, V., Faltus, M. and Tesárek, P. "Recovery of residual anhydrous clinker in finely ground recycled concrete," *Resources, Conservation and Recycling*, **155**, 104640 (2020).

Effect of Treatment on the Mechanical Properties of Additive Manufactured Ti-6Al-4V Parts

Roman Růžek[1,a], Petr Homola[1,b] and Adam Karkulín[1,c]

[1]*Czech Aerospace Research Centre, Aviation Division, Beranových 130, 199 05 Prague, Czech Republic;*

[a]*ruzek@vzlu.cz,* [b]*homola@vzlu.cz,* [c]*karkulin@vzlu.cz*

Abstract: The paper is focused on the evaluation of the influence of surface treatment and machining parameters on the static and fatigue properties of additively manufactured parts. The internal defects and characterization of the specimens were further analyzed using scanning electron microscopy considering surface roughness, possible anomalies in surface chemistry and possible cracks. The influence of selected machining methods (heat treatment, CNC milling, electrolytic polishing, chemical polishing) and operating parameters is discussed in the paper.

Keywords: Experimental; Stress; Analysis

1 Introduction

To reduce fuel consumption and CO_2 emissions in the aerospace sector, innovative solutions are being sought to reduce aircraft weight, including additive manufacturing (AM). Therefore, a high ratio of mechanical properties to weight is required. Due to this parameter, Ti6Al4V alloys are widely used in aircraft structures [1-4]. Nowadays, conventional processes such as casting and CNC machining are commonly used, but they lead to large volume losses, which causes higher costs. Additive manufacturing (AM) offers advantages in terms of weight, lead time, design, and functionality and allows the realization of alternative geometric shapes. Material strength characteristics can be better exploited. However, there are currently technological shortcomings [5] that have led to AM not being approved for applications on parts with high safety requirements, i.e., for use in primary aircraft structures. Therefore, it is necessary to focus on improving the "damage tolerance" parameters of such parts and quality control of printed parts. From this point of view, pores and other defects are particularly important and their occurrence cannot be completely prevented at present. However, the aim is to reduce the occurrence of defects to a minimum. The mechanical properties of the parts produced by AM technology can be further improved in various ways. The test program was designed in accordance with regulation requirements [6,7]. The presented paper is focused on the evaluation of the influence of surface treatment and machining parameters on the static and fatigue properties of additively manufactured parts. The work is a part of the Clean Sky 2 Joint Undertaking (JU) under grant agreement No 101007830 [8].

2 Manufacturing

The AM parts were built from Grade 5 (Ti 6Al 4V) powder in the conditions “new powder” and “recycled powder”, the latter for assessment of the recyclability of used powder, i.e., powder which has been used in former AM processes but has not been fused. The chemical composition of the material is listed in Table 1.

Table 1. Chemical composition of the material (in wt.%).

Al	V	Ti	O	N	H	Fe	C
6.20	3.98	balance	0.09	0.01	0.002	0.03	0.01

Specimens have been built in an upright standing position on a building platform. 6 specimens of each type (round and flat) can be built in one run (totally 20 pieces per processing batch, i.e., 120 test specimens have been produced). AM part on the build plate is shown in Fig.1. The paper deals with round specimens only. Several series of specimens had been manufactured with various post-processing methods – as built, stress relief (SR) treatment at 600 °C for 90 minutes, CNC machining, electropolishing (EP), plasma-electropolishing (PEP), hot isostatic pressing (HIP) at 920 °C and 1000 bar for 2 hours, and heat treatment (HIT) at 800 °C for 90 minutes. Examples of various post-processed round specimens are shown in Fig. 2.

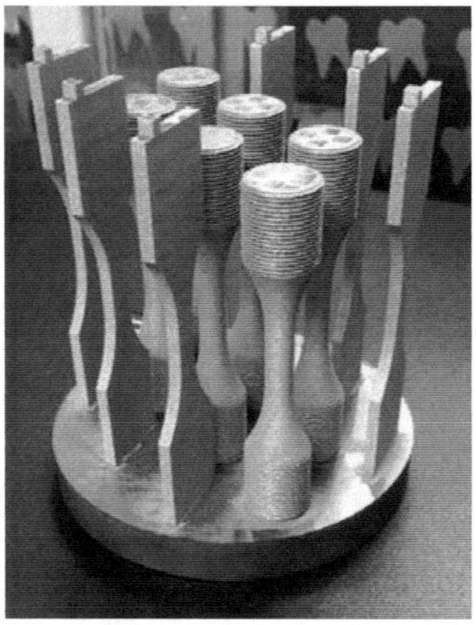

Fig. 1 AM-parts on the build plate.

Fig. 2 Round test specimen post-processing types appearance (from left to right side: as-built, electropolished, HIP and electropolished, CNC milled, HIP and plasma-electropolished, and HIT and plasma-electropolished.

3 Test methods and arrangement

The static tensile tests were carried out according to the EN ISO 6892-1 standard [9] using Sinus (100 kN load cell) and Schenck (250 kN load cell) testing machines using hydraulic mechanical threaded holder fixtures. The static tests were done under displacement control at a test rate of 0.5 mm/min. The strain was measured using one axial extensometer (Instron 2620-604, gauge length of 12.5 mm) attached to the test specimen. The extensometer was always removed during the test after reaching of axial strain value of 1%. All the static properties were calculated using the analysis module of Bluehill 3 software. The tensile test set-up detailed picture is shown in Fig. 3.

The fatigue tests were carried out according to the EN 6072 standard [10] using Sinus (100 kN load cell) testing machine using mechanical threaded holder fixtures. The fatigue tests were done under load control using a harmonic loading (sinusoidal shape of the loading cycle) with a constant load amplitude. Cyclic loading of the specimens was performed at one level of stress ratio $\boldsymbol{R} = -1$ and a nominal test frequency of 10-15 Hz. The fatigue curve was generated by performing a series of experiments at appropriate maximum stress or load levels, such that the desired range from 10^4 to 4.10^6 cycles was achieved. The run-out test specimen was tested at a higher load level in agreement with the EN 6072 standard.

The microfractographic analyses of the fracture surfaces were carried out using the TESCAN 3SBU Scanning Electron Microscope (SEM).

Fig. 3 Round static specimen test set-up – detail with an attached extensometer.

4 Results

Typical stress-strain curves of AM specimens are shown in Fig. 4. Mechanical testing of additively manufactured specimens with and without a post-treatment is summarized in Fig. 5 and Fig. 6. Quasi-static tensile tests performed at room temperature showed a negative effect of higher temperature treatment procedures on the strength properties. The lowest properties exhibited hot isostatic pressured batches (down to 80% of the as-built batch); the highest ones were found in the case of the mechanically milled (CNC) samples.

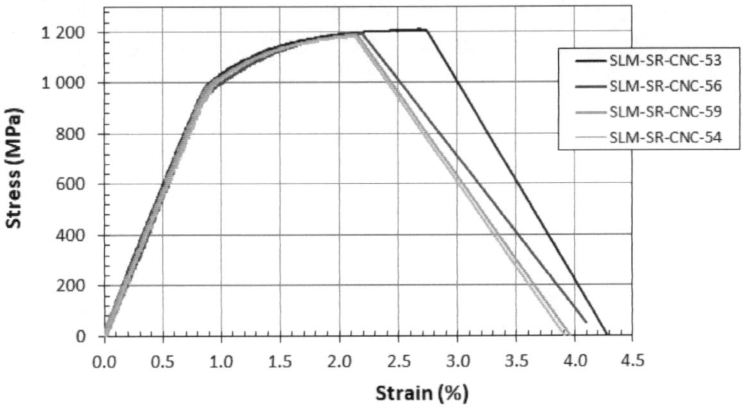

Fig. 4 Typical stress-strain tensile curves of as-built, stress-relieved, and CNC milled (SLM-SR-CNC) batch.

Fatigue test results are summarized and compared in semi-logarithmic coordinates in Fig. 7, Fig. 8, and Fig. 9. The arrows in the Fig.s indicate either samples that have failed outside the critical area in the thread (shorter lives) or completed tests without failure (run out). Tabular comparison of post-treatment effect on the fatigue life at stress level of 400 MPa is presented in Tab. 2. A slightly worse fatigue behaviour could be observed in the case of the electropolished and heat-treated samples. Contrarily, the CNC machined batch shows a significantly improved fatigue behaviour. In the case of the heat-treated batch, no substantial effect on fatigue behaviour in the linear part of the fatigue curve is present, only a lower fatigue limit caused by the heat-treatment process could be observed (a large scatter of the HIP batch results is probably affected by an electro-polishing process). A noticeable influence of plasma-electropolishing could be observed in the case of the HIP batches when compared to standardly electropolished samples, especially in the fatigue limit region (see Fig. 9).

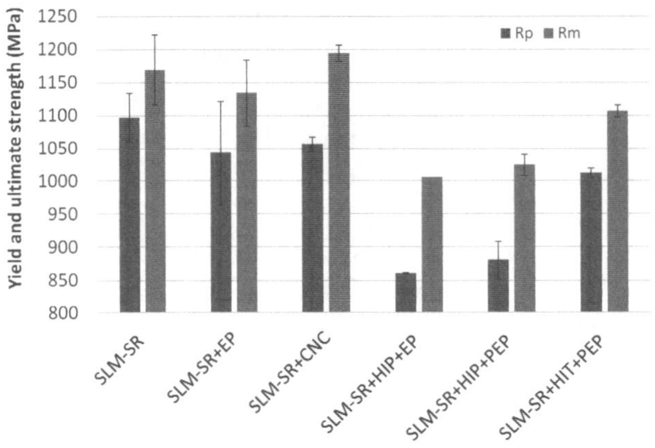

Fig. 5 Comparison of the Yield stress Rp0.2 and Ultimate tensile strength Rm for the as built condition (SLM SR) and post-treated configurations.

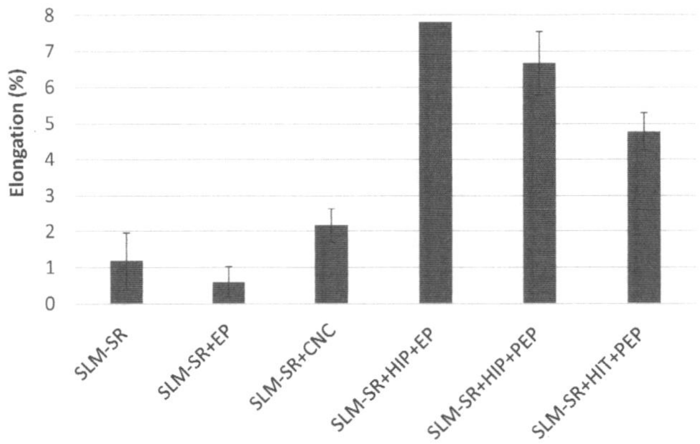

Fig. 6 Comparison of the elongation for the as-built condition (SLM SR) and post-treated configurations.

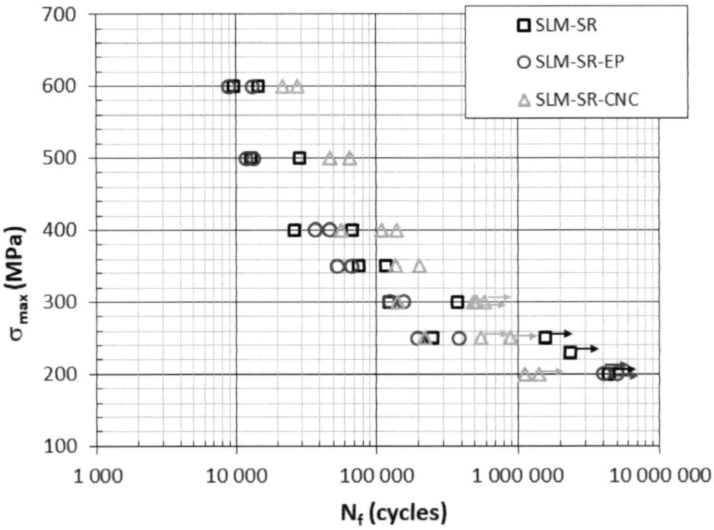

Fig. 7 Fatigue tests – Comparison of the fatigue data of electro/mechanical machining procedures with the as-built condition (SLM SR).

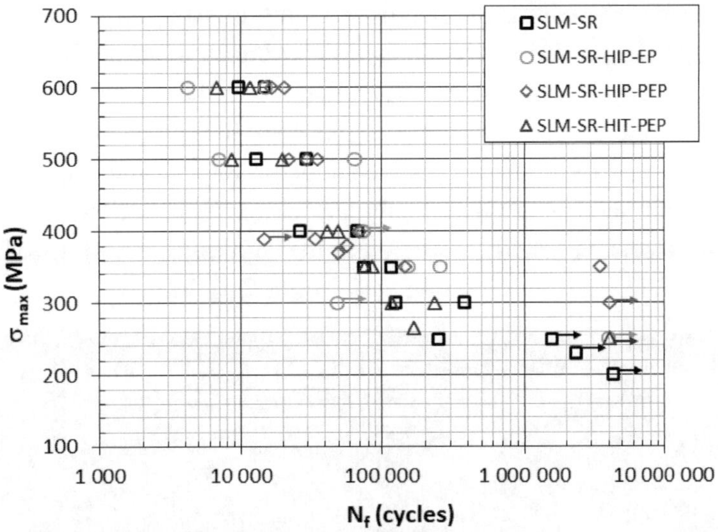

Fig. 8 Fatigue tests – Comparison of the fatigue data of a higher temperature treatment methods with the as-built condition (SLM SR).

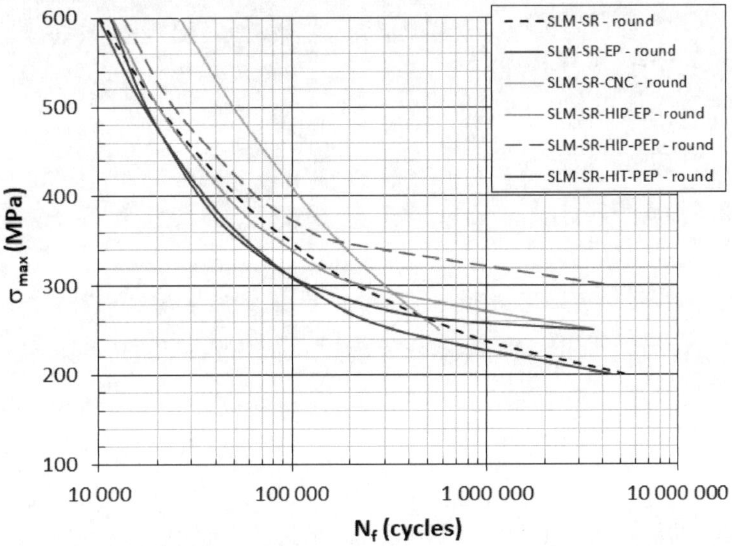

Fig. 9 Fatigue curve comparison (nonlinear 4-parametric regression) for round specimens' batches (R=-1).

Table 2. Effect of post-treatment on the fatigue life at stress level of 400 MPa (comparison with SLM SR).

Batch	Maximum stress (MPa)	Mean fatigue life	% ratio
SLM-SR (original)	400	51 532	–
SLM-SR-EP	400	35 289	68%
SLM-SR-CNC	400	109 147	212%
SLM-SR-HIP-EP	400	45 684	89%
SLM-SR-HIP-PEP	400	64 977	126%
SLM-SR-HIT-PEP	400	33 024	64%

Despite the improved roughness, the electro/mechanical machining procedures caused an exposition of internal subsurface defects on the surface of the samples (open pores by removing up to 0.5 mm surface layer by milling) that acted as a stress concentrator facilitating the initiation of fatigue cracks and deterioration of the fatigue behaviour. Only in the case of the HIP samples, a limited number of open defects in the test specimen's surface have been observed that led to an improved fatigue behaviour connected with a surface initiation type. This resulted even in an order increase in fatigue life. Examples of two fatigue crack initiation modes for as-built and electropolished samples (tested at the same loading level of 400 MPa) in the form of fracture surface micrographs are shown in Fig. 10 and Fig. 11, respectively. In the case of the as-built sample, a typical sub-surface one-site initiation could be observed (Fig. 10), while the electropolished sample exhibited multiple initiations of a fatigue crack from the open pores (Fig. 11) resulting in a half fatigue life (Fig. 8).

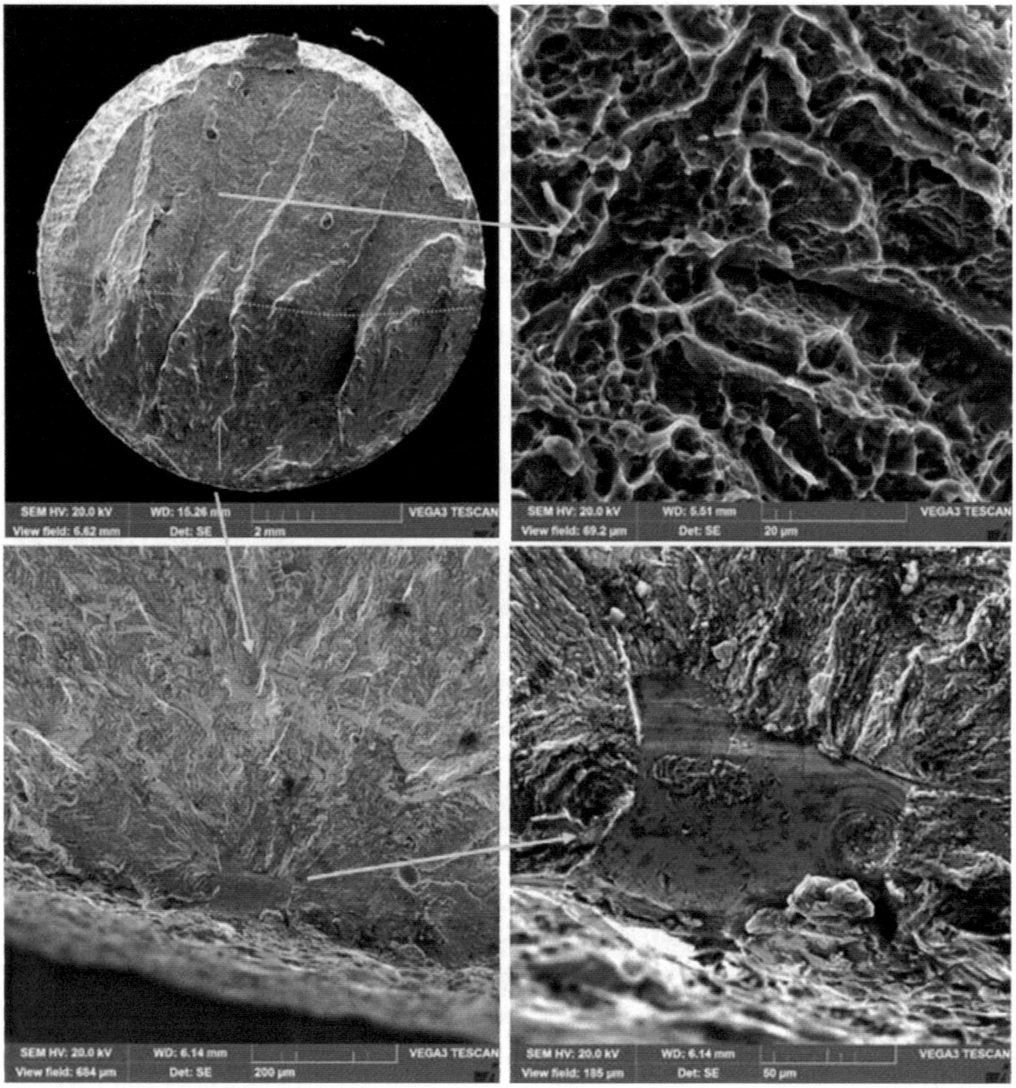

Fig. 10 Fracture surface of as built batch sample tested at the maximum loading level of 400 MPa with the resulting fatigue life of 67 483 cycles (SEM, up to 8000x magnification).

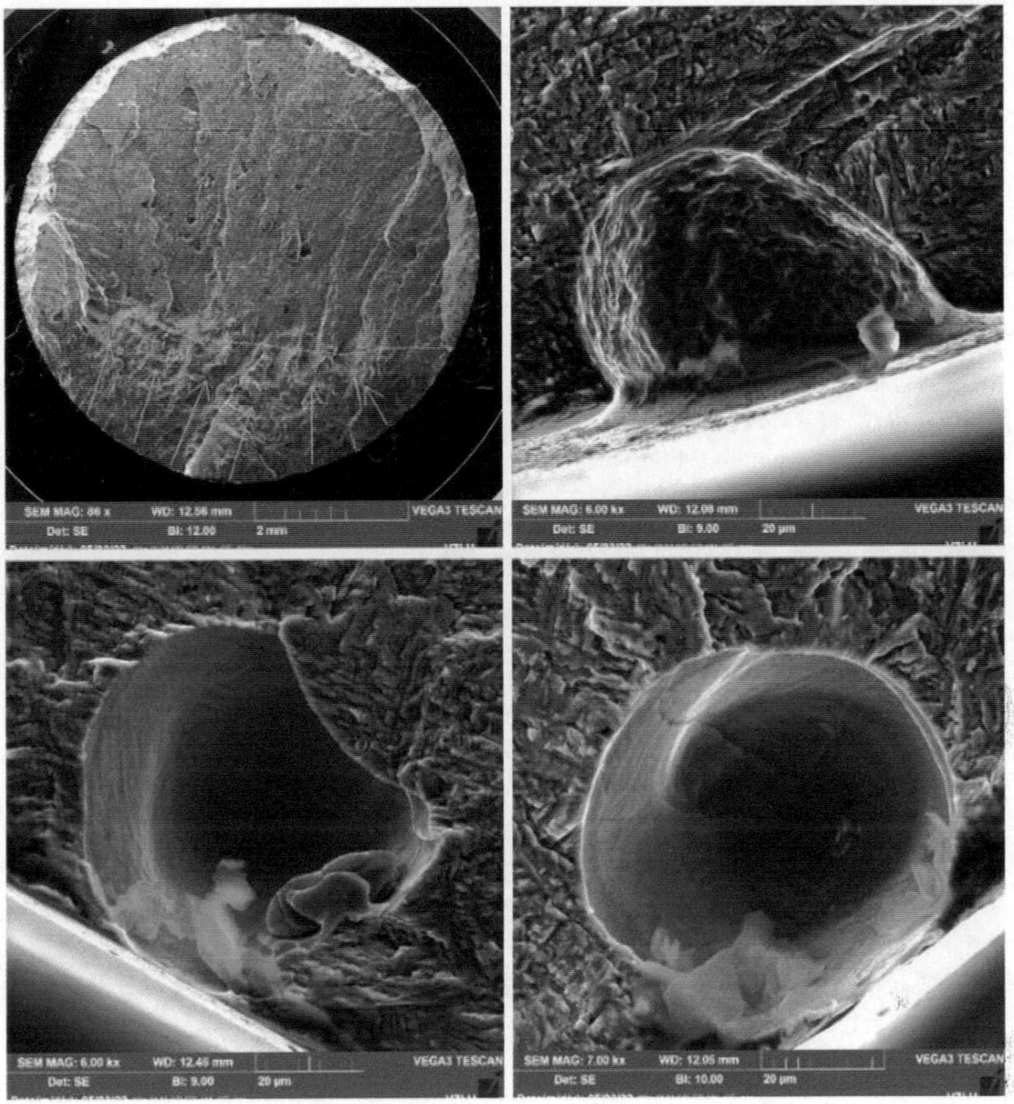

Fig. 11 Fracture surface of electropolished batch sample tested at the maximum loading level of 400 MPa with the resulting fatigue life of 37 372 cycles (SEM, up to 7000x magnification).

Conclusion/Summary

The following conclusion are stated: a negative effect of higher temperature treatment procedures on the strength properties; the lowest properties exhibited hot isostatic pressured batches (down to 80% of the as-built batch); the highest ones were found in the case of the mechanically milled (CNC) samples; a slightly worse fatigue behaviour in case of the electropolished samples (as compared with the as-built batch); improved fatigue behaviour in case of the CNC machined and HIP batches. Fractographic analyses supported the discussion of the results.

Acknowledgements

These results were generated within the framework of the Clean Sky 2 Joint Undertaking (JU) under grant agreement No 101007830. The JU receives support from the European Union's Horizon 2020 research and innovation programme and the Clean Sky 2 JU members other than the Union.

References

[1] Liu, Z., He, B., Lyu, T. and Zou Y.: "A Review on Additive Manufacturing of Titanium Alloys for Aerospace Applications Directed Energy Deposition and Beyond Ti-6Al-4V". *JOM,* 73, 1804–1818 (2021). https://doi.org/10.1007/s11837-021-04670-6.

[2] Salihu S.A., Suleiman Y.I. and Eyinavi A. I.: "Classification, Properties and Applications of titanium and its alloys used in automotive industry- A Review", *American Journal of Engineering Research (AJER)*, 8(8), pp. 92-98 (2019), e-ISSN: 2320-0847.

[3] Shulong Y., Yongyun Z. and Peng Y.: "Chapter 14 - Applications of titanium in the electronic industry" in *Titanium for Consumer Applications*, Ed. Froes F., Quian M and Niinomi M., (Elsevier, pp. 269–278, 2019). ISBN 9780128158203, https://doi.org/10.1016/B978-0-12-815820-3.00019-8.

[4] Veiga, C., Davim J.P. and Loureiro A.J.R.: "Properties and applications of titanium alloys: a brief review", *Rev. Adv. Mater. Sci.* 32 (2), pp. 133–148 (2012).

[5] Tofail S.A.M., Koumoulos E.P., Bandyopadhyay A., Bose S., O'Donoghue L. and Charitidis C.: "Additive manufacturing: scientific and technological challenges, market uptake and opportunities", *Materials Today*, 21 (1), 2018, https://doi.org/10.1016/j.mattod.2017.07.001.

[6] Růžek R. and Homola P. "Certification requirements to additive manufactured parts used in airframe primary structures," in *Proceedings of 58th Int. Conf. of Experimental Stress Analysis*, Sobotín, Czech Republic, 2020 (Technical University of Ostrava, Ostrava, 2020), pp. 432-437.

[7] Metallic Materials Properties Development and Standardization (MMPDS) Handbook. MMPDS-17, Battelle Memorial Institute, 505 King Avenue, Columbus, OH 43201, 2022.Annex: 11th Call for Proposals (CFP11) - List and Full Description of Topics, JTI-CS2-2020-CFP11-AIR-01-46: Evaluation of NDT Techniques for Assessment of Critical Process and Manufacturing Related Flaws and Defects for a Ti-alloy 2019. *Available from* https://3tanium.eu/.

[8] Annex: 11th Call for Proposals (CFP11) - List and Full Description of Topics, JTI-CS2-2020-CFP11-AIR-01-46: Evaluation of NDT Techniques for Assessment of Critical Process and Manufacturing Related Flaws and Defects for a Ti-alloy 2019. *Available from* https://3tanium.eu/.

[9] ČSN EN ISO 6892-1, Metallic materials - Tensile testing - Part 1: Method of test at room temperature (ISO 6892-1:2021), 2021.

[10] EN 6072, Aerospace series - Metallic materials - Test methods – Constant amplitude fatigue testing, 2010.

Design and Behaviour of 3D Printed Joint Connecting Various Polymer Materials

David Rybanský[1,a], Martin Šotola[1,b], René Souček[1,c] and Pavel Maršálek[1,d]

[1]*VSB – Technical university of Ostrava; 17. Listopadu 2172/15, Ostrava-Poruba, Czech Republic;*

[a]*david.rybansky@vsb.cz*, [b]*martin.sotola@vsb.cz*, [c]*rene.soucek.st@vsb.cz*, [d]*pavel.marsalek@vsb.cz*

Abstract: This paper deals with the design and modelling of a joint made by a 3D printing method. The aim of the work is to test and prepare a suitable connection between two printed materials. To determine the best-performing type of joint, Load capacity is used as main criterion. For evaluation, two materials (PA12 and TPU) are used for performing experiments using a universal testing machine. After obtaining data, a mathematical model was developed to correspond with the experiment. A parametric study was performed to evaluate suitable modifications of the design. A computational model was done in the software ANSYS Workbench 2020 R2. The conclusion is the evaluation of the obtained knowledge and comparison of the experimental testing with the FEM analysis. The obtained results can be used for designing biomedicine aids.

Keywords: Joint connection; Load-bearing capacity; Additive manufacturing; Finite Element Method

1 Introduction

Additive manufacturing is becoming more popular for making personalized biomechanical applications such as orthosis or prostheses which have typically complex geometries [1, 2]. The typical material for printing such applications is currently PA12 (Polyamide 12). However, some materials used directly with skin are rough and cause calluses, therefore there is a need for softer materials. These days, manufacturers are experimenting with TPU (Thermoplastic Polyurethane), which could replace foams. Current SLS (Select Laser Sintering) printers cannot print made of two materials.

This paper deals with the design and testing of a joint that can connect two parts made of two materials with different mechanical properties, see Table 1 (the material data were obtained from articles [3, 4]). The main aim of the new joint is to have a strong enough connection of two parts for application and an additional but not mandatory requirement is to be able to dismantle the joint. The first part of the aim is to have a large load-bearing capacity for forced detaching. The second part is due to the ability to wash softer parts of the assembly to minimize health issues.

This work is based on previous results of flexible structures which are used in biomechanical applications [5]. However, the applications were made of one material. due to the higher demand for using multiple materials, this work is aimed for investigating connecting these materials in various places of the part.

Table 1. Material properties of PA12 and TPU.

Material	E [MPa]	ν [-]	σ_Y [MPa]	E_T [MPa]	A [%]
PA12	1 224	0.39	21	334	18
TPU (1301)	60	0.4	5	0.06	250

2 Method

The first step was to design two types of joints for this application base on different concepts [4]. The first joint is a customization of the dovetail joint (see Fig. 1) but applied for polymer materials. This type of joint is typical in wood applications (it is a standardized joint). However, it was necessary to slightly alter it due to the manufacturing process (3D printing), which meant removing sharp edges. The second joint was the annular joint (see Fig. 2), which has no standard. For each joint, three variants were created which different dimensions (see Table 2 and **Table** Table 3). It should be noted that the left part of the joint from Fig. 1 and Fig. 2 was made of PA12, and the right part was made of TPU (1301). Both parts were printed from virgin powder. Depending on the design of the joint, there is a possibility of limited movement (rotation) of connected parts. In our application, it was necessary to restrict this movement.

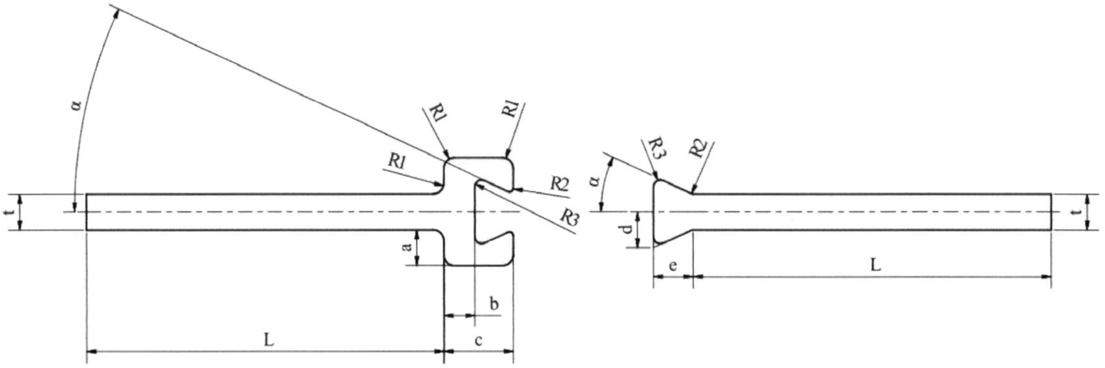

Fig. 1 Scheme of dovetail joint.

Table 2. Dimension of dovetail joint (left part PA12, right part TPU).

Dimensions	*a* [mm]	*α* [°]	*b* [mm]	*c* [mm]	*d* [mm]	*e* [mm]	*L* [mm]	*R1* [mm]	*R2* [mm]	*R3* [mm]	*t* [mm]
Variant 1	1.00	25	0.88	1.95	1.00	1.07	10.0	0.30	0.10	0.20	1.00
Variant 2	1.50	25	0.90	3.02	1.50	2.14	10.0	0.30	0.10	0.20	1.00
Variant 3	1.50	40	0.88	1.95	1.50	1.19	10.0	0.30	0.10	0.20	1.00

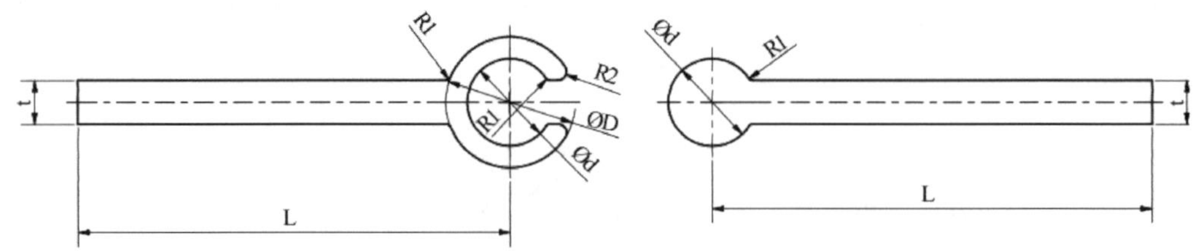

Fig. 2 Scheme of the annular joint (left part PA12, right part TPU).

Table 3. Dimensions of annular joint.

Dimensions	*a* [mm]	*α* [°]	*b* [mm]	*L* [mm]	*R1* [mm]	*R2* [mm]	*t* [mm]
Variant 1	1.07	25	1.00	10.0	0.20	0.10	1.00
Variant 2	2.14	25	1.50	10.0	0.20	0.10	1.00
Variant 3	1.19	40	1.50	10.0	0.20	0.10	1.00

In the second step after manufacturing these variants, experiments were performed using the universal testing machine (TESTOMETRIC M500-50CT) with the force sensor (S2M). The force sensor is in the range of 0 – 100 N. Specimens were thoroughly cleaned and put in jaws of the machine (see the example presented in Fig. 3). The velocity of the jaws was set to v = 5 mm/min. The machine was running until disjointing. From the experiment, we obtained a relation between acting force and displacement. In the beginning, joints had linear characteristics. However, it was not linear in full range. The experiment was performed for each variant using five specimens. The presented results in Fig. 4 represent Variant 2 of the annular joint (specimen O) and Variant 3 of the dovetail joint (specimen L).

Fig. 3 Clamping of the annular joint in the universal testing machine.

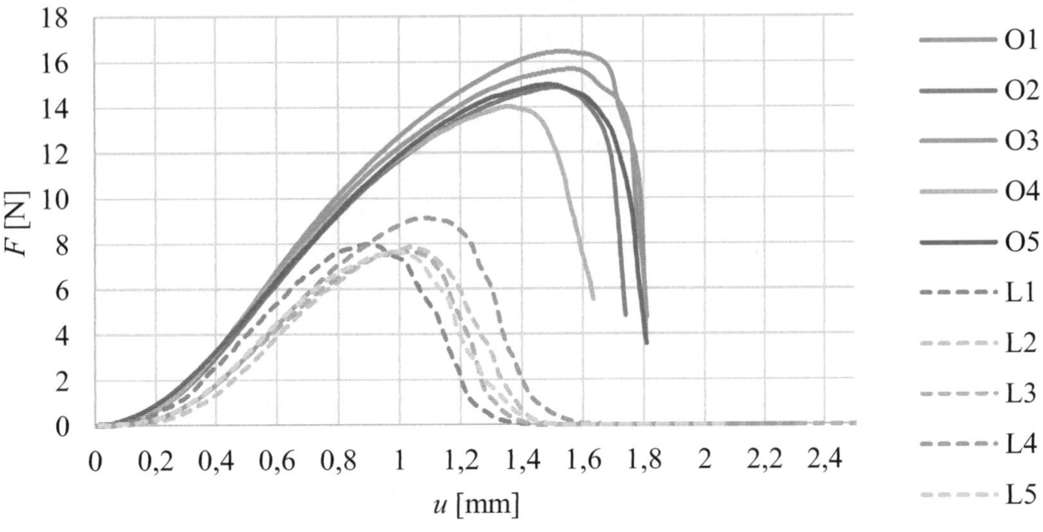

Fig.4 Force displacement diagram of specimens O and specimens L.

As it is apparent from Fig. 4, specimen L had a smaller load-bearing capacity for disjointing than specimen O. For the following study, we focused on the annular join. In the third step, we performed numerical modelling using the finite element method. The numerical model contained multiple nonlinearities, such as material model, contacts, and large deformation, similarly as previous research of load-bearing capacity of wires ropes [6]. Because of these nonlinearities, we chose plane stress analysis, which is a simplified simulation than volume one, but it provided us with enough accurate results for our case. The numerical model used symmetry and had 5 948 nodes and 1 864 elements. Material models were elastic-plastic materials using bilinear isotropic hardening. The used mechanical properties are listed in Table 1. This model aimed to get the same load-bearing capacities as the experiment. For evaluation of difference between numerical modelling and experiment, following expression is used

$$err = \frac{F_{LAB_{ave}} - F_{FEA}}{F_{FEA}} \cdot 100\ (\%), \tag{1}$$

where $F_{LAB_{ave}}$ is average load capacity obtained from experiment and F_{FEA} is load capacity obtained from numerical modelling. Evaluated difference was 7.7 %. Additionally, we obtained a field of stresses (Fig. 5) and mainly plastic strains after disjointing using the finite element method. Comparison of experimental data and of numerical model is presented in Fig. 6.

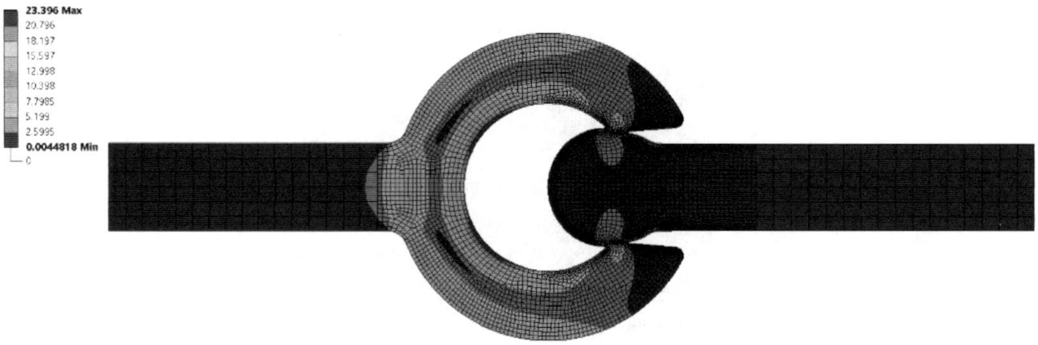

Fig. 5 Field of equivalent von Mises stress during maximal load corresponds to the load-bearing capacity.

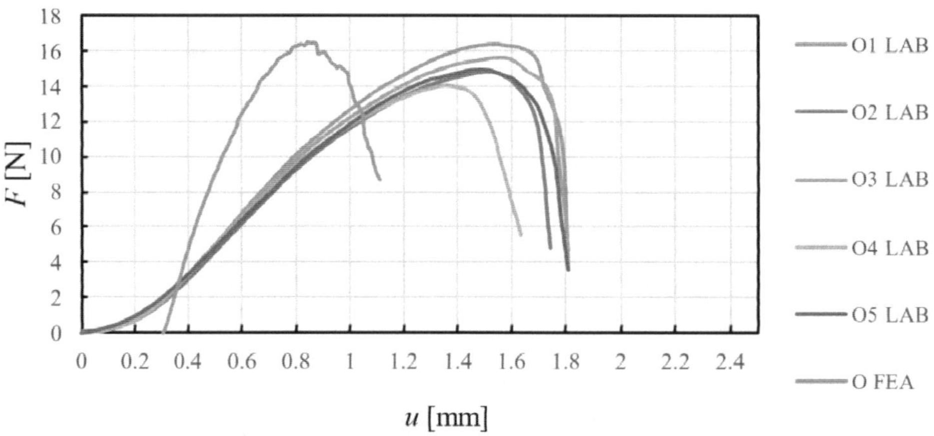

Fig 6 Comparison of experimental data with finite element analysis.

The fourth step was a parametric study of the key parameter. The parametric study dealt with investigating the change in the load-bearing capacity due to resizing the filler's radius ***R1***, see Fig. 7. Radius ***R1*** was assumed as an important geometrical parameter influencing load-bearing capacity. The rest of the dimensions was not the task of the parametric study even though they might affect stiffness or the load capacity. Totally 6 values of the radius were studied. The dimension is limited due to the limitation of manufacturing (a small radius is difficult to manufacture) and larger values are affected by dimension ***d.*** The study resulted in lowering the radius enlarged the load-bearing capacity which is apparent from linear regression in Fig. 7. For the larger load-bearing capacities during disjointing, it is crucial to have it caught by the slot of the tougher part.

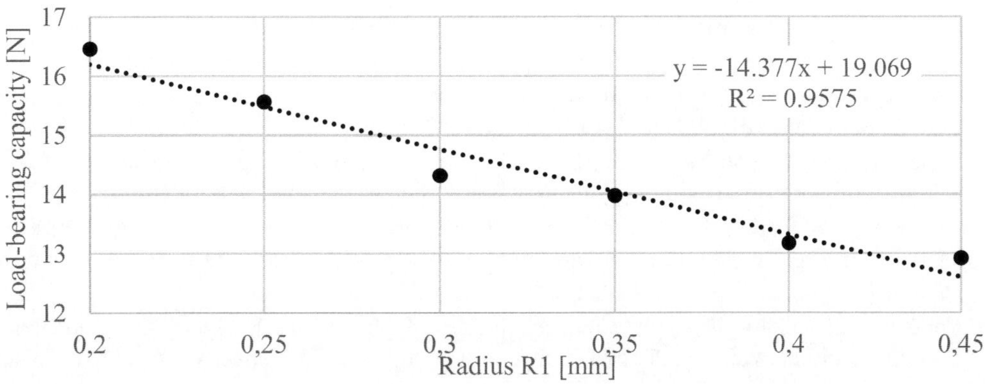

Fig. 7 Parametric study of radius R1.

Conclusions

With the advent of new types of biomechanical application, it is necessary to research possibilities of connecting various 3D printed material. We designed joint connections made of two materials (PA12 and TPU) with main aim to maximized possible load-bearing capacity. In this paper, we presented two types of joints (dovetail and annular). We performed experimental measurement of two possible variants and prepared numerical model which was later used for parametric study.

Comparing the experiment load-bearing capacity with one obtained from the numerical model, the error is 7.7 %. However, numerical modelling had different characteristics than experimental ones, see Fig. 6. Our explanation of this phenomenon was manufacturing inaccuracies, which can be minimized by precious printing. Numerical modelling used simplified analysis which might have affected results. The material models used rather a simple plasticity model. The contact was also simplified since the surface of the part is affected by heat treatment. In this study, we focused only on maximal force before disjointing, therefor model was enough suitable for parametric study. After the parametric study, it was apparent that radius ***R1*** affects load capacity greatly because it might allow a smoother transition of parts (which led to lower values of load capacity).

The presented numerical model is suitable for fast parametric studies but has a lot of shortcomings. For better evaluation of the whole characteristic, one should the material model and prepare a volume analysis of the joint. After treating these shortcomings, a numerical model can be used for preparing a better design of joints.

Acknowledgments

This work was supported by Specific Research „Application of Modern Computational and Experimental Approaches in Applied Mechanics“ (SP2023/027).

References

[1] Marsalek, P., Sotola, M., Rybansky, D., Repa, V., Halama, R., Fusek, M. and Prokop, J. "Modeling and Testing of Flexible Structures with Selected Planar Patterns Used in Biomedical Applications," Materials, 14(1), 140 (2021). ISSN 1996-1944.

[2] Repa, V., Marsalek, P., Prokop, J., Sotola, M., Stareczek, D., Rybansky, D. and Halama, R. "Modelling and testing of 3D printed flexible structures with three-pointed star pattern used in biomedical applications," in Proceedings of Experimental Stress Analysis - 58th International Scientific Conference, EAN 2020, Ostrava, 2020, pp. 451-455.

[3] Lammens, N., De Baere, I. and Van Paepegem, W. "On the orthotropic elasto-plastic material response of additively manufactured polyamide 12," in Proceedings of the 7th bi-annual international conference of Polymers & molds innovations, Ghent, Belgium, 2016.

[4] Stoia, D., Linul, E. and Marsavina, L. "Influence of Manufacturing Parameters on Mechanical Properties of Porous Materials by Selective Laser Sintering," Materials, 12(6), 871 (2019).

[5] Rybansky D, Marsalek P, Sotola M, Hroncek J, Drahorad L, Kusnir O, Prokop J. "Design and Behavior of Lightweight Flexible Structure with Spatial Pattern Reducing Contact Surface Fraction," Polymers. 2023; 15(19):3896.

[6] Hroncek, J., Marsalek, P., Rybansky, D., Sotola, M., Drahorad, L., Lesnak, M. and Fusek M. "Simplified Numerical Model for Determining Load-Bearing Capacity of Steel-Wire Ropes," Materials, 16(10), 3756 (2023).

Splinted and Unsplinted Dental Implants: A QCT/FEA Study

Luboš Řehounek[1,a], Mária Frolo[2,b] and Aleš Jíra[1,c]

[1]*Czech Technical University in Prague, Faculty of Civil Engineering, Department of Mechanics; Thákurova 7, 166 29 Prague 6, Czech Republic;*

[2]*Prague Center of Dental Implantology, Na Okraji 331/45b, 162 00 Prague 6, Czech Republic,*

[a]*lubos.rehounek@fsv.cvut.cz,* [b]*simkova@pcdi.cz,* [c]*jira@fsv.cvut.cz*

Abstract: To properly assess the stress distributions at the peri-implant area of a human maxilla, two variants of implant-supported overdentures (IODs) were evaluated using the Quantitative Computed Tomography-Based Finite Element Analysis (QCT/FEA). The used CT data were acquired from a 76-year-old edentulous male who was predetermined for an anterior maxillary replacement. Between the splinted and unsplinted variant, the bar-supported splinted variant showed favorable stress distributions both in bone and in the bodies of implants.

Keywords: Splinted; QCT/FEA; Implant; Dental; Stress

1 Introduction

Implant-supported overdentures (IODs) are a treatment option that is often used when the patient's bone quality is not sufficient, the lip line is not adequate for good phonetics and aesthetics or there is a buccolingual inclination of the alveolar ridge [1]. The system of connection between the overdenture and the implants includes the unsplinted attachments, such as various balls, locators and double crowns, or splinted attachments with many variants of bar and clip designs [2,3]. In literature, however, there is no clear consensus on which variant is better [4].

2 Materials and Methods

2.1 Data acquisition

We analyzed the maxilla of a 76-year-old male with sufficient dexterity. First, we checked the extraoral and intraoral requirements for an IOD were met [1]. Straumann SLA RN SP Roxolid implants were implanted in the most suitable sites of the patient's maxilla and a 3D X-ray was performed (Fig. 1). Then, a QCT/FEA (Fig. 2, Quantitative Computed Tomography-Based Finite Element Analysis) was performed in Mechanical Finder.

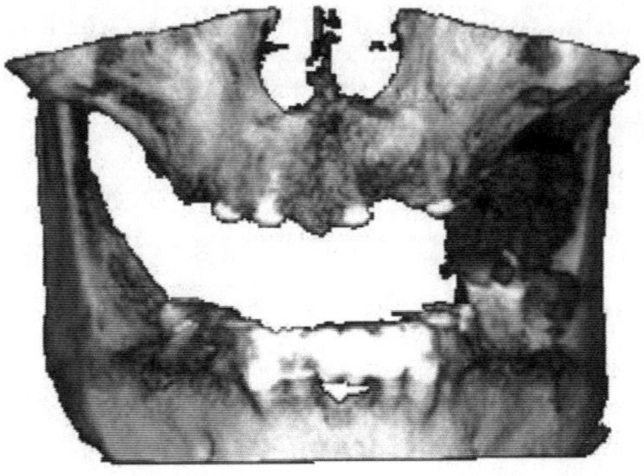

Fig. 1 The 3D X-ray scan of the patient that was obtained with the Planmeca ProMax ®3D Classic.

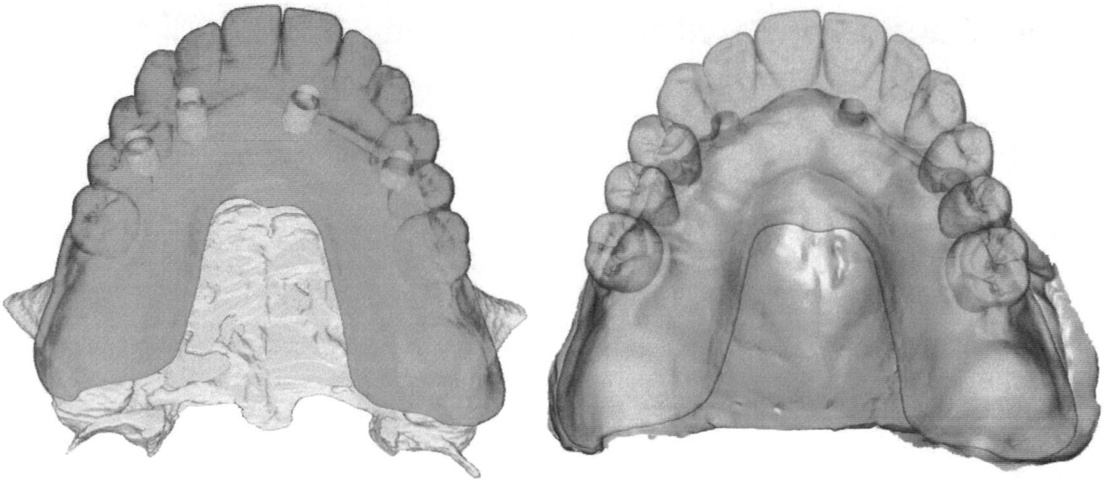

Fig. 2 Left—The QCT/FEA reconstruction with prosthetics, right—the laboratory X-ray model.

2.2 Analyzed implant variants

After the reconstruction of the maxillary model, we modelled two variants of the implant assemblies—a splinted and an unsplinted variant, shown on Fig. 3. The model also contained the overdenture made from PEEK (polyetheretherketone) (Fig 4).

Fig. 3 Left—The splinted design chosen for analysis, right—the unsplinted locator design.

2.3 Boundary Conditions

Simulations were performed in three load cases. The load was always represented by a force slanted in the buccolingual direction. The first load case was loading of both maxillary central incisors by 150 N in the frontal region with the force being slanted by 35°, the second by 600 N in the distal region on both molars and slanted only by 5° and the third by 600 N in the distal region on the first premolar with force components only in the direction of the implant axis. The model was constrained by fixing a section of the cut-off, as shown on Fig. 4.

3 Results

The results shown on Fig. 5 are comparisons of equivalent stress between the splinted and unsplinted variant in bodies of implants. Aside from the first load case (150 N, 35° slanted on incisors), the splints show great contribution towards the overall stiffness of the structure and redistribute stress into the whole assembly—unlike the unsplined variant, where stresses concentrate into individual implants and create large concentrations of stress.

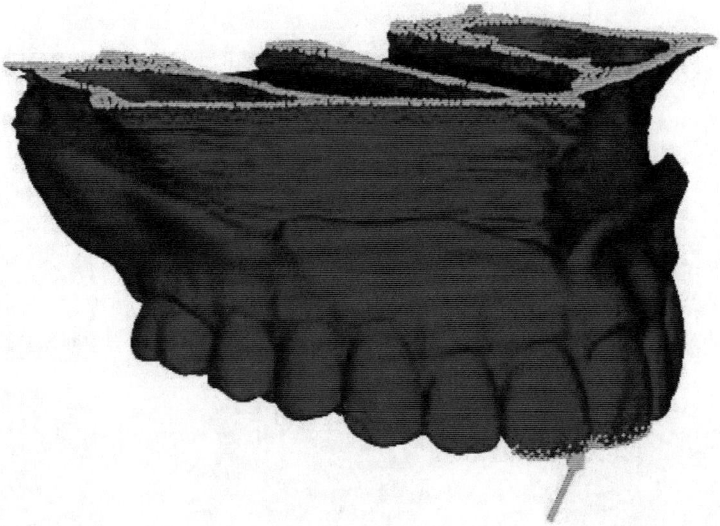

Fig. 4 The complete model with the force from the first load case. Yellow color represents constrained nodes. Green color represents loaded nodes. The model now also contains the prosthetic from PEEK.

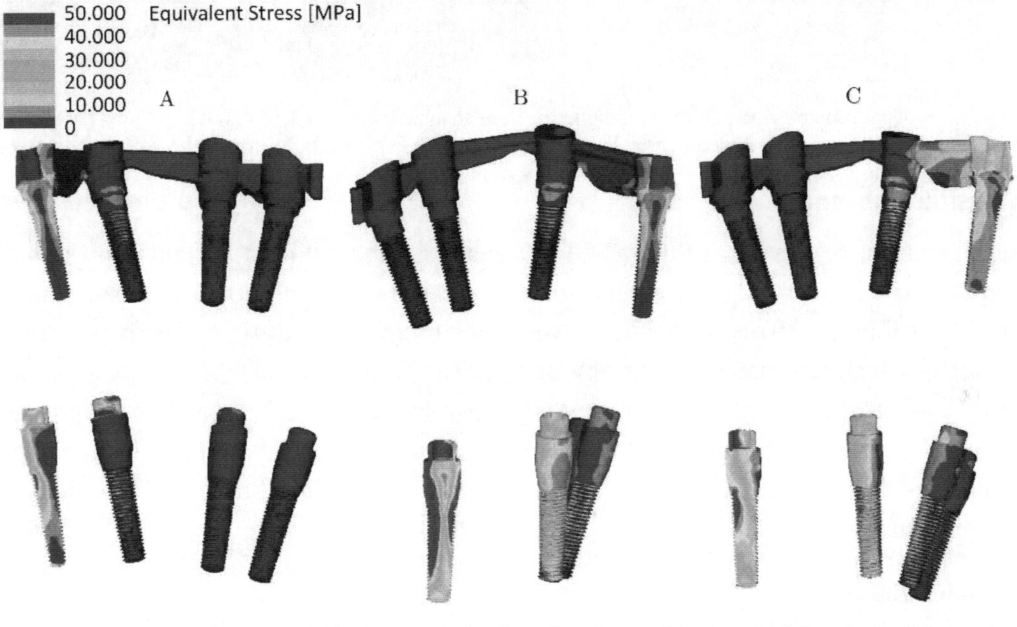

Fig. 5 Isolines of Equivalent (Von-Mises) stress in bodies of implants of both analyzed variants. Column A—1st load case, column B—2nd load case, column C—3rd load case. Top row is the splinted variant, bottom row the unsplinted variant.

The splinting effect also prevents excessive displacements of the implants, which in turn has a positive effect on concentrations of stress in bone (Fig. 6). The splinted variant shows only small areas with concentrations of minimum principal stress. The unsplinted variant, where the stiff implants were displaced into bone, shows large concentrations of stress at the implants' apices. The isolines on Fig. 6 also show the positive effect of splinting on reducing the stresses from flexion (bending).

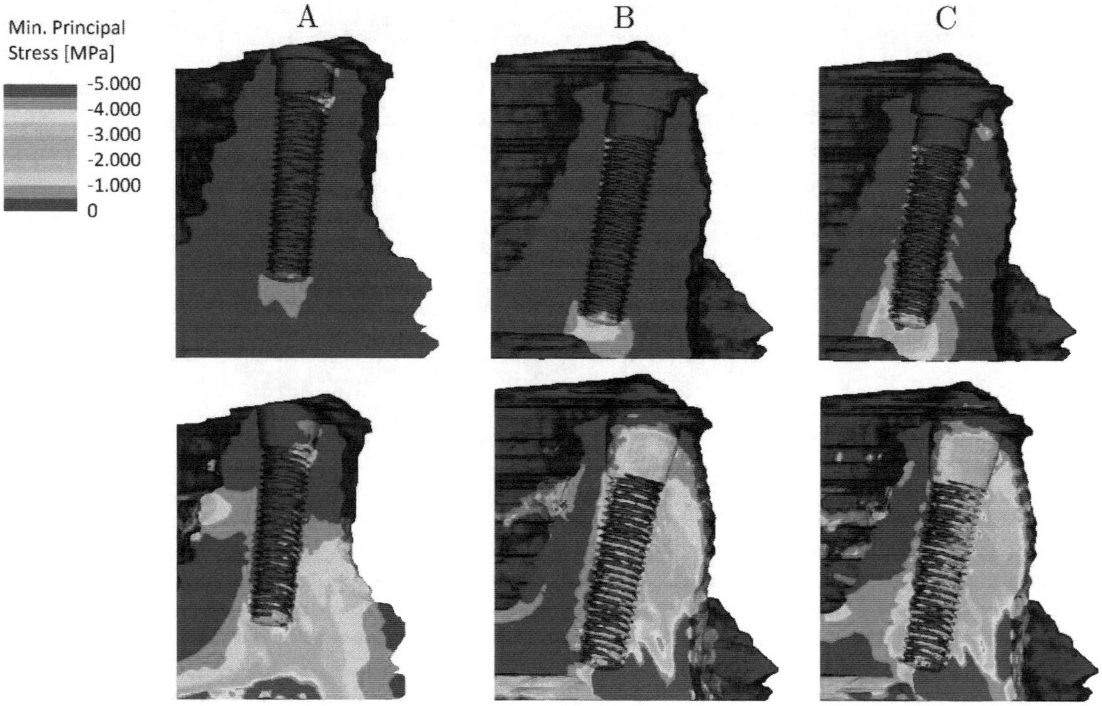

Fig. 6 Isolines of Minimum Principal Stress in bone for both analyzed variants. Column A—1st load case, column B—2nd load case, column C—3rd load case. Top row is the splinted variant, bottom row the unsplinted variant.

Conclusion/Summary

According to the results from the QCT/FEA analyses, the splinted variant shows smaller concentrations of stress in implants and more favorable stress distributions among all load cases. The splinting effect seems to have a positive effect in fixed implant-supported solutions for edentulous patients. This simulation was performed on an X-ray scan, which is arguably harder to work with (weaker imaging capabilities) than a CT scan but can still result in acceptable results if reconstructed properly. Models like these should always include the prosthesis as it provides additional stiffness and load redistribution which helps the unsplinted variant.

Acknowledgements

The financial support provided by the by the Faculty of Civil Engineering, CTU, Prague, project n. SGS23/152/OHK1/3T/11 is gratefully acknowledged.

References

[1] Mericske-Stern R.D., Taylor T.D. and Belser U., "Management of the edentulous patient", Clinical Oral Implants Research: Chapter 7, 11, pp.108-125 (2000).

[2] Ceraulo S., Leonida A., Lauritano D., Baldoni A., Longoni S., Baldoni M. and Caccianiga G., "Proposal for a clinical approach to geriatric patients with anchor need on implant for removable denture: New technique", *Prosthesis*, **2**(3), p.16 (2020).

[3] Amornvit P., Rokaya D., Bajracharya S., Keawcharoen K. and Supavanich W., Management of obstructive sleep apnea with implant retained mandibular advancement device. *World J Dent*, **5**(3), pp.184-189 (2014).

[4] Al Amri M.D., Crestal bone loss around submerged and nonsubmerged dental implants: A systematic review. *The Journal of Prosthetic Dentistry*, **115**(5), pp.564-570 (2016).

Contribution to the Creation of Virtual Models of Rotary Machines

Zdenko Šavrnoch[1,a], Alžbeta Sapietová[1,b], Vladimír Dekýš[1,c] and Barbora Drvárová[1,d]

[1]*University of Žilina; Univerzitná 1, 010 26 Žilina, Slovakia;*

[a]*zdenko.savrnoch@fstroj.uniza.sk*, [b]*alzbeta.sapietova@fstroj.uniza.sk*, [c]*vladimir.dekys@fstroj.uniza.sk*, [d]*barbora.drvarova@fstroj.uniza.sk*

Abstract: This paper presents a review of rotordynamics with a specific emphasis on the Laval rotor. The authors describe the development of a virtual model of the Laval (Jeffcott) rotor using the MSC.ADAMS software. The study focuses on analyzing parameters related to the stability of the rotor's operation, such as the critical angular velocity and the trajectory of the rotor's center of gravity (CoG orbit). These parameters were monitored and evaluated throughout the analysis.

Keywords: Laval's rotor, Jeffcott rotor, rotordynamics, Adams, rotary machines, MATLAB

1 Introduction

Currently, the primary objective of advancements in rotor dynamics is to attain increasingly higher operating speeds. This is because higher rotational speeds in energy machines lead to greater energy generation or transformation, particularly in production settings, resulting in enhanced productivity gains [1]. Although there has been progress in this field in recent years, rotor dynamics remains an extensively researched area, with studies aimed at effectively utilizing existing concepts in new domains and gaining a better understanding of the behaviour of rotating systems. Non-linear and non-stationary rotor dynamics are specific areas that require extensive and in-depth research [2].

2 Rotordynamics

Field of applied mechanics that focuses on studying the behaviour, operation and diagnostics of rotating machinery and equipment [1]. The most basic form of rotating machinery consists of a rotating component (such as a shaft or disk) that is supported by a structure and bearings (called a Laval's rotor). As the rotational speed of the machine increases, the unbalance of the system causes an increase in the magnitude of the oscillations. At critical speeds, the amplitude of these oscillations reaches a maximum value. Examples of rotating machines are steam and gas turbines, crankshafts of internal combustion engines, gearboxes etc.

3 Laval's rotor

When examining the shaft bending and center of gravity (CoG) trajectories of rotating masses, a simple model of an immaterial shaft and point mass representing rotating masses that are attached to it, can be used [3]. If we want to add more complexity to the numerical (or analytical) model, we can include shaft mass in the total mass of the system. The system can be classified as being damped or not. Additional parameters that can be used to describe the system include the method of shaft mounting and deformability of the bodies that form the system [4].

Laval's (Jeffcott) rotor can be assembled in three ways:

- Flexible (compliant) shaft system

- System with rigid shaft and compliant bearings
- System with flexible (compliant) shaft and compliant bearings

4 Computational model of undamped Laval's rotor

System characterized by its placement in rigid rotating supports [3]. A rigid disk is mounted symmetrically on a flexible shaft without any eccentricity. The initial deflection of the center of gravity (CoG) from the central axis (which is the centerline of the rigid rotational support) occurs in the vertical direction of the y-axis and is proportional to the gravity of the system.

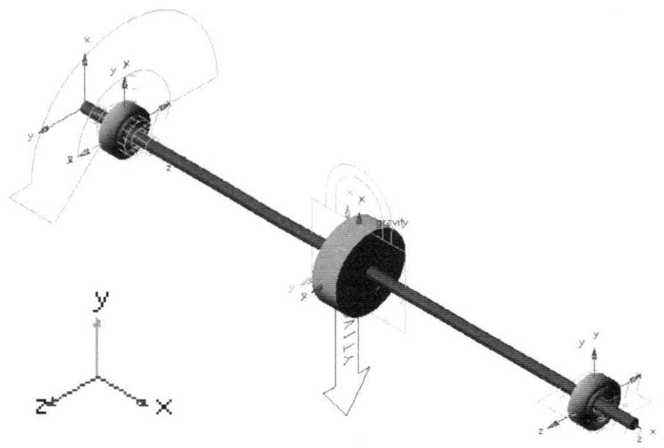

Fig. 1 Disk symmetrically mounted on flexible shaft and rigid supports.

All material constants are based on the material Steel S235 chosen from the material library of MSC.Adams.

Table 1. Other parameters used in the computational model.

Parameters	Denotation	Nominal value	Units
Shaft length	L_h	585	[mm]
Shaft diameter	d_h	10	[mm]
Disk width	b_k	25	[mm]
Disk diameter	d_k	75	[mm]
Shaft mounting distance from disk	l_b	250	[mm]

5 Simulation results

Table 2. Simulation parameters.

End Time	25 [s]
Step size	0.0001 [-] / 10 [kHz]
Solver	WSTIFF/I3

Rotor's angular velocity starts at 130 [rad/s] and its linear ramp is given by the function 130+15·time [rad/s]. Rotor rotates clockwise (CW) and starts at equilibrium.

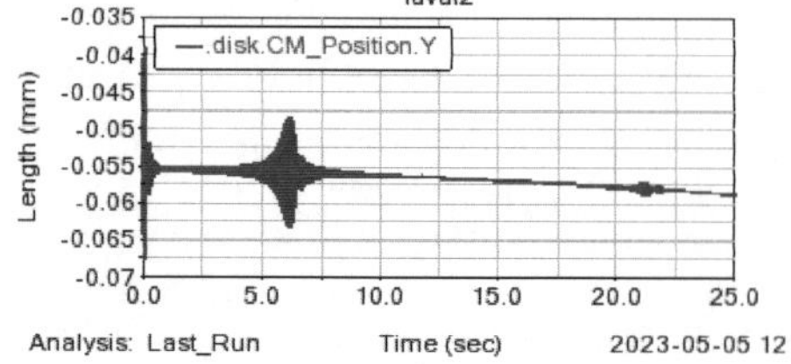

Fig. 2 Deflection in Y axis.

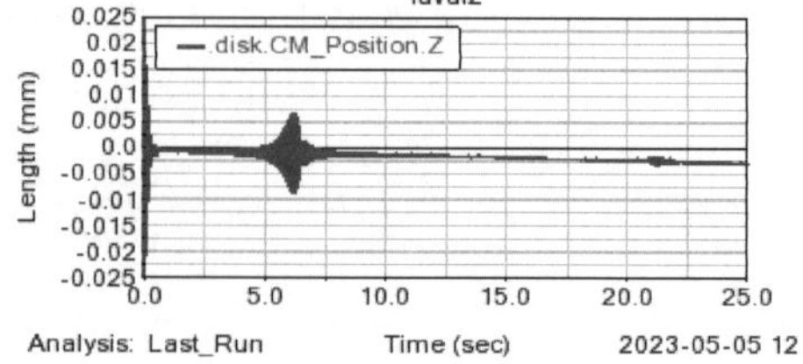

Fig. 3 Deflection in Z axis.

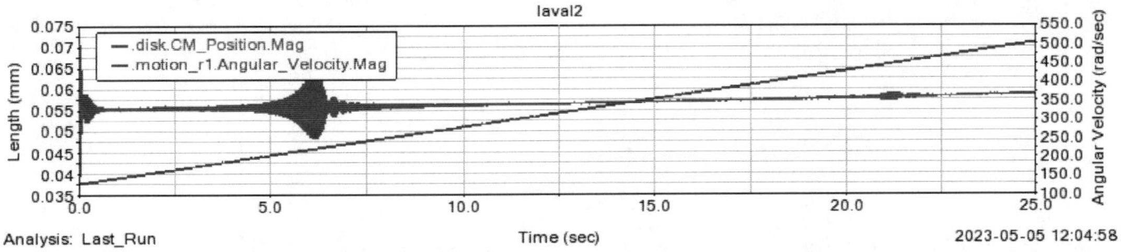

Fig. 4 Deflection magnitude.

Analysis shown the presence of the first critical and the second critical angular velocity in the projected time interval of 25 [s]. Deflection of CoG (Fig. 4) of the first angular velocity ω_{1crit} = 223,7 [rad/s] (located at t_1 = 6,23 [s]) has much higher amplitude than the second critical angular velocity ω_{2crit} = 447,6 [rad/s] (located at t_2 = 21,18 [s]). This difference between the amplitudes is caused by the different shapes of the first and the second modal frequencies (Fig. 5, 6) [5].

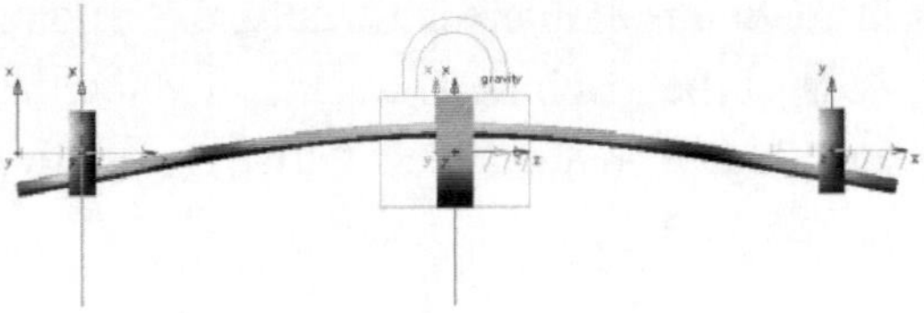

Fig. 5 First mode.

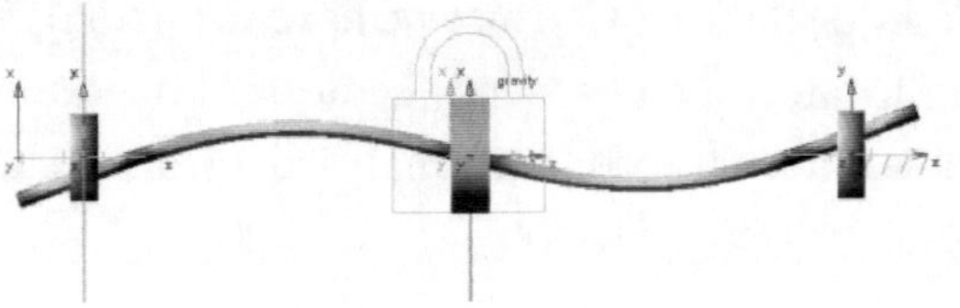

Fig. 6 Second mode.

For a better visualisation of the CoG defflection, it can be projected to the 3D plot (Fig. 7), where the bottom plane axes represent deflection in particular axis of the rotor and the vertical axis represents time or angular velocity (these are interchangable).

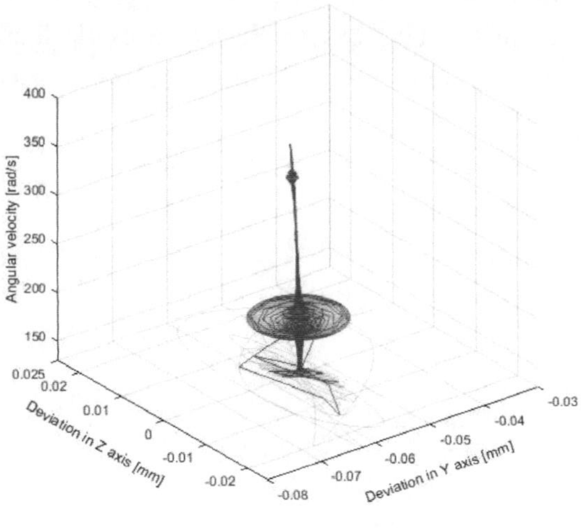

Fig. 7 3D representation of CoG deflection.

For a better display and wider option of diagram modification, were the output data exported from MSC.Adams to MATLAB.

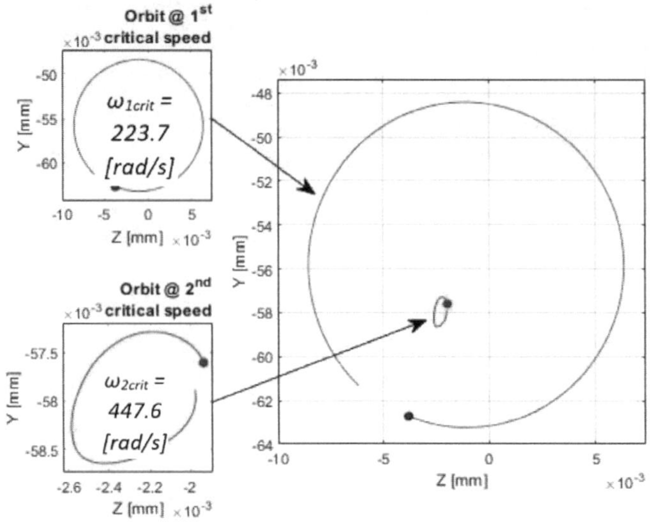

Fig. 8 Square diagrams of system CoG orbit at different speeds and size comparison.

To inspect the CoG orbits we have to cut diagrams in the points of interest. Usually, these points are located at and around the desired operating speeds of the machine or at critical velocities. Therefore, the orbits of analysed rotor model are evaluated at the time of first two critical velocities (Fig. 8).

Summary

The deflection amplitude for both the Y and Z axes steadily increases (as shown in Fig. 4), with the maximum amplitude being reached at 1st critical velocity (as might be seen on Fig. 2,3,4). The rotational supports exhibit equal stiffness in both the Y and Z axes, which causes the first and second critical angular velocity of the rotor to remain the same for both axes. The weight of the disk acting on the flexible shaft causes deflection (as seen in Fig. 2 - Deflection in Y axis) from the initial position of the center of gravity of the bent rotor. Second critical velocity is equal approximately double the first critical velocity. In compare with the circular orbit of 1st critical velocity its orbit has elliptical shape (Fig. 8).

Acknowledgements

This research work was supported by the project KEGA 011ŽU-4/2022.

References

[1] Gash, R. and Pfützner, H. *Dynamika rotorů* (SNTL – Nakladatelství technické literatury, Praha, 1980). Číslo publikace 04-226-80.

[2] Sága, M., Vaško, M., Handrik, M. and Kopas P. "Contribution to Random Vibration Numerical Simulation and Optimisation of Nonlinear Mechanical Systems," *Scientific Journal of Silesian University of Technology – Series Transport*, **103**, pp. 143-154 (2019).

[3] Byrtus, M., Hajžman, M. and Zeman, V. *Dynamika rotujících soustav* (Západočeská univerzita v Plzni, 2010).

[4] Záhorec, O. and Caban, S. *Dynamika* (Olymp, Košice, 2002).

[5] Černohlávek, V., Svoboda, M., Štěrba, J., Chalupa, M. and Sapieta, M. "Analytical and Experimental Solution of Vibrations of a System of Bound Bodies," *Manufacturing Technology*, **20**(6), pp. 699-707 (2020).

Optimization of 3D Printing Parameters to Minimize Residual Stresses in Maraging Steel

Karel Trojan[1,a], Ivana Zetková[2,b], Jiří Čapek[1,c], Petr Thurnvald[2,d], Nikolaj Ganev[1,e] and Miloslav Kepka jr.[2,f]

[1]*Faculty of Nuclear Sciences and Physical Engineering, Czech Technical University in Prague; Trojanova 13, 120 00 Prague, Czech Republic;*
[2]*Regional Technological Institute, Faculty of Mechanical Engineering, University of West Bohemia, Univerzitní 8, 306 14 Pilsen, Czech Republic;*

[a]*karel.trojan@fjfi.cvut.cz,* [b]*zetkova@fst.zcu.cz,* [c]*jiri.capek@fjfi.cvut.cz,* [d]*petrt@fst.zcu.cz,* [e]*nikolaj.ganev@fjfi.cvut.cz,* [f]*kepkami1@fst.zcu.cz*

Abstract: During additive manufacturing using the selective laser melting technology, a complex residual stress distribution is created that can significantly affect the printing itself and also the mechanical properties of the final product. Thus, this research has been carried out to optimize 3D printing parameters to minimize residual in maraging steel. It was found that the preheating temperature of the build platform significantly affects both the residual stresses and microstructure parameters as well as the mechanical properties.

Keywords: Additive Manufacturing; Selective Laser Melting; Maraging steel; Residual stresses; Mechanical properties

1 Introduction

Additive Manufacturing (AM), specifically Selective Laser Melting (SLM) technology, is a promising method of metal powder consolidation and offers tremendous opportunities for parts production. This involves selectively melting parts from a thin flat powder bed in layers using a scanning energy source to produce 3D parts. During the SLM process, the sections are heated according to the scanning strategy, allowing independent heating and cooling of the sections, leading to a more complex residual stress (RS) distribution [1]. The magnitude of these RS may even approach the yield strength of the material. The complexity of the SLM process results in the size and orientation of the RS being highly dependent on laser power, scanning speed, scanning strategy, and other printing parameters.

Fatigue crack initiation and propagation play an important role in fatigue properties, where they are shown to be strongly associated with surface roughness, microstructure parameters (dislocation density, crystallite size, microcracks) and RS [2]. In a study [3], the correlation between the initiation and propagation of surface fatigue cracks with RS was investigated in laser-welded materials using X-ray diffraction, where it was found that not only RS, but also the distribution of microstructure parameters, plays a crucial role in fatigue. In general, the distribution of RS in AM specimens is highly dependent on the printing strategy. The distribution of internal RS is crucial for the deformation of the sample and the formation of fatigue cracks. Therefore, this research was conducted to minimize the RS of 3D printed maraging steel.

2 Experiment

An EOS M290 machine with the printing parameters shown in Table 1 was used to produce 5 mm diameter test specimens of C300 maraging steel (trademark MS1), see Table 2 for chemical composition. Research has been carried out on what changes in the process parameters lead to a reduction in RS after solution annealing heat treatment at 820°C without following machining. The most significant effect was to change the setting of preheat of the building platform from the default 40°C to 120°C. Therefore, two samples were analysed, where the sample marked SP had a preheating temperature of 40°C and the sample P120 a temperature of 120°C.

Table 1. Infill printing parameters.

Parameter	
Laser power	285 W
Laser speed	960 mm/s
Hatch distance	110 μm
Layer thickness	40 μm
Printing strategy pattern	Stripes

Table 2. Chemical composition of C300 maraging steel (trademark MS1) [4].

Element	Composition [wt.%]
Ni	17–19
Co	8.5–9.5
Mo	4.5–5.2
Ti	0.6–0.8
Al	0.05–0.15
Cr, Cu	each ≤ 0.5
C	≤ 0.03
Si, Mn	each ≤ 0.1
P, S	each ≤ 0.01
Fe	Bal.

Macroscopic RS were determined from the centre of the cylindrical test samples using X-ray diffraction. The RS values were calculated from lattice deformations determined on the basis of experimental dependencies of $2\theta(\sin^2\psi)$ assuming a biaxial state of residual stress (θ is the diffraction angle, ψ the angle between the sample surface and the diffracting lattice planes). The diffraction angle was determined as the centre of gravity of the Cr$K\alpha_1\alpha_2$ doublet diffracted by the lattice planes {*211*} of the α-Fe phase. The X-ray elastic constants $\frac{1}{2}s_2 = 5.76\ \text{TPa}^{-1}$, $s_1 = -1.25\ \text{TPa}^{-1}$ were used for the stress calculation. The experimental error is the standard deviation according to the "$\sin^2\psi$" residual stress calculation algorithm. The FWHM value (Full Width at Half Maximum) was determined from the analysed diffraction line {*211*} at a tilt of $\psi = 0°$, the error of determination does not exceed 0.05° 2θ. The depth profile of RS, which was determined in the direction of the axis of the cylindrical test specimen, was obtained by gradual electrolytic removal of the surface layers.

Tensile tests were carried out on the FU250_AH250 tensile testing machine, which is equipped with an extensometer. From the data obtained, the ultimate strength, the yield strength corresponding to 0.2% of plastic strain and the elongation at break were evaluated. All results of mechanical properties shown, see Table 3, are the average of at least two values.

3 Results

From the results of macroscopic RS, obtained from the centre of the cylindrical test samples, it is clear that the depth profile varies significantly from sample to sample, see Fig. 1. Tensile RS still dominate in the near vicinity of the surface for both samples, however, for sample P120 (120°C platform preheating) a rapid decrease of RS can be observed with only compressive residual stresses being described with a maximum at a depth of about 120 μm. On the contrary, the RS for sample SP (default platform preheating, i.e. 40 °C) show a tensile character up to a depth of 170 μm with a maximum of more than 300 MPa. This state of RS is unfavourable as high tensile values reduce the yield stress and promote crack initiation and their growth. However, at greater depths for the SP sample, compressive RS have also been described. The high tensile RS in the surface layers did not affect the strength, as the ultimate tensile strength for the SP specimen is approximately 50 MPa higher than for P120, see Table 3. There could be a greater effect on fatigue life, but the surface tensile RS of specimen P120 probably explains one of the reasons why there was no increase in fatigue life between specimen P120 and SP (results will be the subject of another article). Therefore, further print optimisation will be required to minimize surface RS.

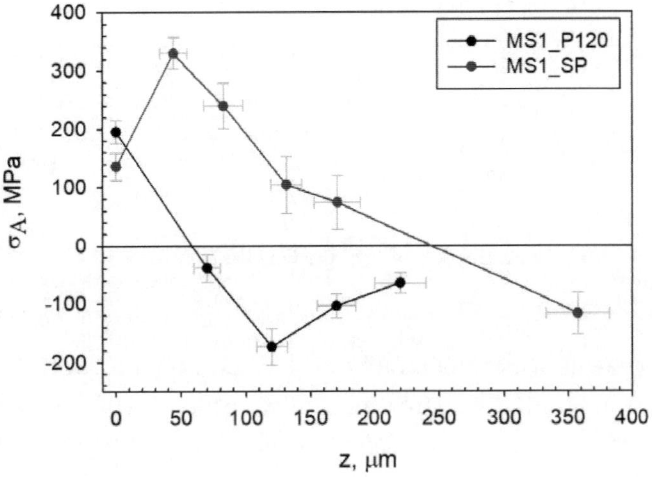

Fig. 1 Depth profile of macroscopic residual stresses.

Fig. 2 shows the depth profile of the FWHM parameter, which is the width at half of the observed diffraction maximum. From the increasing value of FWHM, it could be said that in the analysed volume the crystallites (coherent diffracting domains) are smaller and/or the microstrain (deformation of individual crystallites) is higher. From the comparison, it can be clearly stated that in the surface layers of the P120 sample, the crystallite size is smaller and/or the microstrain value is higher. Thus, if we compare these conclusions with the results of mechanical properties (see Table 3), it is not surprising that the sample P120 shows lower extensibility and higher yield strength, since the value increases with increasing microstrain (which is due to a higher dislocation density) and with decreasing grain size d (according to the Hall-Petch effect following a $d^{-½}$ relationship), since the crystallite size often correlates with the grain size. The values of yield strength and tensile strength are within error identical to those given in the material data sheet of the powder manufacturer (see Table 3), solution heat treatment had no significant effect on the mechanical properties.

Table 3. Comparison of the obtained mechanical properties (solution heat treatment at 820°C) with tabulated values for as-build state [4].

	MS1_SP	MS1_P120	Tabulated values
Yield strength, MPa	879 ± 14	916 ±7	930 ± 150
Ultimate tensile strength, MPa	1120 ± 1	1069 ± 2	1100 ± 150
Elongation at break, %	15	12	12 ± 4

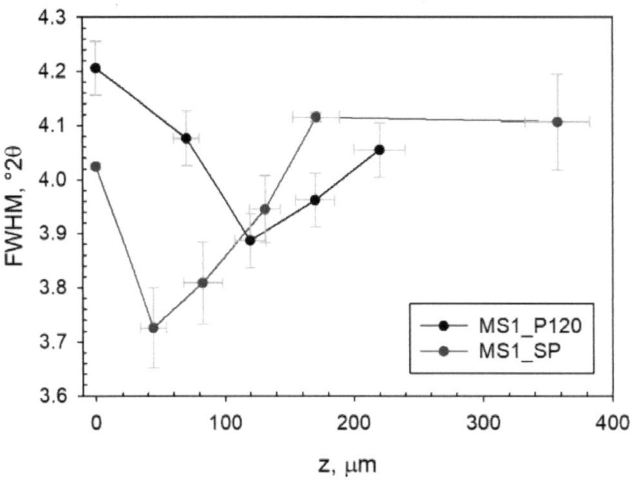

Fig. 2 Depth profile of the FWHM parameter.

Conclusions

It was found that increasing the temperature of the printing platform to 120 °C significantly affected the residual stresses after heat treatment of the printed samples by SLM. The samples with different preheating temperatures also differed in mechanical properties. Further optimization of the printing parameters is needed to minimize the residual stresses also on the surface. This should have a pronounced effect on improving the material properties, especially the fatigue life.

Acknowledgements

The work was supported by Technology Agency of the Czech Republic "Development of "3D print-thermal spray" systems for dynamically and cyclically loaded applications", grant No. TH75020003. The work of CTU staff was supported by the Grant Agency of the Czech Technical University in Prague, grant No. SGS22/183/OHK4/3T/14.

References

[1] Bartlett, J. L. and Li, X. "An overview of residual stresses in metal powder bed fusion," *Additive Manufacturing*, **27**, pp. 131-149 (2019).

[2] Schneller, W., Leitner, M., Pomberger, S., Springer, S., Beter, F. and Grün F. "Effect of post treatment on the microstructure, surface roughness and residual stress regarding the fatigue strength of selectively laser melted AlSi10Mg structures," *Journal of Manufacturing and Materials Processing*, **3**(4), 89 (2019).

[3] Černý, I., Čapek, J., Kec, J., Trojan, K., Ganev, N. and Němeček, S. "Fatigue of Laser-Welded Structures: Role of Residual Stress," in *Comprehensive Structural Integrity*, Aliabadi F. and Soboyejo W., ed. (Elsevier Science, 2023).

[4] EOS GmbH - Electro Optical Systems, "EOS MaragingSteel MS1 – Material data sheet," *Available from* https://www.eos.info/03_system-related-assets/material-related-contents/metal-materials-and-examples/metal-material-datasheet/werkzeugstahl_ms1_cx/ms1/ms-ms1-m280_m290_400w_material_data_sheet_07-22_en.pdf *Accessed:* 2023-05-15.

Mechanical Evaluation of the PEEK Cranial Implants

Patrik Varga[1,a], Teodor Tóth[1,b], Radovan Hudák[1,c], Róbert Huňady[2,d] and Ján Kostka[2,e]

[1]*Department of Biomedical Engineering and Measurement, Faculty of Mechanical Engineering, Technical University of Košice, Letná 1/9, 042 00 Košice-Sever, Slovakia;*
[2]*Department of Applied Mechanics and Mechanical Engineering, Faculty of Mechanical Engineering, Technical University of Košice, Letná 1/9, 042 00 Košice-Sever, Slovakia;*

[a]*patrik.varga@tuke.sk,* [b]*teodor.toth@tuke.sk,* [c]*radovan.hudak@tuke.sk,* [d]*robert.hunady@tuke.sk,* [e]*jan.kostka@tuke.sk*

Abstract: Additive manufacturing and new polymeric progressive biocompatible materials enable the production of individual cranial implants. The advantage of this approach is a reduction in the number of complications during surgery due to inaccuracies in the design and manufacture of the implant. Critical areas of implants are their attachment to the skull and their mechanical properties. The paper deals with the finite element method analysis of force action on a cranial implant made of PEEK. The simulation was made for two methods of fixing the implant to the skull, where in case A, they were used to fix the screw, and in case B, it was only about placing the implant on the defect without fixing it to the bone.

Keywords: Cranial implant; PEEK; Finite Element Analysis

1 Introduction

In the field of medical surgery, skull defects are mostly often caused by congenital deformity, tumors, and traumatic injuries [1, 2]. Cranioplastic reconstruction is one of the more demanding surgeries due to that it involves the brain, nerves, and tissues of the human body [1]. Nowadays, the development of modern additive manufacturing technologies and biocompatible materials makes it possible to design and manufacture different personalized cranial implants. The implementation of implants produced by additive technologies such as material extrusion or selective laser sintering leads to higher production efficiency (lower material consumption), higher precision of implant production, and lower input material costs [3]. Due to suitable mechanical properties and minimal radiodiagnostic disadvantages, it is passing to polymer materials, e.g., polyetheretherketone (PEEK). PEEK is a high-temperature thermoplastic that has similar mechanical properties to cortical bone, so it is suitable for skull bone replacement. Its advantages include high chemical resistance, bioinertness, and translucency [4]. Due to the variety of defects (dimensions, shape, position), it is necessary to carry out a numerical simulation to verify the mechanical characteristics and guarantee the appropriate functionality of cranial implants made of this material [5-7].

2 Material and methods

To perform the simulation, a model of the implant and part of the skull was designed from DICOM data in the software package Mimics Innovation Suite (Materialise, Belgium). The overall methodology of the experiment is shown in Fig. 1.

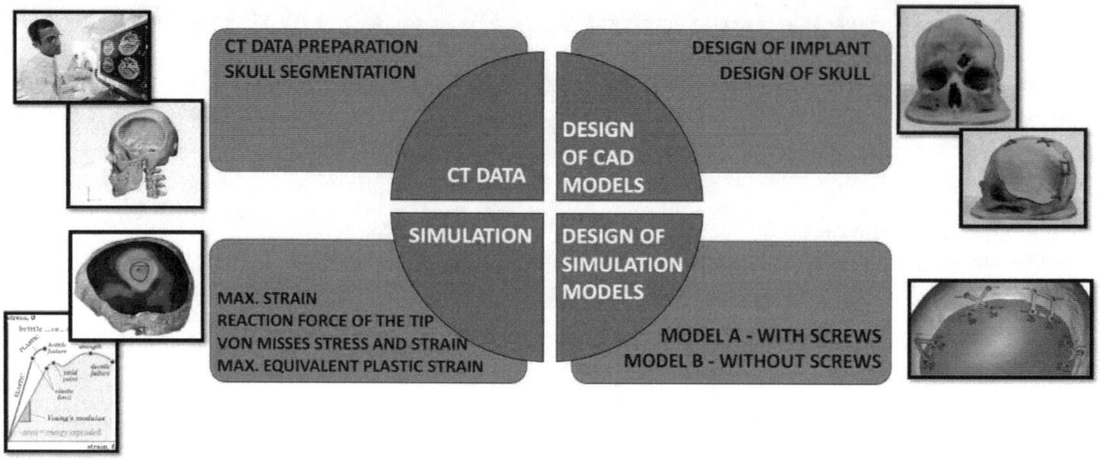

Fig. 1 Experiment plan.

The pressure tip model was designed in the software SolidWorks (Dassault Systèmes, Canada). Nonlinear static analysis of the implant was done in Abaqus (Dassault Systèmes, Canada) with the skull and tip set as perfectly rigid bodies and the implant model as a homogeneously deformable body. Tables 1 and 2 show the mechanical properties and material parameters of the implant used for the finite element method simulation. The analysis was performed for two types of implant fixation – fixed with screws and loose without fixation with screws. The test was completed when a displacement of 5 mm was reached.

Table 1. PEEK mechanical properties.

Elastic modulus PEEK [MPa]	3600
Poisson's ratio μ [-]	0.4
Yield strength [MPa]	132
Yield strength deformation	Approx. 0.06
Ultimate tensile strength deformation	Approx. 0.4

Table 2. Johnson and Cook model parameters.

Parameters	A [MPa]	B [MPa]	n	C	m
Values	132	10	1.2	0.034	0.7

2.1 Boundary conditions of the simulation

The skull was fixed firmly in both cases. The motion of the tip was controlled by a single reference point (RP), and displacement was allowed in the direction of the tip axis. A surface frictional contact between the tip and implant and the skull and implant was defined. The friction coefficient was 0.1. In the first case, the implant was attached to the skull using simplified screw connections that were represented by kinematic couplings and MPC beams (Fig. 2a). In the second case, the implant was positioned loosely on the skull (Fig. 2b), and thus deformation and displacement were limited by the contact regions.

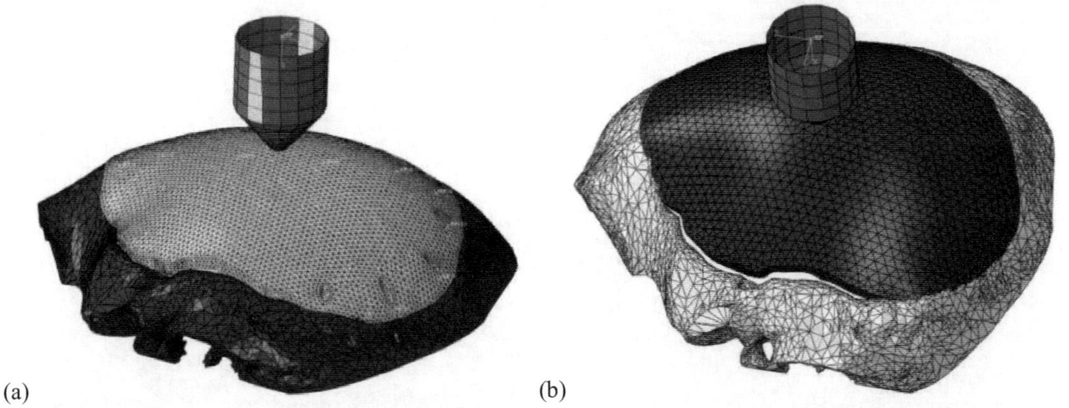

Fig. 2 FE model of the implant with screw holes (a) and without screw holes (b).

The implant mesh was formed by C3D10 quadratic solid elements with an average size of 4 mm. The tip and skull were meshed using Rigid Body elements. Due to the structural response of the implant, nonlinear static analysis was used for the simulation (Table 3).

Table 3. Settings and constrains for simulation elements.

CAD model	
Implant (PEEK)	homogeneous deformable body
Skull	rigid body
Tip	rigid body
Boundary conditions	
Implant	With/without fixation screws
Skull	motionless
Tip	1° degree of freedom
Loading	
Tip	Forced displacement of 5 mm
Surface contact	
Tip/implant	Surface contact, finite sliding, friction $f = 0.1$
Skull/implant	Surface contact, finite sliding, friction $f = 0.1$
Finite Element Mesh	
Implant	C3D10 quadratic solid elements with an average size of 4 mm
Tip/Skull	Discrete Rigid Element
Analysis	
Static	General

3 Results

Fig. 3 shows the course of the reaction force depending on the displacement of the tip for both models. The graph shows that the difference in the course of the force, as well as the maximum force achieved between the implant fixed with screws and without fixation, is negligible.

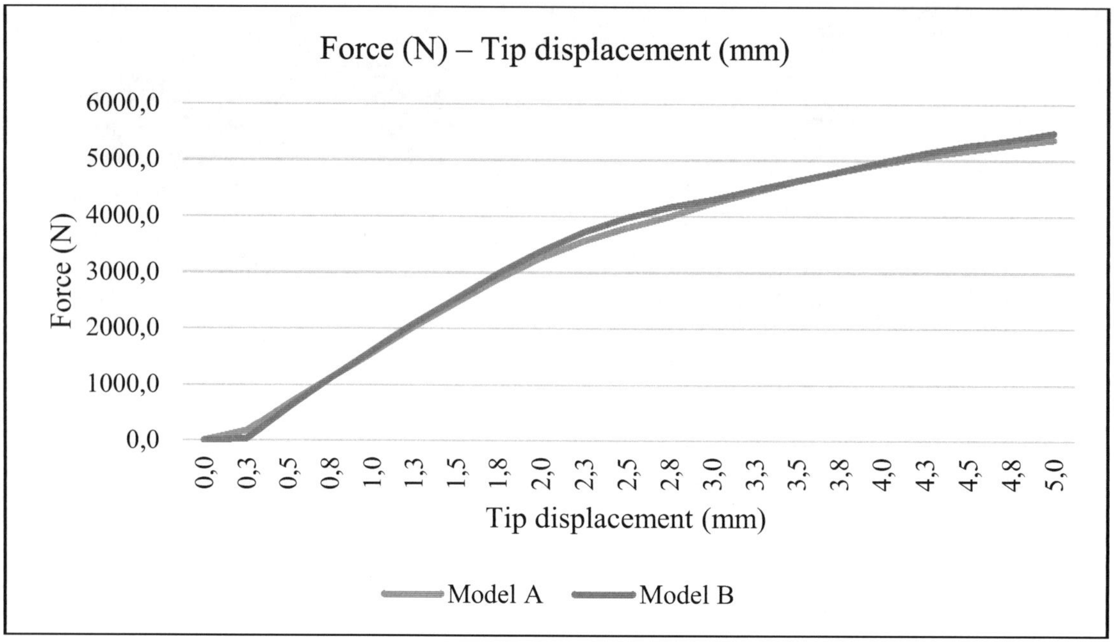

Fig. 3 Reaction force of the tip during loading.

Fig. 4 shows the total strain depending on the tip displacement. The courses of both models are the same at the beginning. For model A, the total strain then has a steeper course and is greater by approximately 0.07, which is probably caused by the screws attaching it to the skull.

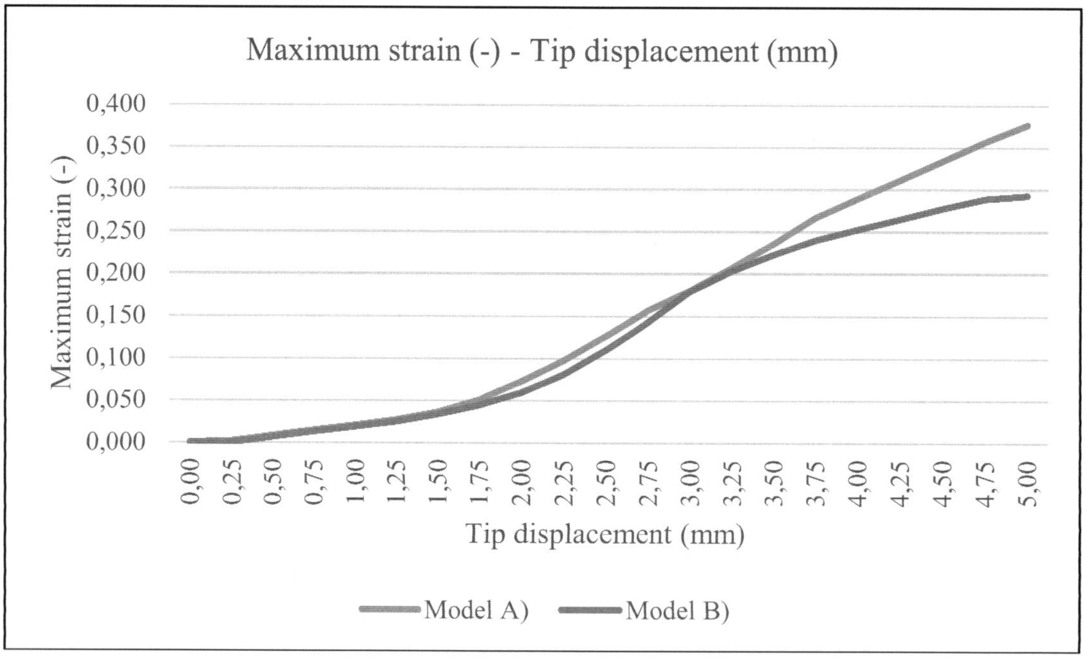

Fig. 4 Dependence of maximum strain on tip displacement.

Fig. 5 shows the maximum equivalent plastic strain at the point of contact depending on the tip displacement. The course of model A has a significantly steeper course, which may be caused by the attached screws and the non-ideal fitting of the contact surface of the implant on the skull.

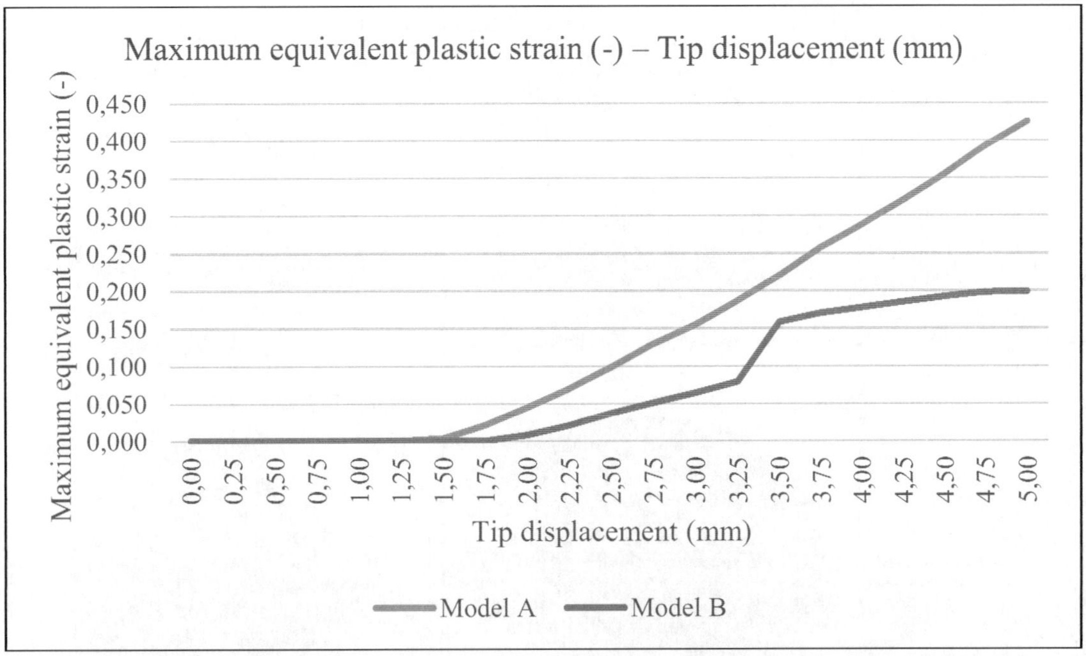

Fig. 5 Dependence of the maximum equivalent plastic strain on tip displacement.

Fig. 6 and Fig. 7 show the von Mises stress distribution and the proportional limit for both models. From the stress distribution, it follows that for model B (Fig. 7), the stress area above the yield strength is larger than for model A (Fig. 6).

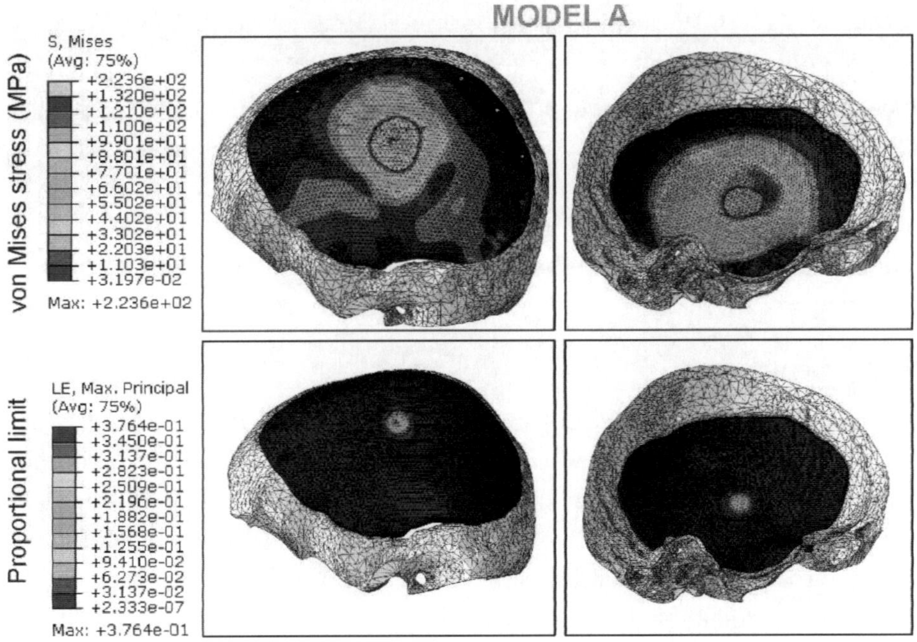

Fig. 6 von Mises stress and proportional limit distribution for Model A.

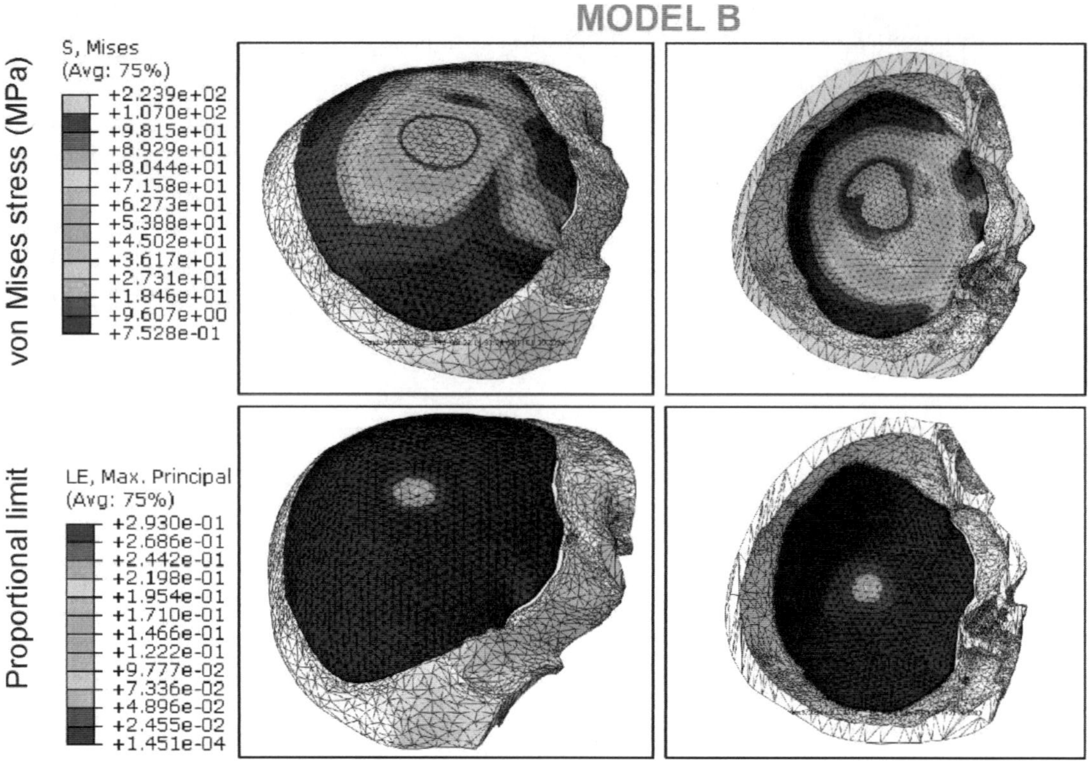

Fig. 7 von Mises stress and proportional limit distribution for Model B.

Conclusions

The results of the simulations show that the course of the force depending on the displacement of the tip differs only to a minimal extent for fixation of the implant with and without screws. The total proportional deformation at the point of action of the tip, depending on the displacement of the tip, is greater by 0.07 for the implant fixed with screws, and its plastic deformation of the implant has a steeper course. During the simulation, the influence of production technology (layering of material) and porosity impact during production were not considered.

Acknowledgements

This publication is the result of the project implementation CEMBAM - Center for Medical Bioadditive Research and Production, ITMS2014+: 313011V358 supported by the Operational Programme Integrated Infrastructure funded by the European Regional Development Fund, KEGA 021TUKE-4/2022 "Implementation of computed tomography in an interdisciplinary technical-natural area", VEGA 1/0516/22 "Research on the impact of composite materials production technology on their mechanical and fatigue properties using numerical and experimental methods" and VEGA 1/0500/20 "Research of mechanical properties of materials with a complex internal structure by numerical and experimental methods of mechanics".

References

[1] Kim, C.N.T., Binh, C.X., Dung, V.T. and Toan, T.V. "Design and mechanical evaluation of a large cranial implant and fixation parts," *Interdisciplinary Neurosurgery*, **31**, 101676 (2023).

[2] Sharma, N., Aghlmandi, S, Dalcanale, F., Seiler, D., Zeilhofer, H.F., Honigmann, P. and Thieringer, F.M. "Quantitative Assessment of Point-of-Care 3D-Printed Patient-Specific Polyetheretherketone (PEEK) Cranial Implants," *Int. J. Mol. Sci.*, **22**, 8521 (2021).

[3] Berretta, S., Evans, K. and Ghita, O. "Additive manufacture of PEEK cranial implants: Manufacturing considerations versus accuracy and mechanical performance," *Materials & Design*, **139**, pp. 141-152 (2018).

[4] Petersmann, S., Smith, A.J., Schäfer, U. and Arbeiter F. "Material extrusion-based additive manufacturing of polyetheretherketone cranial implants: Mechanical performance and print quality," *Journal of Materials Research and Technology*, **22**, pp. 642-657 (2023).

[5] Marcián, P., Narra, N., Borák, L., Chamrad, J. and Wolff, J. "Biomechanical performance of cranial implants with different thicknesses and material properties: A finite element study," *Computers in Biology and Medicine*, **109**, pp. 43-52 (2019).

[6] Garcia-Gonzalez, D., Jayamohan, J., Sotiropoulos, S.N., Yoon, S.H., Cook, J., Siviour, C.R., Arias, A. and Jérusalem, A. "On the mechanical behaviour of PEEK and HA cranial implants under impact loading," *Journal of the Mechanical Behavior of Biomedical Materials*, **69**, pp. 342-354 (2017).

[7] El Halabi, F., Rodriguez, J.F., Rebolledo, L., Hurtós, E. and Doblaré, M. "Mechanical characterization and numerical simulation of polyether-ether-ketone (PEEK) cranial implants," *J Mech Behav Biomed Mater*, **4**(8), pp. 1819-1832 (2011).

Czech Society for Mechanics (CSM)
Dolejskova 5
Praha 8
182 00
Czech Republic

CURRAN ASSOCIATES INC.
proceedings.com

ISBN 978-1-7138-9339-4
9 781713 893394